The Rough G

"The Rough Guide to ...
one of the most complex, politicized and profound ...
our day. It's even-handed and accessible, skipping the jargon and rhetoric to give the reader a solid understanding of the climate challenge and the most promising solutions."

Eileen Claussen, President,
Pew Center on Global Climate Change

"Scientifically up-to-date and clearly written, this courageous book cuts through mystery and controversy to explain climate change for readers who prefer facts."

Richard Somerville, Distinguished Professor,
Scripps Institution of Oceanography

"Much has been published on climate change in recent years, but nothing as clear, succinct, comprehensive and readable as Bob Henson's handy volume. Henson sheds much light on the heat in a balanced, meticulous fashion. His assessment of a staggering number of recent studies and ongoing research is invaluable. This will be the first go-to climate change book on my shelf."

David Laskin, Author of *Braving the Elements*
and *The Children's Blizzard*

"At last, just what we've long needed: an authoritative, no-hysterics guide for climatic crises to come. Bob Henson is one of the world's clearest and most engaging writers on the atmospheric sciences. His new book is a gold mine of information for taxpayers, environmentalists, politicians and anyone else who cares about our climatic tomorrows."

Keay Davidson, Author of *Carl Sagan: A Life*

The first edition of *The Rough Guide to Climate Change* was shortlisted for the 2007 Royal Society Prize for Science Books

Credits

The Rough Guide to Climate Change
Editing & design: Duncan Clark
Picture research & illustration: Robert Henson, Duncan Clark & Hannah Stuart
Proofreading: Serena Stephenson & Stewart Wild
Production: Rebecca Short

Rough Guides Reference
Series editor: Mark Ellingham
Editors: Peter Buckley, Duncan Clark, Matthew Milton, Ruth Tidball, Tracy Hopkins, Joe Staines, Sean Mahoney
Director: Andrew Lockett

Cover images

Front cover, UK edition: bootprints on Lake Austin, Australia (Paul A. Souders/Corbis). **Back cover, UK edition**: iceberg (Peter Nink). **Front cover, US edition**: meltwater lakes and streams on Greenland Ice Sheet east of Kangerlussuaq (James Balog/www.ExtremeIceSurvey.org). **Back cover, US edition**: bootprints on Lake Austin, Australia (Paul A. Souders/Corbis). **Flaps**: Greenland ice sheet (NASA); forest fire (John McColgan, Bureau of Land Management, Alaska Fire Service); European heat wave, July 2003 (NASA); golden toad (Charles H Smith/US Fish and Wildlife Service); New Orleans after Katrina (Jocelyn Augustino/FEMA); arid earth (Patrik Giardino/Corbis); Himalayan melt lakes (NASA)

Publishing Information

This second edition published January 2008 by
Rough Guides Ltd, 80 Strand, London WC2R 0RL
345 Hudson St, 4th Floor, New York 10014, USA
Email: mail@roughguides.com

Distributed by the Penguin Group
Penguin Books Ltd, 80 Strand, London WC2R 0RL
Penguin Putnam, Inc., 375 Hudson Street, NY 10014, USA
Penguin Group (Australia), 250 Camberwell Road, Camberwell, Victoria 3124, Australia
Penguin Books Canada Ltd, 90 Eglinton Avenue East, Toronto, Ontario, Canada M4P 2YE
Penguin Group (New Zealand), 67 Apollo Drive, Mairongi Bay, Auckland 1310, New Zealand

Printed in Singapore by SNP Security Printing Pte Ltd

Typeset in DIN, Myriad and Minion

384 pages; includes index

A catalogue record for this book is available from the British Library

ISBN 13: 9-781-85828-105-6
ISBN 10: 1-85828-105-9

3 5 7 9 8 6 4 2

THE ROUGH GUIDE to

Climate Change

Robert Henson

Edited and designed by
Duncan Clark

www.roughguides.com

Acknowledgements

My deep gratitude goes to the experts who graciously provided interviews and chapter reviews for this book. They include: Sim Aberson (NOAA Atlantic Oceanographic and Meteorological Laboratory); Lisa Alexander (Monash University); David Anderson and Connie Woodhouse (NOAA Paleoclimatology Program); Myles Allen, Ian Curtis, Tina Jardine, Lenny Smith and David Stainforth (Oxford University); Caspar Ammann, Aiguo Dai, Greg Holland, Jim Hurrell, Jeff Kiehl, Joanie Kleypas, Linda Mearns, Jerry Meehl, Susi Moser, Claudia Tebaldi, Tom Wigley and Jeff Yin (NCAR); Vicki Arroyo, Elliot Diringer, Judi Greenwald, Katie Mandes, Truman Semans and John Woody (Pew Center on Global Climate Change); Mustafa Babiker (Arab Planning Institute); Richard Baker (UK Central Science Laboratory); Dan Becker (Sierra Club); Harold Brooks (NOAA Storm Prediction Center); Shui Bin (University of Maryland); Andy Challinor (University of Reading); John Christy (University of Alabama in Huntsville); Hugh Cobb (NOAA National Hurricane Center); Lisa Dilling, Roger Pielke Jr. and Konrad Steffen (University of Colorado); Nikolai Dotzek (Deutsches Zentrum für Luft- und Raumfahrt); David Easterling and Tom Karl (NOAA National Climatic Data Center); Kerry Emanuel (Massachusetts Institute of Technology); Ben Felzer (Marine Biological Laboratory); Jeff Fiedler (UCAR); Hayley Fowler (Newcastle University); Greg Franta and Joel Swisher (Rocky Mountain Institute); Jonathan Gregory (University of Reading/Hadley Centre); James Hansen and Cynthia Rosenzweig (NASA Goddard Institute for Space Studies); Eric Haxthausen (Environmental Defense); Tim Johns, Richard Jones and Jason Lowe (Hadley Centre for Climate Change); Phil Jones and Nathan Gillett (University of East Anglia); David Karoly (University of Oklahoma); David Keith (University of Calgary); Georg Kaser (University of Innsbruck); Paul Kench (University of Auckland); Eric Klinenberg (New York University); Beate Liepert (Lamont-Doherty Earth Observatory); Glenn Milne (Durham University); Andy Parsons (UK Natural Environmental Research Council); Greg Pasternack (University of California, Davis); Rick Piltz (ClimateScienceWatch.org); James Risbey (CSIRO); Alan Robock (Rutgers University); Vladimir Romanovsky (University of Alaska); William Ruddiman (University of Virginia); Stephen Schneider (Stanford University); Peter Schultz (US Climate Change Science Program); Glenn Sheehan (Barrow Arctic Science Consortium); Scott Sheridan (Kent State University); Susan Solomon (NOAA/IPCC); Peter Stott (Hadley Centre, UK Met Office); Jeremy Symons (National Wildlife Federation); David Thompson (Colorado State University); Lonnie Thompson (Ohio State University); David Travis (University of Wisconsin–Whitewater); Chris Walker (Swiss Re); Chris West (UK Climate Impacts Programme); Gary Yohe (Wesleyan University).

Special thanks go to the friends and colleagues who read large portions of the book and provided much-needed moral support along the way: Joe Barsugli, Brad Bradford, Andrew Freedman, Zhenya Gallon, Roger Heape, Theo Horesh, Matt Kelsch, Richard Ordway, John Perry, Catherine Shea, Wes Simmons, Stephan Sylvan and Ann Thurlow.

The enthusiasm of Rough Guides for this project smoothed the way on many occasions. I'm grateful to Mark Ellingham and Andrew Lockett for championing the book and to Richard Trillo, Katy Ball and colleagues for getting the word out about it. The multitalented Duncan Clark provided superb editing, designing and illustration. And every writer should have an agent and supportive friend like Robert Shepard in her or his corner.

No one could ask for a better springboard for writing about climate change than working at the University Corporation for Atmospheric Research, which operates the National Center for Atmospheric Research. It's impossible to adequately convey my appreciation for the opportunity I've had as a writer and editor at UCAR and NCAR to learn about climate change from many dozens of scientists over the last seventeen years. This book couldn't have been written without two intervals of half-time leave made possible by Lucy Warner and Jack Fellows. The opinions expressed herein are my own (as are any errors), and the content does not necessarily reflect the views of my UCAR and NCAR colleagues or the institution as a whole.

The choices we make in the next few years will shape the planet we bequeath to the youth of the new century. This book is dedicated to them, including those in my own circle of life – David, Chloe and Ava Henson, Renée BeauSejour and Natasha Kelsch Mather.

Contents

Part 4: Debates & solutions

From spats and spin to saving the planet

Part 5: What you can do

Reducing your footprint & lobbying for action

Part 6: Resources

Climate change books and websites

Foreword

by James Lovelock

author of *The Revenge of Gaia*

When the Intergovernmental Panel on Climate Change published its third assessment report in 2001 it received only brief attention from the world's media. Yet it was the most thorough statement ever made by climate scientists about Earth and to me the scariest document ever written. It talked laconically but authoritatively about the probability of the world being three degrees Celsius hotter by the end of the century and of a distinct possibility of it being almost six degrees hotter.

Three degrees does not sound like much but it represents a rise in temperature comparable with the global heating that occurred between the last ice age, some 15,000 years ago, and the warmth of the eighteenth century. When Earth was cold, giant glaciers sometimes extended from the polar regions as far south as St Louis in the US and the Alps in Europe. Later this century when it is three degrees hotter glaciers everywhere will be melting in a climate of often unbearable heat and drought, punctuated with storms and floods. The consequences for humanity could be truly horrific; if we fail to act swiftly, the full impact of global heating could cull us along with vast populations of the plants and animals with whom we share Earth. In a worst case scenario, there might – in the 22nd century – be only a remnant of humanity eking out a diminished existence in the polar regions and the few remaining oases left on a hot and arid Earth.

The Rough Guide to Climate Change makes the arcane science of climate change comprehensible and sets the scene for the apocalyptic events that may lie ahead. You may well ask why science did not warn us sooner; I think that there are several reasons. We were from the 1970s until the end of the century distracted by the important global problem of stratospheric ozone depletion, which we knew was manageable. We threw all our efforts into it and succeeded but had little time to spend on climate change. Climate science was also neglected because twentieth-century science failed to recognise the true nature of Earth as a responsive self-regulating entity. Biologists were so carried away by Darwin's great vision that they failed to see that living things were tightly coupled to their material environment and that evolution concerns the whole Earth system with living organisms an integral part of it. Earth is not the Goldilocks planet of the solar system sitting at the right place for life. It was in this favourable state some two billion years ago but now our planet has to work hard, against ever increasing heat from the Sun, to keep itself habitable. We have chosen the worst of times to add to its difficulties.

We are fortunate to have this readable yet accurate book by Robert Henson at the prestigious National Center for Atmospheric Research in Colorado. For me it is a treasured source and map that will guide my thoughts as the climate changes. I recommend it to everyone as a truthful reference point in an ocean of misinformation and special pleading.

James Lovelock, July 2006

Introduction

When strangers meet at a bus stop or in a coffee shop, weather is the universal icebreaker. Yesterday's sweltering heat, the storm predicted for this weekend: it's all fair game. Even longer-term climate shifts find their way into chit-chat. "It used to snow harder when I was a kid" is a classic example – and one explicable in part by the fact that any amount of snow looks more impressive from a child's height.

Today, however, such clichés have an edge to them, because we know that humans play a role in determining the course of climate. When we hear about Arctic tundra melting or a devastating hurricane, we're now forced to consider the fingerprints of humanity – and that's going well beyond small talk. Indeed, climate change is as much a divider as weather has traditionally been a unifier. Weather has always seemed to transcend politics, but human-induced climate change is wedded to politics: it's an outgrowth of countless decisions made by local, regional and national governments, as well as individuals and corporations. Sadly, it's also become – particularly in the US – a polarized subject, linked to other issues so frequently that it often serves as shorthand for one's entire world view.

It might come as a surprise, then, how much of the basic science behind global climate change is rock-solid and accepted by virtually all parties. A handful of sceptics aside, the debate between experts these days revolves around interpretation. Just how warm will Earth get? Which computer projections for the year 2050 are likely to be the most accurate? How should we go about trying to reduce the blanket of greenhouse gases that's getting thicker each year? How can we best adapt to unavoidable changes? These are difficult questions – but they're about the *nature* of global climate change, not its mere existence.

Ever since Louis Agassiz introduced the notion of ice ages in the 1830s, we've known that prehistoric climates differed markedly from the present. And it's been clear for many decades that carbon dioxide and other greenhouse gases keep our planet warmer than it would otherwise be. But it's taken a long time to truly connect the dots, to fully understand that we can push our planet's climate into dangerous, uncharted territory just by the way we go about living our lives.

Although more and more people recognize the risks of climate change, not everyone is yet convinced of the danger, in part thanks to a battlefield of rhetoric created by those with ideological axes to grind. Energy and auto corporations, heavily invested in the status quo, have also contributed to the spin. All of this makes it difficult for the public – not to mention journalists and politicians – to separate fact from fiction, thus making a solution seem unlikely or even impossible.

However, if global warming is one of the most daunting challenges humanity has faced, it's also a unique opportunity. Fossil fuels do more than power our cars and homes. They fuel crisis – from instability in the oil-drenched Middle East to smog-choked skies across the great cities of the developing world. But the era of cheap, plentiful oil is drawing to an end, and the difficult steps needed to deal with global warming could hasten the world's transition to cleaner, sustainable forms of energy. And, as many who have written on this subject point out, we could emerge from that transition with new ways of achieving global unity on other tough issues. For a topic that often seems shrouded in layers of grey, there may be a few silver linings after all.

How this book works

Whether you're alarmed, sceptical or simply curious about climate change, this book will help you sort through the many facets of this sprawling issue. **The Basics** lays out some key questions and answers, explains how global warming actually works, and examines the sources of the greenhouse gases heating up our planet. **The Symptoms** provides an in-depth look at how climate change is already affecting life on Earth – from rising sea levels to polar bears stranded on tenuous ice – and how these changes may play out in the future.

The Science describes how the global warm-up has been measured and puts the current climatic changes in the context of Earth's distant history and future. It also takes a look at the computer models that tell us what we can expect over the next century. How society uses such scientific findings, of course, is shaped by the political and media landscape. **Debates & Solutions** surveys the global-warming dialogue and explores the ways in which we might be able to eliminate or reduce the threat of climate change. These include political agreements, such as the Kyoto Protocol, as well as cleaner energy sources and sci-fi-esque geo-engineering schemes.

For solutions on an individual or family scale, turn to **What You Can Do**, which provides tips on reducing your carbon footprint at home and on the road. Finally, **Resources** provides details of other books on the subject and a list of recommended websites.

PART 1

THE BASICS

Global warming in a nutshell

Climate change: a primer

Key questions and answers

Before exploring the various aspects of climate change in depth, let's quickly answer some of the most frequently asked questions about the issue. The following fifteen pages will bring you up to speed with the current situation and the future outlook. For more information on each topic, follow the reference to the relevant chapter later in the book.

The big picture

Is the planet really warming up?

In a word, yes. Independent teams of scientists have laboriously combed through more than a century's worth of temperature records (in the case of England, closer to 300 years' worth). These analyses all point to **a rise of more than 0.7°C (1.3°F)** in the average surface air temperature of Earth over the last century (see the graph on the inside front cover). The chapter Keeping Track (p.171) explains how this average is calculated. The chart overleaf shows how warming since the 1970s has played out regionally.

In recent years global temperatures have spiked dramatically, reaching a new high in 1998. An intense **El Niño** (see p.118) early that year clearly played a role in the astounding warmth, but things haven't exactly chilled

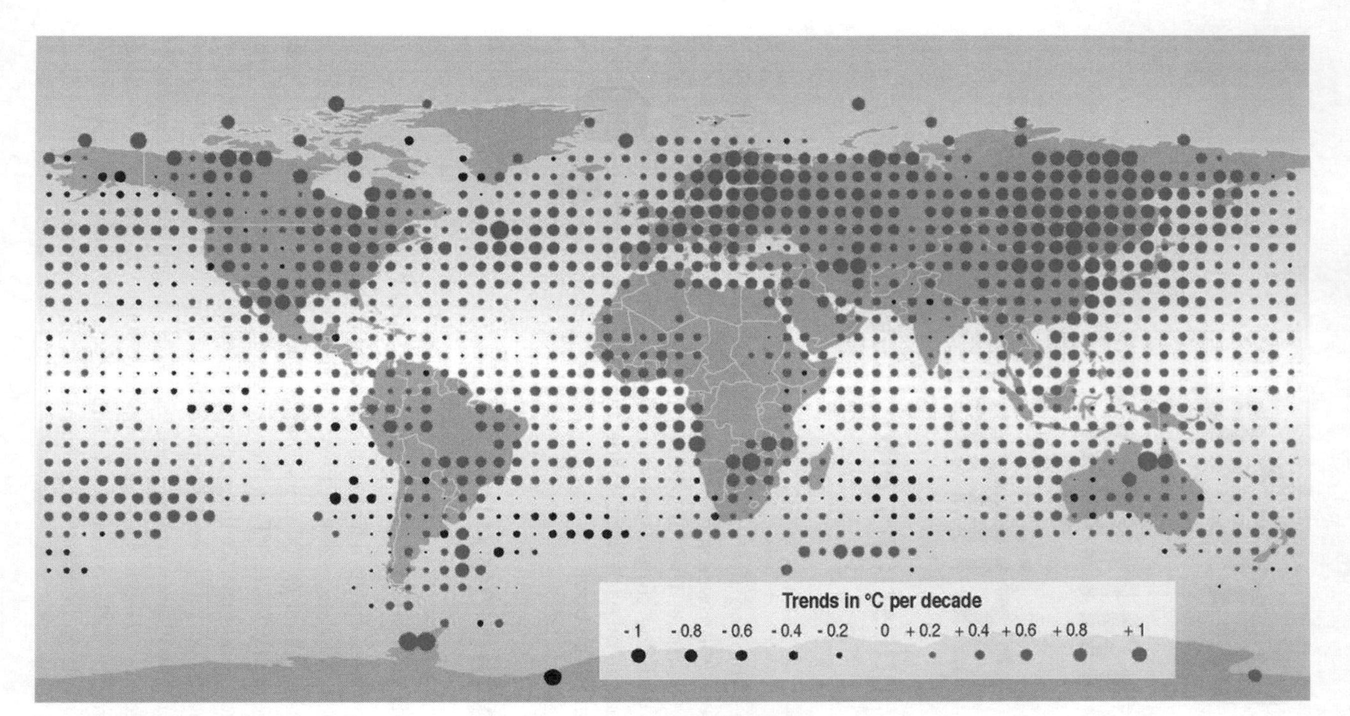

View of a warming world: a depiction of temperature change by region, 1976–2000
IPCC

down since then. The first six years of the twenty-first century, along with 1998, were the hottest on record – and quite possibly warmer than any others in the past millennium (see p.220).

Apart from what temperatures tell us, there's also a wealth of circumstantial evidence to bolster the case that Earth as a whole is warming up.

▶ **Ice** on land and at sea is melting dramatically in many areas outside of interior Antarctica and Greenland. Montana's Glacier National Park is expected to lose its glaciers by 2030. Arctic sea ice has lost nearly half its average summer thickness since 1950, and by mid-century the ice may disappear completely each summer, perhaps for the first time in more than a million years. The warmth is already heating up international face-offs over shipping, fishing and oil-drilling rights in parts of the Arctic once written off as inaccessible.

▶ **The growing season** has lengthened across much of the Northern Hemisphere. The most common species of Japan's famed sakura (cherry blossoms) now blooms five days earlier on average in Tokyo than it did fifty years ago. At some higher latitudes, the growing season is now more than two weeks longer than it was in the 1950s – hardly a crisis in itself, but a sign that temperatures are on the increase.

▶ **Mosquitoes, birds and other creatures** are being pushed into new territories, driven to higher altitudes and latitudes by increasing warmth. The range of twelve bird species in Britain shifted north in the 1980s and 1990s by an average of 19km (12 miles). And Inuits in the Canadian Arctic report the arrival over the last few years of barn swallows, robins, black flies and other previously unseen species. (As we'll see later, however, not all fauna will migrate so successfully.)

But don't many experts claim that the science is uncertain?

There is plenty of uncertainty about details in the global-warming picture: exactly how much it will warm, the locations where rainfall will increase or decrease, and so forth. Some of this uncertainty is due to the complexity of the processes involved, and some of it is simply because we don't know how individuals, corporations and governments will change their greenhouse emissions over time. But there's near-unanimous agreement that global climate is already changing and that fossil fuels are at least partly to blame.

Climate change or global warming?

The phrases that describe climate in transition have a history of their own. Early in the twentieth century, researchers preferred **climatic change** or **climate change** when writing about events such as ice ages. Both terms are nicely open-ended and still used often. They can describe past, present or future shifts – both natural and human-produced – on global, regional or local scales.

Once scientists began to recognize the specific global risk from human-produced greenhouse gases, they needed a term to describe it. In 1975 Wallace Broecker, of New York's Lamont-Doherty Earth Observatory, published a breakthrough paper in the journal *Science* entitled, "Climatic Change: Are We on the Brink of a Pronounced Global Warming?" By the early 1980s the phrase **global warming** – without the "a" in front – was gaining currency among scientists. Meanwhile, the term **global change** emerged as a way to embrace all modes of large-scale human tampering with the planet. When 1988's watershed events arrived (see p.250), the global-warming label broke into headlines worldwide and became standard shorthand among media and the public.

Of course, the planet as a whole *is* warming, but many scientists avoid that term, preferring 'global change' or more specifically **global climate change**. One of their concerns is that global warming could be interpreted as a uniform effect – an equal warming everywhere on the planet – whereas in fact a few regions may cool slightly, even as Earth, on average, warms up.

Politicians hoping to downplay the reality of global warming gravitate towards 'climate change' for entirely different reasons. US political pollster and consultant Frank Luntz has reportedly advised clients that 'climate change' sounds less frightening to the lay ear than 'global warming'. Scary or not, a number of other surveys support the idea that 'global warming' gets people's attention more quickly than the less ominous (though more comprehensive) 'climate change.' And a few activists and scientists, including the Gaia theorist James Lovelock, now favour **global heating** – a phrase that implies humans are involved in what's happening.

The uncertainty that does exist has been played both ways in the political realm. Sceptics use it to argue for postponing action, while others point out that many facets of life require acting in the face of uncertainty (buying insurance against health or fire risks, for example).

Is a small temperature rise such a big deal?

While a degree or so of warming may not sound like such a big deal, the rise has been steeper in certain locations, including the Arctic, where small changes can become amplified into bigger ones (see p.75). The warming also serves as a base from which heat waves become that much worse – especially in big cities, where the **heat-island effect** comes into play. Like a thermodynamic echo chamber, the concrete canyons and oceans of pavement in a large urban area heat up more readily than a field or forest, and they keep cities warmer at night. During the most intense hot spells of summer, cities can be downright deadly, as evidenced by the hundreds who perished in Chicago in 1995 and the thousands who died in Paris in 2003 (see p.47).

How could humans change the whole world's climate?

By adding enormous quantities of carbon dioxide and other **greenhouse gases** to the atmosphere over the last 150 years. As their name implies, these gases warm the atmosphere, though not literally in the same way a greenhouse does. The gases absorb heat that's radiated by Earth, but they release only part of that heat to space, which results in a warmer atmosphere (see p.24).

The amount of greenhouse gas we add is staggering – in carbon dioxide alone, the total is more than thirty billion metric tonnes per year, which is more than four metric tonnes per person per year. And that gas goes into an atmosphere that's remarkably shallow. If you picture Earth as a soccer ball, the bulk of the atmosphere would be no thicker than a sheet of paper wrapped around that ball.

Even with these facts in mind, there's something inherently astounding about the idea that a few gases in the air could wreak havoc around

Earth's atmosphere seems huge, but it's actually extremely thin – like a piece of paper wrapped around a soccer ball
NASA

the world. However, consider this: the eruption of a single major volcano – such as Krakatoa in 1883 – can throw enough material into the atmosphere to cool global climate by more than 1°C (1.8°F) for over a year. From that perspective, it's not so hard to understand how the millions of engines and furnaces spewing out greenhouse gases each day across the planet, year after year, could have a significant effect on climate. (If automobiles spat out chunks of charcoal every few blocks in proportion to the invisible carbon dioxide they emit, the impact would be more obvious.) Yet many people respond to the threat of global warming with an intuitive, almost instinctive denial.

When did we discover the issue?

Early in the twentieth century, the prevailing notion was that people could alter climates locally (for instance, by cutting down forests and ploughing virgin fields) but not globally. Of course, the ice ages and other wrenching climate shifts of the past were topics of research. But few considered them an immediate threat, and hardly anyone thought humans could trigger worldwide climate change. A few pioneering thinkers saw the potential global impact of fossil-fuel use (see p.27), but their views were typically dismissed by colleagues.

Starting in 1958, precise measurements of carbon dioxide confirmed its steady increase in the atmosphere. The first computer models of global climate in the 1960s, and more complex ones thereafter, supported the idea floated by mavericks earlier in the century: that the addition of greenhouse gases would indeed warm the climate. Finally, global temperature itself began to rise sharply in the 1980s, which helped raise the issue's profile among media and the public as well as scientists.

Couldn't the changes have natural causes?

The dramatic changes in climate we've seen in the past hundred years are not proof in themselves that humans are involved. As sceptics are fond of pointing out, Earth's atmosphere has gone through countless temperature swings in its 4.5 billion years. These are the results of everything from cataclysmic volcanic eruptions to changes in solar output and cyclic variations in Earth's orbit (see p.196). The existence of climate upheavals in the past raises the question asked by naysayers as well as many people on the street: how can we be sure that the current warming isn't "natural" – ie caused by something other than burning fossil fuels?

That query has been tackled directly over the last decade or so by an increasing body of research, much of it through the **Intergovernmental Panel on Climate Change (IPCC)**, a unique team that draws on the work of more than one thousand scientists. We'll refer often throughout this book to the IPCC's work (see p.287 for more on the panel itself). Back in 1995, the IPCC's Second Assessment Report included a sentence that made news worldwide:

> "The balance of evidence suggests a discernible human influence on global climate."

By 2001, when the IPCC issued its third major report, the picture had sharpened further:

> "There is new and stronger evidence that most of the warming observed over the last 50 years is attributable to human activities."

And in its fourth report (2007), the IPCC spoke even more strongly:

> "Human-induced warming of the climate system is widespread."

To support claims like these, scientists call on results from two critical types of work: **detection** and **attribution** studies. Detection research is meant to establish only that an unusual change in climate has occurred. Attribution studies try to find the likelihood that humans are involved.

One way to attribute climate change to greenhouse gases is by looking at the signature of that change and comparing it to what you'd expect from non-greenhouse causes. For example, over the past several decades, Earth's surface air temperature has warmed most strongly near the poles and at night. That pattern is consistent with the projections of computer models that incorporate rises in greenhouse gases. However, the pattern agrees less well with the warming that might be produced by other causes, including natural variations in Earth's temperature and solar activity.

As computer models have grown more complex, they've been able to incorporate more components of climate. This allows scientists to tease out the ways in which individual processes helped shape the course of the last century's warm-up. One such study, conducted at the US National Center for Atmospheric Research, examined five different factors: volcanoes, sulphate aerosol pollution, solar activity, greenhouse gases and ozone depletion. Each factor had a distinct influence. The eruption of Mount Pinatubo in 1991 helped cool global climate for several years. Sulphate pollution (see p.190) peaked in the middle of the twentieth century, between World War II and the advent of environmentalism, and it may have helped produce the mid-century cool-down already discussed. Small ups and downs in solar output probably shaped the early-century warming and perhaps the mid-century cooling as well. However, the Sun can't account for the pronounced warming evident since the 1970s. The bottom line is that the model couldn't reproduce the most recent warming trend unless it included greenhouse gases.

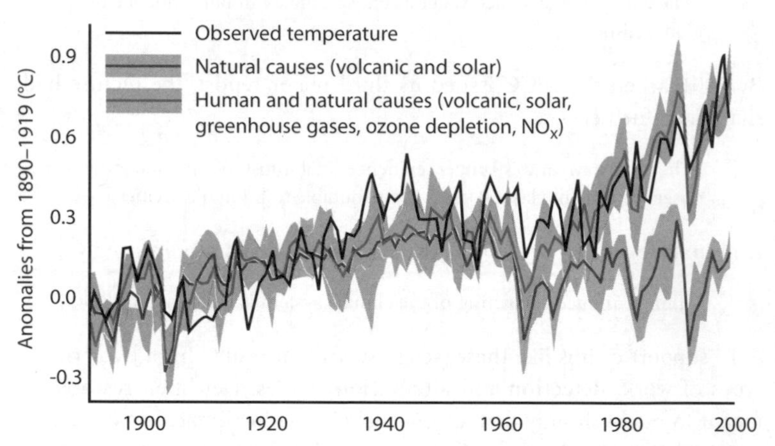

The red and grey shaded areas show the departure in global temperature from the 1890–1919 average, as produced by four computer model simulations. The red and grey lines show averages of the four models. The grey runs factor in natural agents of climate change only; the red runs include both human and natural factors. The black line shows the temperature measured in the real world.

Couldn't some undiscovered phenomenon be to blame?

Although many people would love to find a "natural" phenomenon to blame for our warming planet, such as the relationship between clouds and cosmic rays (see p.259), it's growing extremely unlikely that a suitable candidate will emerge. Even if it did, it would beg a difficult question: if some newly discovered factor can account for the climate change we've observed, then why aren't carbon dioxide and the other greenhouse gases producing the warming that basic physics tells us they should be?

And there's another catch. Any mystery process could just as easily be a cooling as a warming agent, and if it were to suddenly wane, it could leave us with even greater warming than we imagined possible. Trusting in the unknown, then, is a double-edged sword. As such, most scientists in the trenches trust in Occam's razor, the durable rule credited to the medieval English logician and friar William of Occam: "One should not increase, beyond what is necessary, the number of entities required to explain anything."

How do the rainforests fit into the picture?

The destruction of rainforests across the tropics is a significant contributor to climate change, accounting for roughly a fifth of recent human-produced CO_2 emissions. Tropical forests hold nearly half of the carbon present in vegetation around the world. When they're burned to clear land, the trees, soils and undergrowth release CO_2. Even if the land is eventually abandoned, allowing the forest to regrow, it would take decades for nature to reconcile the balance sheet through the growth of replacement trees that pull carbon dioxide out of the air. In addition to the CO_2 from the fires, bacteria in the newly exposed soil may release more than twice the usual amount of another greenhouse gas, nitrous oxide, for at least two years. Brazil's National Institute for Amazon Research estimates that deforestation puts four times more carbon into the atmosphere than the nation's fossil-fuel burning does.

Rainforests also cool the climate on a more local level, their canopy helping to trap moisture and allow it to slowly evaporate, providing a natural air-conditioning effect. When the rainforest has been slashed and burned over large areas, hotter and dryer conditions often set in, although the exact strength of this relationship is difficult to quantify. Across eastern Brazil, where nearly 20% of the Amazonian rainforest has been destroyed, 2005 saw the region's worst drought in a century, perhaps related to changes in the nearby Atlantic and to rain-suppressing smoke from fires as well as to the deforestation itself.

By contrast, in mid-latitude and polar regions, forests actually tend to warm the climate (see p.354).

Was Hurricane Katrina related to global warming?

It's impossible to tie any single weather event, including Katrina, to global warming. Several hurricanes of comparable strength have been observed across the Atlantic over the last century. The horrific damage in New

Orleans was the result not only of Katrina's strength but also the storm's track, the weakness of levees and many other factors. That said, the waters of the Gulf of Mexico that fuelled Katrina were at near-record warmth. It appears that the tropics are part of a global trend towards ocean warming that goes hand in hand with atmospheric warming, and several studies have found an increase in hurricane intensity since the 1970s. See p.131 for more on Katrina and other hurricanes and p.106 for details on oceanic changes.

Whatever happened to "global cooling"?

The planet did cool slightly from the 1940s to the 1970s, mainly in the Northern Hemisphere. However, despite a flurry of 1970s media reports on an imminent ice age (see p.248), there was never anything approaching a scientific consensus on that issue. And while a slight decrease in the amount of sunlight reaching Earth – or "global dimming" – has been measured over the last few decades, it's not been enough to counteract the overall warming.

And the ozone hole?

There are a few links between ozone depletion and global warming, but for the most part they're two separate issues. The world community has already taken steps to address the Antarctic ozone hole, which is expected to disappear by the end of the twenty-first century. See p.29.

The outlook

How hot will it get?

According to the 2007 IPCC report, the global average temperature is likely to rise anywhere from 1.1°C to 6.4°C (2.0–11.5°F) by 2080–2099, relative to 1980–1999. This range reflects uncertainty about the quantities of greenhouse gases we'll add to the atmosphere in coming decades and also about how the global system will respond to those gases. Some parts of the planet, such as higher latitudes, will heat up more than others. The warming will also lead to a host of other concerns – from intensified rains to melting ice – that are liable to cause more havoc than the temperature rise in itself.

Is global warming necessarily a bad thing?

Whether climate change is bad, good or neutral depends on your perspective. Some regions – and some species – may benefit, but many more will suffer intense problems and upheavals. And some of the potential impacts, such as a major sea-level rise, increased flooding and droughts, more major hurricanes and many species being consigned to extinction, are bad news from almost any perspective. So while it may be a bit of a reach to think in terms of "saving the planet" from global warming, it's perfectly valid to think about preserving a climate that's sustaining to as many of Earth's residents as possible.

Perhaps the more pertinent question is whether the people and institutions responsible for producing greenhouse gases will bear the impacts of their choices, or whether others will – including those who had no say in the matter. Indeed, people in the poorest parts of the world – such as Africa – will generally be least equipped to deal with climate change, even if the changes are no worse there than elsewhere. Yet those regions have released only a small fraction of the gases that are causing the changes.

Will anyone be killed or displaced?

Quantifying the human cost of climate change is exceedingly difficult. Weather-related disasters kill thousands of people each year, regardless of long-term changes in the climate. Many of the projected impacts of global warming on society are the combined effects of climate change and population growth (some claim the latter is far more important than the former). For this reason, it's hard to separate out how much of the potential human suffering is due to each factor.

In the decades to come, the warming of the planet and the resulting rise in sea level will likely begin to force people away from some coastlines. Low-lying islands are already vulnerable, and entire cities could eventually be at risk. The implications are especially sobering for countries such as **Bangladesh**, where millions of people live on land that may be inundated before the century is out.

Another concern is moisture – both too much and too little. In many areas rain appears to be falling in shorter but heavier deluges conducive to flooding. However, **drought** also seems to be becoming more prevalent. (See p.59 for more on this seeming paradox.) Changes in the timing of rainfall and runoff could complicate efforts to ensure clean water for growing populations, especially in the developing world.

Warming temperatures may also facilitate the spread of vector-borne diseases such as **malaria** and dengue fever (see p.156). The World Health Organization estimates that in 2000 alone, more than 150,000 people died as a result of direct and indirect climate-change impacts.

Will agriculture suffer?

That depends on where the farming and ranching is done. Global agricultural productivity is predicted to go *up* over the next century, thanks to the extra CO_2 in the atmosphere and now-barren regions becoming warm enough to bear crops. However, the rich world looks set to reap the benefits: crop yields in the tropics, home to hundreds of millions of subsistence farmers, are likely to drop. See p.162.

And wildlife?

Because climate is expected to change quite rapidly from an evolutionary point of view, we can expect major shocks to some ecosystems – especially in the Arctic – and possibly a wholesale loss of species. According to a 2004 study led by Chris Thomas of the University of Leeds and published in the journal *Nature*, climate change between now and 2050 may commit as many as 37% of all species to eventual extinction – a greater impact than that from global habitat loss due to human land use. Similar figures emerged from the 2007 IPCC report, which pegs the percentage of plant and animal species that are at risk from a temperature rise of 1.5–2.5°C (2.7–4.5°F) at 20–30%.

Will rising seas really put cities such as New York and London under water?

Not right away, but it may be only a matter of time. In its 2007 report, the IPCC projects that sea level will rise anywhere from 180 to 590mm (7–23") by 2090–2100. This range is smaller than in the IPCC's 2001 report, but it excludes some key uncertainties about how quickly warming will melt land-based ice. While the new IPCC figures don't signal a catastrophic sea-level rise this century, hurricanes and coastal storms on top of that rise could still cause major problems.

There's also the chance that sea-level rise over the next few decades and beyond could surprise us. The last few years have seen glaciers accelerating their seaward flow in many spots along the margins of Greenland and

West Antarctica. Computer models don't depict the dynamics behind this speed-up very well, so it's not explicitly included in the IPCC projection, but the report does note the added risk at hand. If emissions continue to rise unabated through this century, the Greenland and/or West Antarctica ice sheets could be thrown into an unstoppable melting cycle that would raise sea level by more than 7m (23ft) each. This process would take some time to unfold - probably a few centuries, although nobody can pin it down at this point - but should it come to pass, many of the world's most beloved and historic cities would be hard-pressed to survive.

Will the Gulf Stream pack up, freezing the UK and northern Europe?

The Gulf Stream and North Atlantic Drift bring warm water (and with it warm air) from the tropical Atlantic to Northern Europe. This helps keep the UK several degrees warmer than it would otherwise be. Although this system is unlikely to pack up entirely, there is a possibility that it could be diminished by climate change. The reason is that increasing rainfall and snow-melt across the Arctic and nearby land areas could send more freshwater into the North Atlantic, pinching off part of the warm current. The best estimate is that the flow might weaken by 10–50% over the next century or so. That's probably not enough to offset global warming completely for the UK or northwest Europe, although it could certainly put a dent in it. In any case, the impacts would be much smaller - and would take much longer to play out - than the scenario dramatized in the film *The Day After Tomorrow.* See p.275.

What can we do about it?

What's the Kyoto Protocol?

It's a United Nations-sponsored agreement among nations to reduce their greenhouse-gas emissions. Kyoto emerged from the UN Framework Convention on Climate Change, which was signed by nearly all nations at the 1992 mega-meeting popularly known as the Earth Summit. The framework pledges to stabilize greenhouse-gas concentrations "at a level that would prevent dangerous anthropogenic interference with the climate system". To put teeth into that pledge, a new treaty was needed, one with binding targets for greenhouse-gas reductions. That treaty was finalized in Kyoto, Japan, in 1997 after years of negotiations.

From the start, the chances that the Kyoto Protocol would become international law were tenuous. The US and Australia indicated early on that they wouldn't ratify it, citing the absence of binding targets for developing countries. But the protocol itself required ratification by enough industrialized countries to represent 55% of the developed world's CO_2 output. With the US and Australia out of the picture, virtually every other first-world country would have to ratify the treaty, a process that took seven uncertain years. Finally, Russia's decisive vote in late 2004 brought Kyoto into force the following year. As of mid-2007, 172 states had ratified the treaty (see map).

The Kyoto World

- Signed and ratified (or ratification pending)
- Signed, no intention to ratify
- No position

Under Kyoto, industrialized nations have pledged to cut their yearly emissions of carbon, as measured in six greenhouse gases, by varying amounts, averaging 5.2%, by 2012 as compared to 1990. That equates to a 29% cut in the values that would have otherwise occurred. However, the protocol didn't become international law until more than halfway through the 1990–2012 period. By that point, emission amounts had risen substantially in many countries: over 20% in Canada, for instance. And in some countries exempt from the Kyoto rules, particularly China, emission levels are skyrocketing.

Will Kyoto make a difference?

It appears that few if any of the world's big economies will meet their Kyoto targets by 2012. Even if they did, it would only make a tiny dent in the world's ever-increasing output of greenhouse gases. Reducing greenhouse gas emissions by a few percent over time is akin to overspending your household budget by a decreasing amount each year: your debt still piles up, if only at a slower pace.

The century-long lifespan of atmospheric CO_2 means that the planet is already committed to a substantial amount of greenhouse warming. Even if we turned off every fuel-burning machine on Earth tomorrow, climate modellers tell us that the world would warm at least another 0.5°C (0.9°F) as the climate adjusts to greenhouse gases we've already emitted. The bottom line is that we won't come close to keeping greenhouse heating in check until changes in technology and lifestyle enable us to pull back far beyond our current emission levels, or unless we find some safe method to remove enormous amounts of carbon from the atmosphere, or both. That's a tall order – but if we're determined to reduce the risk of a wide range of climate impacts, we have no choice but to fulfil it.

Will we reach a "tipping point"?

The effects of climate change aren't expected to be strictly linear. A 4°C warming could be more than twice as risky as a 2°C warming, because of **positive-feedback processes** that tend to amplify change and make it worse. The challenge is to identify the points at which the most dangerous positive feedbacks will kick in. For instance, scientists consider it likely that the Greenland ice sheet will begin melting uncontrollably if global temperatures climb much more than 2°C (3.6°F). Because of the implications for coastal areas, as noted above, this is a particularly worrisome threshold.

Since each positive feedback has its own triggering mechanism, there is no single temperature agreed upon as a tipping point for Earth as a whole. However, scientists, governments and activists have worked to identify useful targets. One goal adopted by the European Union, as well as many environmental groups, is to limit global temperature rise to 2°C (3.6°F) over pre-industrial levels. But that ceiling looks increasingly unrealistic – we're already close to 40% of the way there, and only the lower fringes of the latest IPCC projections keep us below the 2°C threshold by century's end.

Another approach is to set a stabilization level of greenhouse gases – a maximum concentration to be allowed in the atmosphere – such as 500 parts per million, as compared to 270–280ppm in pre-industrial times and about 380ppm today. You can learn more about these and other goals in The Predicament, p.278.

Which countries are emitting the most greenhouse gases?

For many years the United States was in first place, with 30% of all of the human-produced greenhouse emissions to date and about 20% of the current yearly totals – despite having only a 5% share of global population. However, China is now taking the lead. Its emissions are *much* lower per capita, but due to its growing population and affluence, China will overtake the US as the world's leading greenhouse emitter by 2008. As shown in the diagram on p.294, the world's industrialized countries varied widely in how much they have increased or decreased their total emissions since 1990. Some of the decreases were due to efficiency gains, while others were due to struggling economies.

Does the growth of China and India make a solution impossible?

Not necessarily. Although its growth in its coal production is hugely worrisome, China is already making progress on vehicle fuel efficiency and other key standards. And because so much of the development in China and India is yet to come, there's a window of opportunity for those nations to adopt the most efficient technologies possible. At the same time, the sheer numbers in population and economic growth for these two countries are daunting indeed – all the more reason for prompt international collaboration on technology sharing and post-Kyoto diplomacy.

If oil runs out, does that solve the problem?

Hardly. It's true that if oil resources do "peak" in the next few years, as some experts believe, we're likely to see economic downswings, and those could reduce oil-related emissions, at first over periods of a few years and eventually for good. The same applies to natural gas, although that peak could arrive decades later. Then, the question becomes what fuel sources the world will turn to: coal, nuclear, renewables or some combination of the three. If the big winner is coal – or some other, less-proven fossil source such as shale or methane hydrates – it raises the potential for global warming far beyond anything in current projections.

Even if renewables win the day later in this century, we're still left with the emissions from today's stocks of oil, gas and coal, many of which would likely get burned between now and that eco-friendly transition. With this in mind, research has intensified on **sequestration** – how carbon might be safely stored underground. The idea appears promising, but big questions remain. See p.310.

Won't nature take care of global warming in the long run?

Only in the *very* long run. The human enhancements to the greenhouse effect could last the better part of this millennium. Assuming that it takes a century or more for humanity to burn through whatever fossil fuels it's destined to emit, it will take hundreds more years for those greenhouse gases to be absorbed by Earth's oceans.

There are few analogies in the geological past for such a drastic change in global climate over such a short period, so it's impossible to know what will happen after the human-induced greenhouse effect wanes. All else being equal, cyclical changes in Earth's orbit around the Sun can be expected to trigger an ice age sometime within the next 50,000 years, and other warmings and coolings are sure to follow, as discussed in the box on p.224. In the meantime, we'll have our hands full dealing with the next century and the serious climate changes that our way of life may help bring about.

The greenhouse effect

How global warming works

Imagine our planet suddenly stripped of its atmosphere – a barren hunk of rock floating in space. If this were the case, then Earth's near-ground temperature would soar by day but plummet by night. The average would be something close to a bone-chilling -18°C (0°F). In reality, though, Earth's surface temperature averages a much more pleasant 14.4°C (57.9°F). Clearly, there's something in the air that keeps things tolerably warm for humans and other living things. But what?

One of the first people to contemplate Earth's energy balance was the French mathematician and physicist Joseph Fourier. His calculations in the 1820s were the first to show the stark temperature contrast between an airless Earth and the one we actually enjoy. Fourier knew that the energy reaching Earth as sunlight must be balanced by energy returning to space, some of it in a different form. And though he couldn't pin down the exact process, Fourier suspected that some of this outgoing energy is continually intercepted by the atmosphere, keeping us warmer than we'd otherwise be.

"Remove for a single summer-night the aqueous vapour from the air which overspreads this country, and you would assuredly destroy every plant capable of being destroyed by a freezing temperature."

John Tyndall, 1863

Venus: a cautionary tale

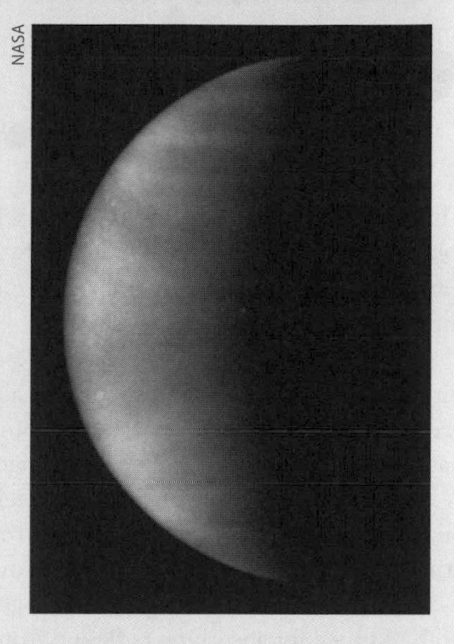
NASA

Scientists don't have the luxury of an extra Earth for experimentation. However, the nearest planet to us serves as an example of what can happen when the greenhouse effect runs amok. Many scientists once saw Venus as a mysterious sibling, one whose climate might be mild enough to support human or humanlike life. But although the planet is named for the Roman goddess of love and beauty, it's actually a rather harsh place. That's because of carbon dioxide, which makes up less than 0.04% of Earth's atmosphere but more than 96% of Venusian air. Factor in the planet's location – about 25% closer to the Sun than Earth is – and you end up with a surface air temperature in the neighbourhood of 460°C (860°F), hot enough to melt lead. (All that CO_2 is invisible, by the way; it's a dash of sulphuric acid that produces Venus's legendary cloak of haze.)

Carl Sagan, the US astronomer and science popularizer, built his early career on Venus, as it were, studying its sizzling atmosphere. In the early 1960s Sagan drew on radio observations and simple mathematical modelling to explain how the planet's dense, superheated atmosphere could be produced by what he called a "runaway greenhouse effect". Sagan's findings led to his concern about the fate of Earth's own atmosphere: in 1984 he was the first witness in then-senator Al Gore's landmark congressional hearings on global warming.

Harking back to experiments by others on how a glass box traps heat, Fourier likened the atmosphere to a hothouse (or greenhouse). *Voilà* – the concept of the **greenhouse effect** was born. It's been with us ever since, even though it's a flawed analogy. The atmosphere doesn't imprison the air the way a glass box does. Instead, it absorbs **infrared radiation** rising from Earth's Sun-warmed surface. All else being equal, the more greenhouse gas there is, the less radiation can escape from Earth to space, and the warmer we get (but there are a few twists along the way, as we'll see).

The diagram overleaf shows what happens to the sunlight that reaches our planet:

▶ **About 30%** gets reflected or scattered back to space by clouds, dust or the ground, especially from bright surfaces like ice.

▶ **More than 20%** is absorbed in the atmosphere, mainly by clouds and water vapour.

▶ **Almost 50%** gets absorbed by Earth's surface – land, forests, pavement, oceans and the rest.

The incoming radiation from the intensely hot Sun is mostly in the visible part of the spectrum – which is why you shouldn't stare at the Sun. Earth, being much cooler than the Sun, emits far less energy, most of it at infrared wavelengths we can't see.

Some of Earth's outgoing radiation escapes through the atmosphere directly to space. Most of it, though, is absorbed en route by clouds and greenhouse gases (including water vapour) – which in turn radiate some back to the surface and some out to space. Thus, Earth's energy budget is maintained in a happy balance between incoming radiation from the Sun and a blend of outgoing radiation from a warm surface and a cooler atmosphere (an important temperature distinction, as we'll see below).

The air's two main components, nitrogen (78%) and oxygen (20%), are both ill suited for absorbing radiation from Earth, in part because of their linear, two-atom (diatomic) structure. But some other gases have three or more atoms, and these branched molecules capture energy far out of proportion to their scant presence. These are the **greenhouse gases**, the ones that keep Earth inhabitable but appear to be making it hotter.

Most greenhouse gases are well mixed throughout the **troposphere**, the lowest 8–16km (5–10 miles) of the atmosphere. (Water vapour is the big exception; it's much more concentrated near ground level.) As mountain climbers well know, the troposphere gets colder as you go up, and so these greenhouse gases are cooler than Earth's surface is. Thus, they radiate less energy to space than Earth itself would. That keeps more heat in the atmosphere and thus helps keep our planet livable.

All well and good – but the more greenhouse gas we add, the more our planet warms. As carbon dioxide and other greenhouse gases accumulate, they block each other's ability to radiate to space, causing the atmosphere to heat up further. The diagram on p.24 shows how this works in more detail. Once the extra gases get the ball rolling, a series of atmospheric readjustments follows. These are largely **positive feedbacks** that amplify the warming. More water evaporates from oceans and lakes, for instance, which roughly doubles the impact of a carbon-dioxide increase. Melting

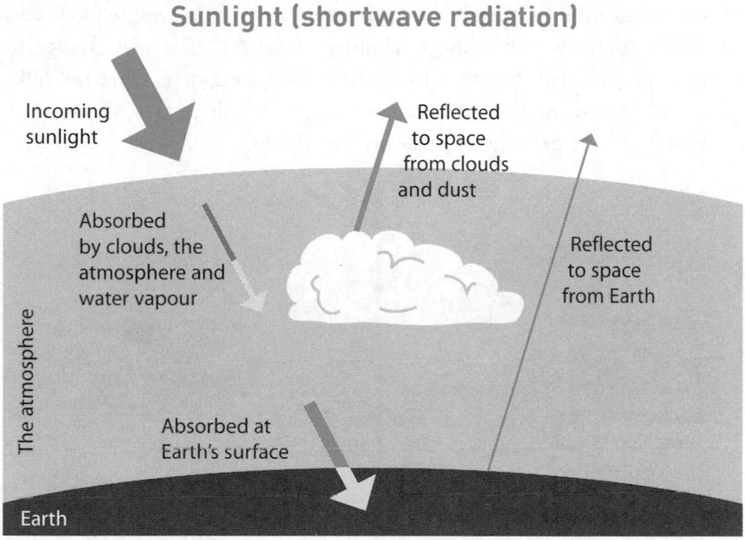

Sunlight (shortwave radiation)

Incoming sunlight
Reflected to space from clouds and dust
Absorbed by clouds, the atmosphere and water vapour
Reflected to space from Earth
The atmosphere
Absorbed at Earth's surface
Earth

Incoming and outgoing radiation
The above diagram shows sunlight entering the atmosphere. The light absorbed (as opposed to reflected) is given out as infrared radiation, as shown in the diagram below. The widths of the arrows in both diagrams reflect the warming power of each process, relative to the total sunlight coming in.

Infrared (longwave radiation)

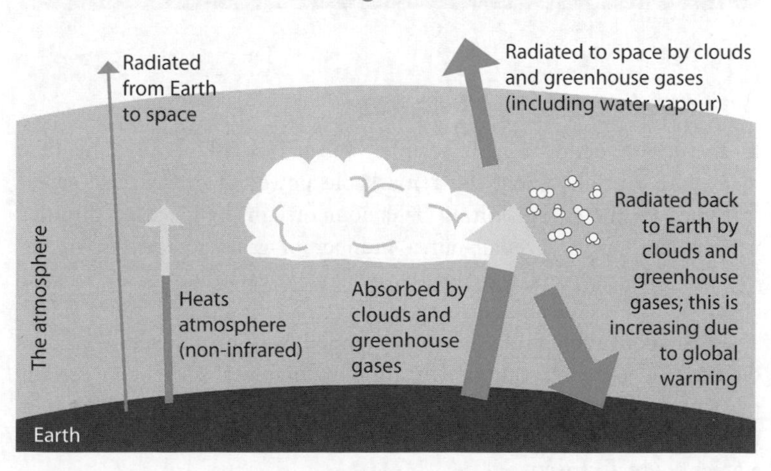

sea ice reduces the amount of sunlight reflected to space. Some feedbacks are less certain: we don't know whether cloud patterns will change to enhance or diminish the overall warming. And, of course, we're not talking about a one-time shock to the system. The whole planet is constantly readjusting to the greenhouse gases we're adding.

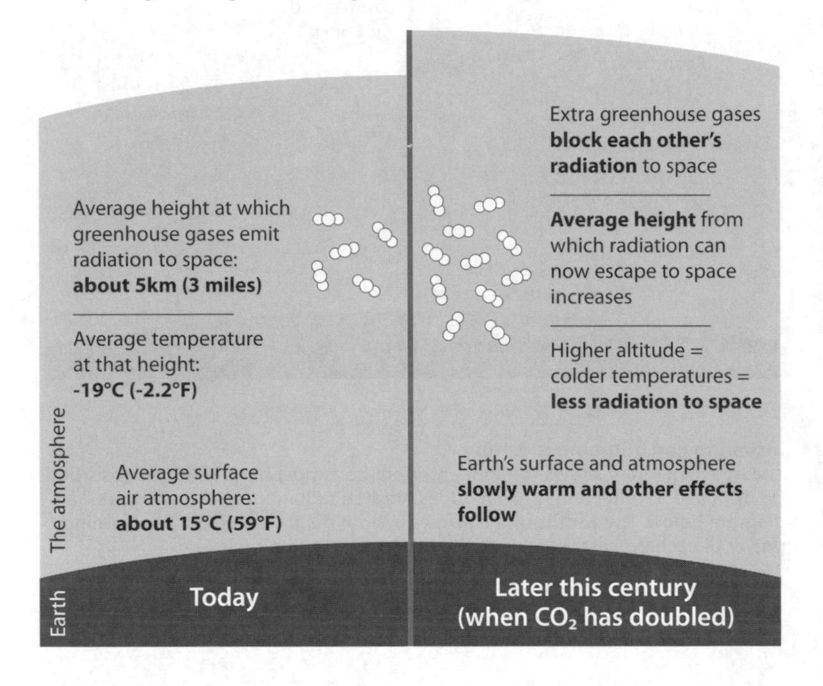

A rogue's gallery of gases

In the 1860s, eminent Irish scientist **John Tyndall** became the first to explore and document the remarkable power of greenhouse gases. Intrigued by the shape-shifting behaviour of light as it passes through various substances, Tyndall put a number of gases to the test in his lab, throwing different wavelengths of light at each one to see what it absorbed. Almost as an afterthought, he tried coal gas and found it was a virtual sponge for infrared energy. Tyndall went on to explore carbon dioxide and water vapour, which are both highly absorbent in certain parts of the infrared spectrum. The broader the absorption profile of a gas – the more wavelengths it can absorb – the more powerful it is from a greenhouse perspective.

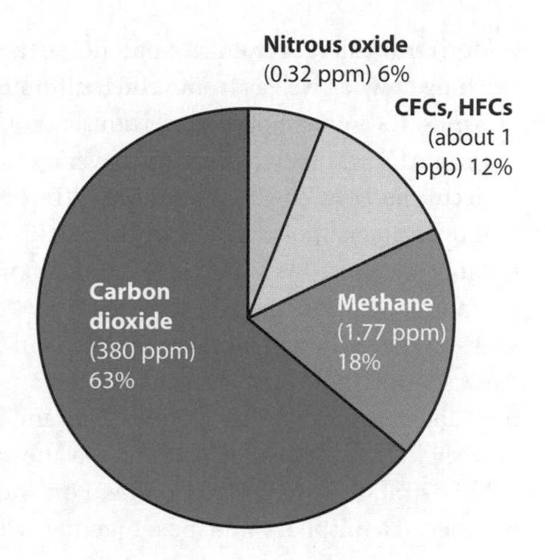

The prevalence of human-produced greenhouse gases and their relative importance in affecting Earth's radiative balance in 2005, according to the 2007 IPCC report. The percentages depend on each gas's prevalence and its molecular ability to trap energy. The numbers total less than 100% due to rounding and other trace gases. Not shown are tropospheric ozone (produced indirectly) and water vapour.

Here are the main greenhouse gases of concern. The diagram shows the prevalence of each gas and how much of an impact it's had in enhancing the overall greenhouse effect.

▶ **Carbon dioxide (CO_2)**, the chief offender, accounts for about 380 of every million molecules in the air, or 380 **parts per million (ppm)**. That number has been climbing by 1–3ppm, or about one-quarter to three-quarters percent per year. The worldwide **emissions** of CO_2 are increasing at several percent per year, but that annual ramp-up becomes a smaller percentage when it joins the large amount of CO_2 already in the air.

Both a pollutant and a natural part of the atmosphere, carbon dioxide is produced when fossil fuels are burned as well as when people and animals breathe and when plants decompose. Plants and the ocean soak up huge amounts of carbon dioxide, which is helping to keep CO_2 levels from increasing even more rapidly (see p.35). Because the give-and-take among these processes is small compared to the atmospheric reservoir of CO_2, a typical molecule of carbon dioxide stays airborne for over a century.

Atmospheric levels of CO_2 held fairly steady for centuries – at around 270–280ppm – until the Industrial Revolution took off. In past geological eras, the amount of CO_2 has risen and fallen in sync with major climate changes, although there's chicken-or-egg uncertainty in whether CO_2 led or lagged some of these transitions (see p.208).

▶ **Methane** emerges from rice paddies, peat bogs and belching cows as well as from vehicles, homes and factories. It's a greenhouse powerhouse: though it stays in the air for less than a decade on average, a methane molecule absorbs 20–25 times more infrared energy in that time than a molecule of carbon dioxide does over roughly a century. As such, methane's total impact on the current greenhouse effect is thought to be roughly a third as big as carbon dioxide's, even though it makes up less than 2ppm of the atmosphere. After soaring in the last few decades, the amount of atmospheric methane rose more fitfully in the 1990s and since 2000 has increased only slightly. A 2006 study by the US National Oceanic and Atmospheric Administration pinned the slowdown on widespread drought in Northern Hemisphere wetlands, together with sagging economies across eastern Europe and Russia in the 1990s.

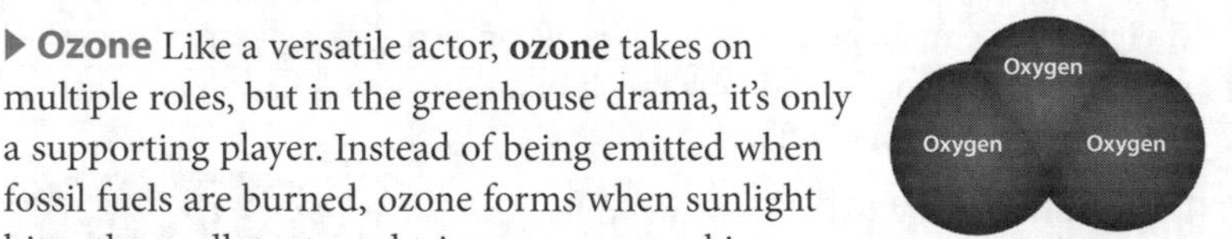

▶ **Ozone** Like a versatile actor, **ozone** takes on multiple roles, but in the greenhouse drama, it's only a supporting player. Instead of being emitted when fossil fuels are burned, ozone forms when sunlight hits other pollutants and triggers ozone-making reactions. Its presence can spike to unhealthy levels when stagnant air takes hold for days near ground level, where people, animals and plants live and breathe. Fortunately, ozone survives only a few days in the **troposphere**. This makes it hard to assess its global concentration, though the limited data that exist indicate an average of around 34 parts per billion. There's apparently been little change in tropospheric ozone amounts since the 1980s, but models hint at a global increase of about 30% since the Industrial Revolution.

Higher up, a protective natural layer of ozone warms the stratosphere while helping to shield Earth from ultraviolet light. The human-induced hole in this layer of "good" ozone has led to record-low temperatures in the lower stratosphere (see box overleaf).

▶ **Water vapour** isn't a very strong greenhouse gas, but it makes up for that weakness in sheer numbers. If you're on a warm tropical island with 100% relative humidity, that means the balmy breezes are wafting as much water vapour as they can – perhaps 3% or more of the air as a whole. The global average is much less, and it varies greatly by location and time of year. It appears to be increasing by perhaps 1% a decade, far more slowly than carbon dioxide.

Water vapour works to propagate itself and boost global warming through an interesting twist. As global temperatures rise, oceans and lakes release more water vapour, obeying a well-known law of thermodynamics. The extra water vapour, in turn, adds to the warming cycle. This is one of several positive feedbacks critical to the unfolding of future climate change.

▶ **A few other gases** – extremely scant but extremely powerful – make up the rest of the greenhouse palette. **Chlorofluorocarbons** (CFCs) and related compounds increased rapidly until they were identified as a key player in stratospheric ozone depletion (see box). Under the Montreal Protocol, they've begun to level off. Along with helping to destroy "good" ozone, they are also powerful, long-lived greenhouse gases – another good reason we're phasing them out. **Nitrous oxide** is also an industrial byproduct, showing up at only about 300 parts per billion (ppb), but with about 300 times the effect of CO_2, molecule for molecule, over its century-long lifespan in the atmosphere.

Greenhouse pioneers

Virtually no one at the peak of the Victorian age reckoned that burning coal or oil would tamper with our climate. But it was becoming clear that great changes in climate had occurred before. Chief among those were the ice ages, which coated North America and Eurasia with kilometre-thick sheets of ice as far south as modern-day Germany and Illinois. Fossils proved that this ice cover had persisted well into the era of early human civilization, some 12,000 years ago. What made the climate plummet into such frigidity and then recover? And could it happen again?

In the mid-1890s, the Swedish chemist **Svante Arrhenius** took a look at how carbon dioxide might be implicated in ice ages. His unexpected results provided our first real glimpse of a greenhouse future. Scientists already knew that volcanoes could spew vast amounts of greenhouse gas, so Arrhenius wondered if a long period of volcanic quiet might allow carbon dioxide levels to draw down and perhaps help plunge Earth into an ice age. He set out to calculate how much cooling might result from a halving of CO_2.

"By the influence of the increasing percentage of carbonic acid [CO_2] in the atmosphere, we may hope to enjoy ages with more equable and better climates, especially as regards the colder regions of the Earth..."

Svante Arrhenius, 1908

Is the ozone hole linked to global warming?

The saga of ozone depletion in the stratosphere is conflated with global warming in the minds of many. Both topics came into public view during the 1980s, often lumped together under the heading of **global change**. There are several links between the two – just enough to cause confusion – but at heart they're two distinct issues.

Ozone, a greenhouse gas, is a pollutant at ground level, harmful when we breathe it. However, the ozone layer that sits within the **lower stratosphere** (especially at about 25–40km/15–25 miles high) is a godsend. Even though the ozone makes up only a tiny fraction of the stratospheric air, it intercepts much of the **ultraviolet light** that can produce sunburns and skin cancer, damage our eyes, and cause other kinds of trouble for people and ecosystems.

A dramatic seasonal depletion in this layer of ozone was found over Antarctica in 1985. Shortly thereafter, scientists identified the three factors that conspire to form the ozone hole, which waxes and wanes during each Southern Hemisphere spring. The first ingredient is a special type of cloud, a **polar stratospheric cloud**, that only forms when winter temperatures fall below about -80°C/-112°F at high altitudes and latitudes. Also needed are **chlorofluorocarbons (CFCs)**, used since the 1920s in spray cans, air conditioners and many other places. CFCs are heavier than clean air, but they mix easily through the atmosphere; once lofted into the stratosphere, they can remain there long enough to do damage. (CFCs are also greenhouse gases themselves; they account for perhaps 10% of the human-enhanced greenhouse effect to date.)

The final protagonist is **sunlight.** As the six-month Antarctic night comes to an end each September, round-the-clock sunshine helps break down the CFCs. This releases **chlorine**, and the chlorine uses the surface of the polar stratospheric cloud to break down ozone into **oxygen**. A single molecule of chlorine can destroy many ozone molecules over a few weeks. During that time, about half of all the ozone through the depth of the Antarctic atmosphere typically vanishes, with near-complete ozone loss in parts of the lower

When he included the role of water vapour feedback, Arrhenius came up with a global temperature drop of around 5°C (9°F).

Soon enough, a colleague inspired him to turn the question around: what if industrial emissions grew enough to someday double the amount of carbon dioxide in the air? Remote as that possibility seemed, Arrhenius crunched the numbers again and came up with a similar amount, this time with a plus sign: a warming of about 5°C (9°F).

Amazingly, this century-old calculation isn't too far off the mark. Scientists now use a **doubling of CO_2** as a reference point for comparing computer models and a benchmark for where our atmosphere might stand by around 2050 or a little afterwards, if present trends continue.

stratosphere. By December of each year, though, the stratosphere has warmed up, the clouds disappear, and the ozone hole fills in.

Fortunately, the ozone hole has never extended much beyond Antarctica, although it has encroached on southern Chile. Southern Australia and New Zealand, while outside the hole per se, have seen ozone reductions of more than 10% at times. As far as the Arctic goes, its wintertime vortex is less stable than its Antarctic counterpart, which limits the growth of polar stratospheric clouds and helps keep a bona fide ozone hole from forming. Still, the springtime depletion over the Arctic can be as high as 60% in some years. A weaker but broader and more persistent **ozone depletion** of some 5–10% extends across much of the globe. Some of this worldwide depletion is likely due to the yearly dispersal of the Antarctic ozone hole and the mixing of that ozone-depleted air around the globe.

Since ozone absorbs sunlight, its depletion can have a **cooling** effect. This helps explain why, even as Earth's surface air temperatures reach record highs, record lows are being notched up in the stratosphere (see p.182). The resulting changes in air circulation over Antarctica sometimes extend to the surface, where interior temperatures have been cooling even as readings near the coast warm up (see p.90).

Unlike global warming, there's an end to ozone depletion in sight, at least on paper. The 1987 **Montreal Protocol**, orchestrated by the United Nations and ratified with amazing speed, called for CFCs to be replaced by substitutes such as **halochlorofluorocarbons**, which have shorter lifetimes and are far less likely to break down in a way that damages ozone. Chlorine concentrations have stabilized in the stratosphere and may already be going down. According to a special IPCC report in 2005, the ozone layer should be rebuilding over the next several decades, although we can expect a few ups and downs along the road to recovery. Indeed, in 2005 and 2006 the ozone hole covered more area than the preceding decade's average, though 2007's results were more variable.

Most models peg the likely warm-up from doubled CO_2 at somewhere between 1.5°C and 4.5°C (2.7–8.1°F). All in all, Arrhenius's initial forecast was quite impressive for the pre-computer era.

In his own time, though, Arrhenius was a lone voice. He himself far underestimated the pace of global development, figuring it would take until the year 4000 or so for his projected doubling to occur. Even then, he figured, a little global warming might not be such a bad thing. Europe was just emerging from the hardships of the Little Ice Age, which had much of the continent shivering and crops withering from about 1300 to 1850. Arrhenius figured a little extra CO_2 might help prevent another such cold spell.

Another greenhouse pioneer came along in the 1930s, by which time the globe was noticeably heating up. **Guy Stewart Callendar**, a British engineer who dabbled in climate science, was the first to point to human-produced greenhouse gases as a possible cause of already observed warming. Callendar estimated that global temperatures might rise around 1.0°C (1.8°F) by 2200. Like Arrhenius, he wasn't too perturbed about this prospect.

In any case, many scientists discounted the message of Arrhenius and Callendar because of the fact that CO_2 and water vapour absorbed energy at overlapping wavelengths. Laboratory tests seemed to show that the two components of the atmosphere were already doing all the absorbing of infrared energy that they could: enlarging the atmospheric sponge, as it were, could have only a minuscule effect. Only after World War II did it become clear that the old lab tests were grievously flawed because they were carried out at sea level. In fact, carbon dioxide behaved differently in the cold, thin air miles above Earth, where it could absorb much more infrared radiation than previously thought. And CO_2's long lifetime meant that it could easily reach these altitudes. Thus, the earthbound absorption tests proved fatefully wrong, one of many dead ends that kept us from seeing the power of the greenhouse effect until industrialization was running at full tilt.

The tale told by a curve

If there's one set of data that bears out the inklings of Arrhenius and Callendar, it's the record of CO_2 collected atop Hawaii's Mauna Loa Observatory since 1958 (see graph opposite). **Charles Keeling** convinced the Scripps Institution of Oceanography to fund the observing site as part of the International Geophysical Year. Because of CO_2's stability and longevity, Keeling knew that the gas should be well mixed throughout Earth's atmosphere, and thus data taken from the pristine air in the centre of the Pacific could serve as an index of CO_2 valid for the entire globe. After only a few years, Keeling's sawtoothed curve began to speak for itself (see graph). It showed a steady long-term rise in CO_2, along with a sharp rise and fall produced each year by the wintering and greening of the

"Charles Keeling was a stickler for precision… He brought a clarity to the problem that altered our perception of global change."
James Hansen, NASA

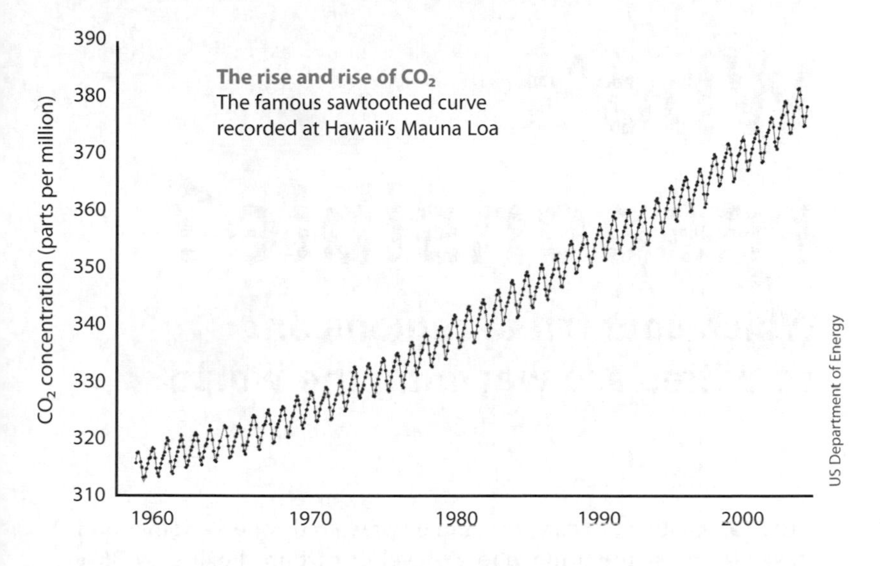

Northern Hemisphere (whose land area and plant mass far outstrip those of the Southern Hemisphere).

With Keeling's curve as a backdrop, a growing number of scientists began to wonder whether human-induced warming might take shape much sooner than Callendar or Arrhenius predicted, as population grew and industrialization proceeded. In the 1960s, climate scientists devised their first primitive renditions of Earth's atmosphere in **computer models**. Like the first crude motion pictures, even these early models had a compelling quality, a whiff of the plausible. One model created by Syukuro Manabe and Richard Wetherald at the US Geophysical Fluid Dynamics Laboratory showed in 1967 that a doubling of CO_2 could produce warming of 2.0°C (3.6°F).

The increasingly complex models developed since the 1960s have refined the picture, but they haven't changed it all that much (see p.227). The best-guess warming for a doubling of CO_2 is in the ballpark of 3.0°C (5.4°F). The Mauna Loa data continue to show carbon dioxide concentrations rising inexorably. And despite ratification of the Kyoto Protocol, the world remains perilously far from even beginning to stabilize CO_2 levels in the atmosphere. Although Arrhenius foresaw a century ago that humans could accentuate the natural greenhouse effect, he would doubtless be amazed at how quickly the process is unfolding and the far-reaching effects it's having on society.

Who's responsible?

Which countries, regions and activities are warming the world?

It took an awful lot of coal, gas and oil passing through engines and stoves to boost the atmosphere's level of carbon dioxide by 35% over the last century and a half. There's been able assistance from other heat-trapping greenhouse gases, such as methane, as well as from deforestation and other changes in land use. Still, most of the human-produced global warming to date appears to be due to CO_2.

Where did it all come from? And what are the main sources today? Asking these questions isn't merely an exercise in finger-pointing. In order to get a handle on where global emissions are headed, it's critical to know what parts of the world and what economic sectors are contributing the most to the current situation.

When it comes to the greenhouse effect, one nation's emissions are everyone's problem. Ordinary pollution disperses in a few days and tends to focus its worst effects in the region where it was generated, plus some areas downwind. Carbon dioxide is much longer lived: once added to the air, it stays there for more than a century. That gives the natural mixing processes of the atmosphere time to shake and stir CO_2 evenly around the globe. Some of the countries producing minimal amounts of greenhouse gases, such as tiny island nations, are among the most vulnerable to the climatic fallout from emissions produced largely by a few big countries.

Scientists can't yet draw a straight line from emissions to a specific climate impact, but in some cases they can now characterize the extent to which greenhouse warming hiked the odds of a particular climate event, such as the 2003 European heat wave (see p.45). It's not a stretch to imagine courts in the future using such findings, along with the tally of who emits what and where, in order to assign partial responsibility for major climate shocks.

Most of the data below are drawn from analyses by the **Intergovernmental Panel on Climate Change** (IPCC) and the **Pew Center on Global Climate Change** (working with the **World Resources Institute**), as well as the **US Carbon Dioxide Information Center.**

How much greenhouse gas is in the air right now?

The global atmosphere currently carries about **3000 gigatonnes** of carbon dioxide, which includes about 800 gigatonnes of carbon. (A gigatonne is a billion metric tonnes.) That's by far the largest presence of any human-produced greenhouse gas. The atmosphere also holds about four gigatonnes of carbon in the form of methane, which is a much stronger but shorter-lived greenhouse gas.

How much are we adding each year?

As of 2006, the most recent year of data at press time, humans were putting close to 31 gigatonnes of carbon dioxide into the atmosphere per year – a stout increase from around 15 gigatonnes in 1970 and about 26 gigatonnes as recently as 2002. This is mainly from burning fossil fuels,

Comparing greenhouse gases

Because greenhouse gases vary so widely in the power of their climatic effects (see p.24), researchers often rely on a unit called **global warming potential**. The global warming potential of a gas is a measure of its contribution per unit mass to greenhouse warming in the atmosphere over a given time span as compared to that of carbon dioxide. Methane, for example, is shorter-lived than CO_2 but much more powerful in its ability to trap heat in the atmosphere. Thus, over a century's time, methane's global warming potential is estimated to be somewhere around 25 (a number that's not yet set in stone), compared to a value of 1 for carbon dioxide. These figures can be multiplied by the prevalence of each gas to produce a **carbon equivalent** that enables all emissions to be considered as a group. By using carbon equivalents, researchers get a better sense of the impact of the atmosphere's total greenhouse-gas burden. Some studies use **carbon dioxide equivalent,** a number obtained by multiplying the carbon equivalent by 44/12 (the ratio of the molecular weight of carbon dioxide to that of carbon).

though about 3% of the total comes from cement production. Almost half of this yearly output remains in the atmosphere, while the rest is absorbed (more on this later).

The global **emission rate** of CO_2 doesn't go up every year. During economic slowdowns, the rate can level off or even decline slightly, as it did in 1992 and 1993. Even in those years, though, we're still adding lots of carbon dioxide to the atmosphere. Indeed, the **total amount** of CO_2 in the air – as seen in the graph on p.31 – has risen every year since measurements began in the late 1950s.

To put these figures in perspective, consider **per-person emissions**. Each human on Earth is responsible, on average, for putting about 4.6 metric tonnes of carbon dioxide into the air each year. That's 4600kg, or about 10,100lb. In other words, someone who weighs 68kg (150lb) puts the equivalent of her weight in carbon dioxide into the atmosphere every six days. Of course, the real figures vary enormously based on where and how you live. For most people in Europe and North America, the average is much higher, as shown by the chart later in this chapter. For more on the "carbon footprints" of individuals, see p.336.

Human-produced methane emissions are harder to quantify, but they could represent another 0.3 to 0.45 gigatonnes per year, with natural processes adding almost that much methane each year. Interestingly, the total amount of methane in the atmosphere hasn't changed much at all in in the new millennium. Recent research points to widespread episodes of drought and weak Eurasian economies as factors.

What happens to the greenhouse gas we put into the air?

It's easy to assess how much CO_2 is burned each year, at least in principle. And it's also fairly straightforward to measure the amount of greenhouse gas in the air, since most of it is well mixed across the globe. From these two factors, we know that about 45% of the carbon that enters the atmosphere each year due to human activity stays there, adding to airborne CO_2 for a century or more (though individual molecules cycle through the system more quickly). The other 55% is absorbed by the ocean and taken up by trees, crops, soils and the like.

The rest is either absorbed by the ocean or taken up by land-based ecosystems – trees, crops, soils and the like. Plants take up carbon dioxide when they photosynthesize and return it to the soil and the atmosphere when they die and decompose.

Will Earth's carbon sink get stopped up?

One of the critical issues in climate-change science is why and how Earth's land areas are able to soak up roughly 25% of our carbon emissions. Several possibilities are on the table. One is that the CO_2 we're adding to the atmosphere may be stimulating plants to grow more vigorously. Many laboratory tests support this notion; after all, CO_2 is a fertilizer. Even so, we can't count on this process indefinitely, because plant growth may run up against other constraints, such as the supply of other nutrients or changes in land use. Indeed, climate change itself could turn the tide by reducing water supply, raising temperatures beyond an optimal level, or increasing the risk of fires – which themselves would release even more CO_2.

Reforestation is also likely to have an important effect. Many parts of eastern North America that were deforested and then farmed for decades are now tree-covered once more. That growing forest is pulling additional CO_2 from the air, but this effect will slow down as the trees reach maturity. On the other side of the equation, the massive deforestation of Earth's tropics (see p.11) is releasing vast amounts of CO_2, which only adds to the carbon puzzle. A 2007 study led by Britton Stephens of the US National Center for Atmospheric Research found that northern land areas were taking up less CO_2 than earlier believed, implying that intact tropical forests might be a net sink for carbon in spite of deforestation. On top of these regional differences, the normal respiration of Earth's plant life may go through large multiyear changes related to climate patterns such as El Niño. Researchers are still working on ways to measure the year-to-year and decade-to-decade variability.

The bottom line is that, one way or another, Earth's land areas as a whole appear to be serving as a sink for carbon right now. Happily, this takes a substantial edge off the impact of our greenhouse emissions. That could change, though. "Will the terrestrial carbon sink continue to operate the same way in the future, as climatic changes become larger and larger?" asks Jonathan Foley of the University of Wisconsin in a 2004 review of the topic. Some studies, including one led by Atul Jain at the University of Illinois in 2005, indicate that deforestation and other land-cover changes could be turning the global biosphere from a sink into a source – just what we don't need.

This simplified picture gets a bit more complex when you consider that Earth's vegetation evolves from year to year and decade to decade. El Niño and other atmospheric cycles can spawn drought over huge areas. The dry conditions make for struggling plants that absorb less CO_2, and they allow for more deforestation and large-scale fires that add CO_2 to the atmosphere. And recent work hints that the absorption of CO_2 by plants and oceans could decrease in the long term as the atmospheric load increases. It's very hard to measure and separate out all these effects; for one thing, what we add is folded into a much larger exchange of carbon that takes place naturally. Over the last few years, the balance has played out as follows:

▶ **The atmosphere** retains about 45% of each year's added CO_2 when averaged across several years, though in a given year the percentage can range anywhere from 30% to 80% as land absorption rises and falls.

▶ **Oceans** absorb close to 25%.

▶ **Land-based ecosystems** take up the rest of the atmosphere's carbon – around 30% on average. There are large variations from year to year due to climate cycles and changes in land use, and scientists aren't yet sure which regions account for most of the absorption (see p.38).

Because most of Earth's plant life is north of the Equator, the northern spring produces a "breathing in" effect (a dip in CO_2 levels), with a "breathing out" (a rise in CO_2) each northern autumn. These ups and downs show up as the sawtooths atop the steady multiyear rise in CO_2 visible on the graph on p.31.

How much do different activities contribute?

The IPCC's 2001 report broke down global emissions of carbon dioxide into four main sectors that account for virtually all of it.

▶ **Industry:** more than 40%

▶ **Buildings (homes, offices and the like):** about 31%

▶ **Transportation:** around 22%

▶ **Agriculture:** about 4%

By and large, the world's most technologically advanced nations have become more energy efficient in recent years, with **industrial emissions** actually declining a few percent in some developed countries since 1990.

Aviation: taking emissions to new heights

George Hall/Corbis

The advent of low-cost flying has sent millions more people hopping from country to country over the last decade, especially across Europe. In many developed countries, aviation is the fastest-growing transportation segment, and it threatens to counteract progress in reducing greenhouse-gas emissions on other fronts.

Part of the reason is pure volume. In 2006 the UK released its first inventory of aviation's impact on the nation's greenhouse emissions. The numbers were as jolting as a rough take-off. Since 1990 the emissions based on fuel use at the UK's international flight terminals have more than doubled, topping nine million tonnes of carbon equivalent in 2005. That's equal to about half of the emissions produced by all UK housing. To make matters worse, it appears likely that contrail clouds generated by aircraft also contribute to warming, perhaps boosting aviation's greenhouse impact several times over, though the numbers are still being researched (see p.188).

International airplane and ship traffic aren't considered in the Kyoto Protocol, so their emissions won't play a role in whether the UK or other nations meet their Kyoto requirements (although aviation may be factored into the European Union's emissions trading scheme before 2012). All told, the UK's greenhouse emissions rose by just under 1% from 1990 to 2004. But when shipping and aviation are excluded – as they are in the Kyoto Protocol – the emissions fell by 4%. Of course, as Kevin Anderson of the Tyndall Centre observed in *The Guardian*, "The atmosphere doesn't care where the carbon comes from."

It's a good thing that aviation is a fairly small piece of the greenhouse emissions picture for the time being, because it may be one of the most troublesome ones to address in the long term. Electric cars are already taking to the streets, and fuel-cell vehicles could become viable later on, but there's no obvious way to wean aircraft off fossil fuels in the next several decades. Biofuels will be tested soon, but these are unlikely to make up more than a sliver of airlines' carbon footprint.

Rough Guides encourages you to "fly less and stay longer," and to make smart choices when you do fly. See p.348 for details on how you can minimize the greenhouse impact of your own flights.

Counterbalancing this progress is the explosion of industry in the world's up-and-coming economies, such as China's and India's. Again, there's lots of variety in how energy efficient each developing nation is striving to be. In sum, the global CO_2 emissions from industry are climbing a bit more slowly (less than 1% a year) than the average emissions for all sectors combined (between 1% and 2%).

The picture is a bit less positive when it comes to **buildings**. The emissions produced by heating, cooling and powering homes have been rising by close to 2% a year, thanks in large part to bigger homes stuffed with more energy-hungry devices. The average American family in a new home has more than four times the living space per person than in 1950. In many parts of the world, there are plans to reverse the growth in household emissions – the UK, for example, is aiming for a 60% reduction by 2050 (see p.302) – but it remains to be seen whether such targets will be met. In the meantime, industrial construction is tending to become more efficient more quickly than the household sector, with companies motivated by the potential for long-term savings.

Among the four sectors, **transport** is where the most trouble appears to be brewing. Emissions from this sector are climbing at well over 2% a year, stoked by the growth of gas-guzzling sport utility vehicles and the ever-longer distances many people are driving and flying (see box on p.37). These perks of middle-class life aside, it's in the developing world where transport-related emissions are growing the most quickly, as many millions of people across China, India and other newly prospering nations take to the roads for the first time ever in their own cars. Right now, these aren't necessarily the most energy-efficient vehicles, although China is in the process of dramatically tightening its emission and fuel-economy standards beyond those in many Western nations. Still, every new gasoline-powered car emits at least some greenhouse gas, and the world's highways could carry up to 700 million additional vehicles by the year 2020.

For more on how personal choices in home life and transportation feed into the global picture, see p.335.

Which countries are most responsible?

Establishing which countries are most responsible for climate change is more complicated than simply totting up the amount of fossil fuels that each nation burns each year. First, there's the fact that a country's total greenhouse-gas emissions may be relatively large even if its per-capita emissions – ie its emissions per person – are quite small, as is the case with China. Then there's the issue of greenhouse gases emitted in the past (some of them before global warming was a known problem), many of which still remain in the atmosphere.

The question of how countries are using – or abusing – their landscapes also needs to be considered. Countries undergoing major deforestation, such as Brazil and Indonesia, would rank significantly higher in the list shown on p.41 of nations with the highest total emissions if the greenhouse-boosting effects of forest destruction were taken into account. Indonesia may even make it into the top five. To add another level of complexity, there's the claim that emissions figures are currently rigged in favour of nations that tend to import, rather than export, goods (see box below).

Outsourcing emissions

The nation-by-nation figures cited in this section look only at the direct emissions of greenhouse gases within those countries. But there's an important side effect of globalization to be considered: the shift it produces in the balance of greenhouse emissions. When a country imports consumer goods, should the emissions produced by the manufacture of those goods be assigned to the destination country rather than the supplier? If they were, the United States would leap even further ahead of the rest of the world as a greenhouse emitter, because so many of its household products are made in other countries, particularly China. One preliminary study led by Shui Bin (US Pacific Northwest National Laboratory) and Robert Harriss (Houston Advanced Research Center) estimated that US emissions would have been 6% higher in 2003 if the products imported by Americans had been made in the US.

Shifting emissions isn't the main purpose of US trade with China, of course, because the US is not a party to the Kyoto Protocol. But what if a country whose emissions are limited by Kyoto did decide to transfer its greenhouse-intensive industry to a nation unfettered by the protocol? The risk of this so-called **carbon leakage** has been studied in some depth. Economists are still tussling over how big a concern it is. A 2005 study by Mustafa Babiker of the Massachusetts Institute of Technology argues that carbon leakage could be substantial, perhaps enough in some cases to counteract any benefits from Kyoto. Babiker's model suggests that in some circumstances the total emissions (direct and outsourced) could rise by 30%.

Other studies show that there could be a much smaller amount of leakage, with the overall benefits of Kyoto more than making up for the outsourced emissions. These brighter scenarios typically assume that most countries, even those outside the Kyoto framework, will be market-motivated to adopt energy-efficient practices. In any case, Kyoto, while an important first step in addressing the global warming problem, clearly isn't going to solve it (see p.280).

When it comes to overall greenhouse gas output, however, two countries stand above the rest. The **United States** is responsible for around 30% of the **cumulative CO_2** emissions to date (that is, all emissions since the industrial era began) and continues to generate about 20% of annual greenhouse-gas emissions – roughly four times its share, considering that the US represents about only 5% of world population. Some of America's outsized emission rate is clearly due to a lack of emphasis on energy efficiency and a focus on economic growth as opposed to environmental virtue. There are also historical factors in the mix that are difficult to change: a car-loving and car-dependent culture, an economy built on once-vast reserves of fossil fuels, and the simple fact that the US is large in both population and land area. The country now has more than fifty years' worth of suburban development that virtually forces millions of Americans to drive to work, school and just about anywhere else outside the home.

Quickly outpacing the US in terms of current emissions is **China**. With its billion-plus citizens, the country's emissions per person are still comparatively low. But as the nation continues to industrialize, its total emissions have been climbing at a staggering pace – close to 10% annually. Only a few years ago, it was thought that China might surpass the US in

America's vast areas of suburbia are an important factor in its profligate energy consumption
Bob Henson

total emissions by the year 2025. But that point was actually reached in 2006, according to the Netherlands Environmental Assessment Agency (or it will be reached by 2008, estimates the International Energy Agency). In recent years the annual *increase* in China's emissions has run higher than Britain's *total* emissions.

The **United Kingdom**, for its part, has produced 6% of cumulative emissions, thanks to its head start as the birthplace of the Industrial Revolution. In the course of a major industrial transformation, the UK has cleaned up its act significantly. It now generates only 2% of the yearly global total of greenhouse gases. Still, that's twice its rightful share, considering that the UK houses only 1% of Earth's population.

Below is a list of the world's top twenty greenhouse emitters, measured in three different ways: percentage of all global emissions, emissions per capita, and carbon intensity (see overleaf). For **total emissions**, the list is a mixed bag, with the contributions split almost evenly between developing and developed nations. Note that China now matches the US in total emissions, though this isn't reflected in the 2004 data used for this chart.

Percentage of global CO_2 emissions *(fossil fuels, cement and gas flaring only)*		***Emissions per capita*** *(tonnes of CO_2 emissions from fossil fuel use, per person)*		***Carbon intensity*** *(tonnes of carbon emitted in CO_2 per millions of dollars in GDP/PPP)*	
United States	20.9	Qatar	21.6	Ukraine	483
China	17.3	Kuwait	10.1	Russia	427
Russia	5.3	UAE	9.3	Saudi Arabia	260
India	4.6	Aruba	8.3	Poland	230
Japan	4.3	Luxembourg	6.8	Iran	223
Germany	2.8	Trinidad/Tobago	6.8	China	201
Canada	2.2	Brunei	6.6	South Africa	200
United Kingdom	2.0	Bahrain	6.5	Australia	193
South Korea	1.6	United States	5.6	South Korea	185
Italy	1.6	Canada	5.5	Canada	172
Mexico	1.5	Norway	5.2	United States	162
South Africa	1.5	Dutch Antilles	5.1	Turkey	149
Iran	1.5	Australia	4.4	Indonesia	127
Indonesia	1.3	Falkland Islands	4.1	Mexico	125
France	1.3	Faroe Islands	3.9	Pakistan	112
Brazil	1.1	Estonia	3.8	Germany	111
Spain	1.1	Oman	3.7	United Kingdom	110
Ukraine	1.1	Saudi Arabia	3.7	EU (collectively)	107
Australia	1.1	Gibraltar	3.9	Japan	104
Saudi Arabia	1.1	Kazakhstan	3.6	Spain	104

Columns 1 & 2 show 2004 data from the US Carbon Dioxide Information Analysis Center. Column 3 shows 2000 data presented in the 2004 Pew Center report, "Climate Data: Insights and Observations".

The **emissions per capita** column tells a different story, with the list topped by the tiny oil-producing nations of **Qatar**, **United Arab Emirates**, **Kuwait** and **Bahrain**. These countries have so few residents that their national contributions to the global greenhouse effect remain small, but because they are heavy producers and consumers of oil, they have a high per-capita emissions rate. Otherwise, industrialized nations lead the way, with the sprawling, auto-dependent trio of Australia, the US and Canada clustered behind the oil producers.

Carbon intensity: an easy way out?

George W. Bush's administration in the US found itself under intense pressure in 2002 to ratify the Kyoto Protocol, which was then making its way through the world's legislatures. Instead, Bush steered America away from Kyoto and towards a different way of assessing progress on climate change. His plan emphasized greenhouse gas intensity, aka **carbon intensity**. This is a measure of how much fossil fuel it takes to produce a certain amount of economic output. Thus, carbon intensity is not the actual amount of carbon emitted, but a number pro-rated by the Gross Domestic Product (GDP) or purchasing power parity (PPP). For example, if GDP and emissions both climbed 3% in a given year, the carbon intensity would remain unchanged even though the actual emissions had risen.

The Bush administration called for an "ambitious but achievable" reduction in carbon intensity of 18% by the year 2012. Environmentalists pointed out that the US carbon intensity dropped 17.4% from 1990 to 2000 without any special attempt to reduce it. Thus, they claimed, the plan offered little more than business as usual. The picture is similar elsewhere. Globally, carbon intensity (looking only at CO_2) dropped by 13% from 1990 to 2000, even as total emissions grew. In China, a booming economy helped reduce carbon intensity by 47% at the same time that CO_2 emissions climbed by 39%. Despite its shortcomings, the concept of carbon intensity is still widely used. As part of their 2007 Sydney Declaration, the 21-nation Asian–Pacific Economic Cooperation group held back from any direct targets for emission reduction, but agreed on an "aspirational" goal of reducing energy intensity by at least 25% by 2030, as compared to 2005 values.

In the long run, then, intensity is a useful way of gauging the impact of greenhouse-gas reductions on the economy. But when it comes down to effects on the physical world, a molecule of gas is still a molecule of gas.

PART 2

THE SYMPTOMS

What's happening now, and what might happen in the future

Extreme heat

Too hot to handle

The super-hot European summer of 2003 ended not with a bang, but with a simmer – a relentless barrage of heat that produced a slow-motion catastrophe. How could a simple hot spell become the deadliest weather event in modern European history? And what did this portend? The unsettling questions hung in the air long after the heat abated. In at least one study, a group of climate scientists confirmed the suspicions held by many: it's unlikely the great heat wave of 2003 would have played out as it did without a helping hand from fossil fuels.

Heat waves themselves are nothing new, of course. Many parts of the US Great Plains have yet to top the all-time records they set during the 1930s Dust Bowl, including a reading of 49.5°C (121°F) as far north as North Dakota. However, climate-modelling studies and our understanding of the greenhouse effect both indicate that the next few decades could bring hot spells that topple many longstanding records across mid-latitude locations. And a heat wave doesn't have to bring the warmest temperatures ever observed to have catastrophic effects. All you need is a long string of hot days combined with unusually steamy nights. Stir in a complacent government, cities that were built for cooler times and a population that can't or won't respond to the urgency, and you have a recipe for the kind of disaster that 2003 brought to Europe.

2003: a record of records

Extremes in summertime heat are one of the long-anticipated outgrowths of a warming planet. Part of this is due to simple mathematics. If the average temperature goes up, the deviations from that average should rise as well, so the most intense spikes in a hot summer ought to climb accordingly. Computer models and recent data bear out this intuitive concept.

We don't know with certainty that temperatures will become more variable as the climate warms, but some simulations of greenhouse-warmed climate do show an increase in heat extremes, especially across interior areas. In other words, even as the typical summertime temperatures rise, warm departures from that norm (both night and day) could become even larger. This is especially true if the bulk of North America and Eurasia undergoes a shift towards drier summers, as many models are indicating.

The notion of larger upward spikes atop a warming climate is a real cause for concern – and there's evidence that just such spikes occurred in 2003. But the peak temperature on a single day, however high it may be, isn't the ideal measure of a heat wave's ferocity. When it comes to the toll on people, animals and plants, the **duration** of a heat wave and the warmth of the **night-time lows** are the real killers. A weeklong stretch of severe heat can be far more deadly than just two or three days of it. If a heat wave is defined at a certain location by a certain number of days with readings above, say, 35°C (95°F), then a warmer climate would likely bump more days above such a threshold.

There's a different connection between global warming and steamy nights. As noted in the Floods & Droughts chapter (see p.58), higher temperatures have boosted **atmospheric water vapour** on a global scale. When there's more water vapour in the air, nights tend to stay warmer.

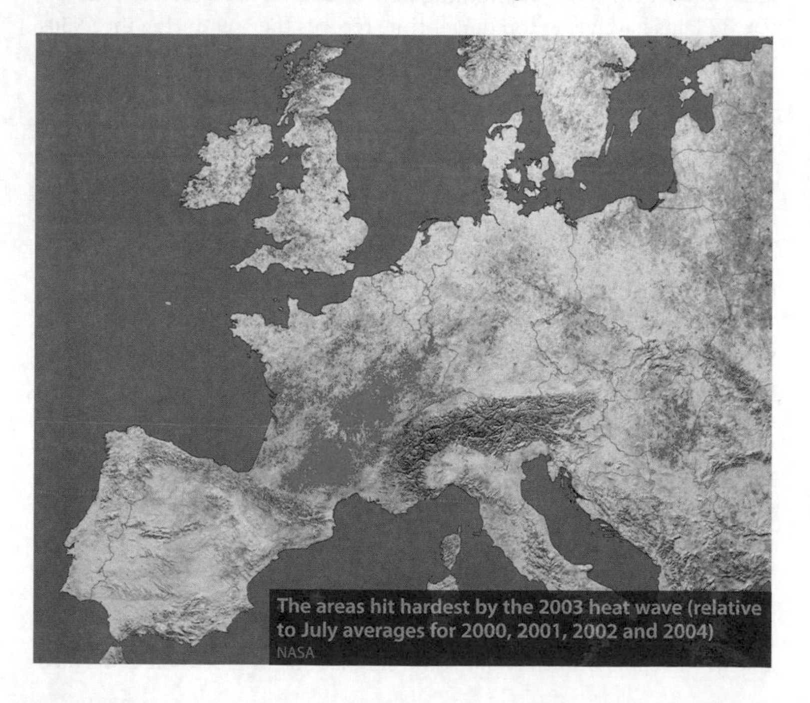

The areas hit hardest by the 2003 heat wave (relative to July averages for 2000, 2001, 2002 and 2004)
NASA

And cool nights act as a critical safety valve, giving people a physiological and psychological break from the intense daytime stress of a heat wave. If the nights warm up too much, that safety valve goes away.

"I hate every single second of this weather."

UK resident Pauline Smith quoted by the BBC during the 2003 heat wave

The 2003 disaster in Europe brought home these points vividly and poignantly. The stage was set in May, as sea-surface temperatures across the Mediterranean soared to record levels. One stretch of serious heat spanned the latter part of June, and temperatures soared again during the last three weeks of July. These were major heat waves in their own right, causing hundreds of deaths and plenty of suffering. But the truly exceptional period was the first two weeks of August. Day after day, large swaths of the continent topped 35°C (95°F) and found themselves hard-pressed to cool below 20°C (68°F). Such temperatures might not sound like a big deal to an American from the Sunbelt. But most of Europe's urban areas lack the ubiquitous air conditioning of the hottest US cities, and certainly most Europeans aren't accustomed to dealing with weather like this.

Among the history-making statistics from 2003:

▶ The 10th of August was **London's** hottest day on record, with highs of 37.9°C (100.2°F) at Heathrow Airport and 38.1°C (100.6°) at the Royal Botanic Gardens. Not to be outdone, Faversham, Kent, climbed to 38.5°C (101.3°F) that same afternoon. It was the first day in nearly 300 years of record-keeping that any place in Britain had topped the 100°F mark.

▶ All-time national records were also broken with 40.4°C (104.7°F) in **Roth, Germany;** 41.5°C (106.7°F) in **Grono, Switzerland**; and a scorching 47.3°C (117.1°F) in **Amareleja, Portugal**.

▶ Germany notched its warmest night on record, with the weather station at **Weinbiet**, a mountain near the Rhine, recording nothing lower than 27.6°C (81.7°F) on the 13th.

Numbers can't do full justice to the impact of this atmospheric broadside on society. In **England,** railroad tracks buckled, tube stops resembled ovens, and the London Eye and its glassed-in Ferris-wheel pods had to shut down. Schools and offices closed for "heat days" in parts of Germany. But perhaps no place suffered so acutely as did **Paris**. With little air conditioning and block after block of tall, closely-packed stone buildings, the French capital – ill prepared for weather that was more suited to Paris,

Texas – became a literal heat-trapping machine. The city ended up falling a whisker short of its all-time high of 40.4°C (104.7°F), which was set in 1947, but it did endure nine consecutive days that topped 35°C (95°F). Even worse for many were the nights of surreal heat. On 12 August, the temperature never dropped below 25.5°C (77.9°F), making it Paris's warmest night on record. The early-August heat was so sustained that it kept the temperature within the intensive-care unit of one suburban Paris hospital, which lacked air conditioning, at or above a stifling 29°C (84°F) for more than a week.

The human cost of heat

Although it was clear in early August 2003 that people were dying in large numbers across Europe, it took several weeks to start getting a handle on the heat's human cost. By September more than 20,000 Europeans were listed as casualties of the summer of 2003, including at least 13,000 who died in France during that awful first half of August. The summer's toll continued to mount over time: some 900 dead in England, more than 1300 in Portugal, as many as 8600 in Spain, up to 4600 in the Netherlands, more than 1000 in Germany and nearly that many in Switzerland. It hardly made the news when, in 2005, authorities in Italy abruptly raised their nation's toll from 8,000 to 20,000, pushing the continent-wide toll well above 50,000. No heat wave in global history has produced so many documented deaths. Even the horrific heat-driven fires across Greece in summer 2007, which killed at least 67 people, fell far short of 2003's human toll.

How could thousands of heat victims go uncounted for more than a year? The answer lies partly in the way mortality statistics are compiled and partly in the way heat waves kill. Many people die indirectly in heat waves – from **pollution**, for instance (see box), or from pre-existing conditions that are exacerbated by the heat – so it's not always apparent at first that weather is the culprit. And unlike a tornado or hurricane, a hot spell doesn't leave a trail of photogenic carnage in its wake. People tend to die alone, often in urban areas behind locked doors and closed windows. The piecemeal nature of heat deaths over days or weeks makes it hard to grasp the scope of an unfolding disaster until the morgue suddenly fills up.

Furthermore, it often takes months or even years for countries to collect and finalize data from the most far-flung rural areas, where many older Europeans live. That was the case in Italy: together, the major cities of **Bologna**, **Milan**, **Rome** and **Turin** reported about 3000 heat-related

Pollution: heat's hidden partner in crime

Some victims of heat waves die not because the air is so warm but because it's so dirty. The sunny, stagnant conditions prevalent during heat waves make an ideal platform for the sunlight-driven processes that create **ozone** – which is a godsend in the stratosphere (see p.26) but a dangerous pollutant at ground level. Ozone irritates the lungs and makes them more vulnerable to other chemicals. Moreover, the relative calm of a heat wave allows tiny bits of heavy metals, sulphates, nitrates and other substances to accumulate in the air. These are often grouped into the classes **PM10** (particulate matter smaller than 10 microns or 0.0004 inch) and **PM2.5** (particles smaller than 2.5 microns). The smallest of these particulates can sneak past the body's natural respiratory filters, causing a variety of lung problems and raising the risk of heart attacks.

After focusing on other pollutants for decades, scientists have only recently learned how deadly ozone and fine particulates can be. The World Health Organization (WHO) has estimated that mortality goes up by 0.3% during low-level ozone episodes. The WHO's Global Burden of Disease project estimates that 100,000 deaths a year in Europe may be related to fine particulates, with US fatalities estimated in the tens of thousands annually. The growing megacities of the developing world are at particular risk. One study led by US scientists Mario and Luisa Molina showed that a 10% reduction in fine particulates in Mexico City's air could save roughly one thousand lives a year.

Several studies have linked a substantial fraction of 2003's heat-wave deaths to ozone and particulates. A group of British epidemiologists and atmospheric scientists concluded that 21–38% of the UK deaths classified as heat-related could instead be attributed to ozone and PM10. Writing in the *Swiss Medical Weekly*, analysts at the Universities of Basel and Bern connected ozone to 13–30% of Switzerland's 2003 heat fatalities. And three scientists with the Dutch government suggest that around 40% of that country's heat-attributed deaths were triggered in roughly equal numbers by ozone and PM10. If there's an upside to these numbers, it's that a concerted effort to reduce ozone and particulate pollution might help save many of the people who die in the worst heat waves.

Road pollution in Mexico City
Stephanie Maze/Corbis

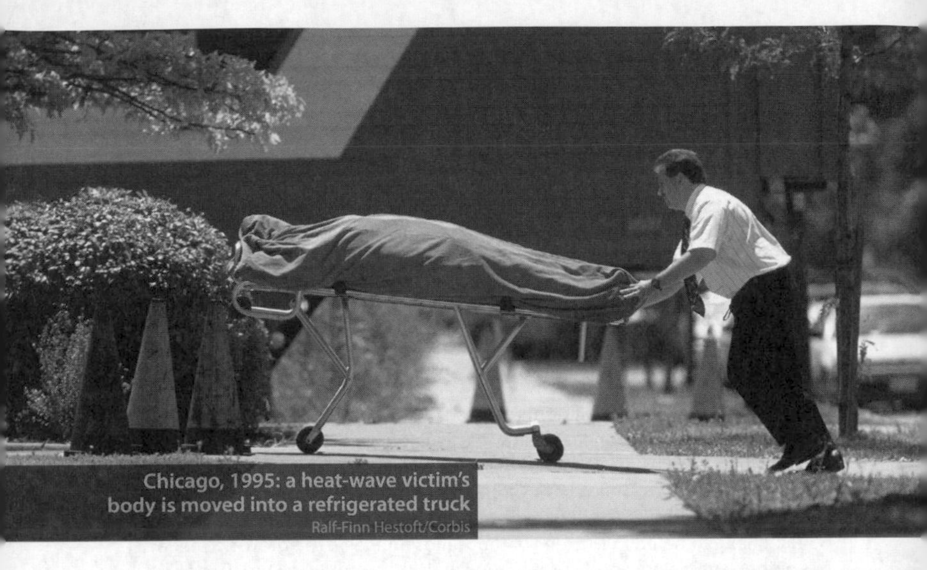

Chicago, 1995: a heat-wave victim's body is moved into a refrigerated truck
Ralf-Finn Hestoft/Corbis

deaths early on, but figures for the rest of the country – where the bulk of the deaths occurred, bucking the usual urban trend – were much slower to arrive. (Spain also reported more heat-related deaths in rural than in urban areas.)

On top of this, heat is a selective killer. It targets the very young and especially the very old, people whose metabolisms can't adjust readily to temperature extremes. In Rome, more than half of the 700 deaths attributed to 2003's heat were among people older than 85. Because the elderly die so disproportionately in heat waves, it's tempting to assume that weather is simply claiming those fated to die shortly anyhow. If that were the case, however, then you'd expect mortality to dip below average in the months following a heat wave. Careful studies of major heat disasters have, in fact, shown that such dips typically account for only about 20% to 30% of the spikes above average, or what epidemiologists call the **excess deaths**, that were observed in the preceding heat. It thus seems that most victims of heat wave, even the elderly, die well before their time.

In France, the situation was exacerbated by local culture. The worst of the heat struck during the August vacations, when much of professional Paris heads out of town. Left behind to fend for themselves were many thousands of elderly Parisians, the ones most susceptible to heat-related health problems and least able to seek refuge in a cool location. As many as half of France's fatalities occurred at rest homes, few of which had air conditioning.

The nights Chicago fried

A little more wind in the Windy City might have been appreciated in the second week of July 1995. During four days of searing conditions, more than 700 people died from heat-related causes. The days were hot indeed – ranging from 37 to 41°C (98 to 106°F) at the in-town Midway Airport – but in this case, it truly was the humidity as much as the heat. Tropical amounts of moisture in the air, coupled with the heat-island effect (see p.175), helped produce overnight lows of 27°C and 29°C (81° and 84°F) on two consecutive nights. As in Paris, many of the city's older multistorey buildings lacked air conditioning. When children and teens sought relief by opening fire hydrants, police put a stop to it.

Chicago's heat-emergency plan sat on the shelf, and the city's mayor and other key officials remained on vacation until the disaster became dire. Then things really heated up – at least politically, according to New York University sociologist Eric Klinenberg, author of *Heat Wave: A Social Autopsy of Disaster in Chicago*. Mayor Richard Daley acknowledged the heat but asked people not to blow it out of proportion. The city's commissioner of human services blamed victims for not taking care of themselves. "They were often interpreted as individual failure", says Klinenberg of the Chicago heat deaths. "In Europe, the heat wave was immediately framed as a political event."

Chicago did learn from its calamity. In 1999, when another heat wave struck, the city's action plan included not only the usual warnings but also free bus rides to "cooling centers." Crews of city workers phoned and checked in person on elderly people living alone. Those steps helped keep the death toll to 110, a number that Klinenberg still finds too high. As global warming unfolds, he says, "we know that more heat waves are coming." And, he adds, dying from heat is not a nice way to go. "If you look closely at the police reports, or the medical autopsies, they're just horrific. These are isolated, lonely, painful deaths."

In Chicago, densely settled neighbourhoods with busy streets and public spaces, such as the heavily Latino Little Village, fared much better than areas where people were more disconnected from neighbours and had few places to gather, such as the African-American neighbourhood of North Lawndale. As Klinenberg sees it, heat waves are "invisible disasters that kill largely invisible people. Perhaps that's the reason we don't care enough about them."

Could a better **warning system** have saved lives in Europe? That's been the case in the United States, where most parts of the country can easily top 35°C (95°F) in a typical summer. After years of perfunctory heat advisories, issued by the US National Weather Service (NWS) but seldom heeded, the country was shocked into awareness by a 1995 heat wave in Chicago that killed more than 700 people (see box above). That same year, Laurence Kalkstein of the University of Delaware teamed up with the NWS to launch a new watch-warning system for extreme heat. The scheme also included intensive **neighbourhood interventions**, such as block captains assigned to check on neighbours. It's been estimated that Philadelphia's heat plan saved more than one hundred lives in four years.

As it happened, Kalkstein and Italian colleagues had just launched a US-style warning system in Rome when the 2003 heat wave struck. The city ended up declaring heat emergencies on eighteen days that summer. Even with its new plan in place, hundreds more Romans died than the scientists expected. In a post-mortem analysis, the team found that the heat was worse than forecast and, indeed, transcended anything they'd considered in planning their warning system.

The future of summer sizzle

On a broader scale, the European disaster of 2003 was just a blip – albeit a spectacular one – in a long-term warming trend that spans most of the world, particularly large parts of North America and Eurasia. While the planet as a whole has warmed nearly 0.8°C (1.4°F) since 1900, the rise has generally been greater over the Northern Hemisphere continents (especially at higher latitudes) than over the tropics. That simplified picture doesn't hold quite as well, though, when you go beyond the average temperature and look at various types of heat extremes, some of which have increased in the tropics as well as mid-latitudes.

It's taken a long while for scientists to verify how heat waves have changed globally over the last few decades, not to mention how they might evolve in the future. Why is this so? After all, the average worldwide temperature has been tracked for decades (see p.172), and individual stations keep close tabs on daily weather. Between the local and the global, though, it's surprisingly difficult to assess regional trends in temperature and precipitation extremes.

Nations have traditionally shared more general data, such as the average high and low temperatures for each month at major reporting sites. But knowing that the average August high in a given city rose, say, 1°C in fifty years doesn't tell you whether stretches of the most intense heat are becoming more frequent, intense or longer-lived. To find that out, you need a consistent way to measure heat extremes, and you need day-by-day data extending over decades. That's been hard for researchers to obtain. Some nations balk at releasing such data – considering it valuable intellectual property – and some aren't set up to provide the data easily.

Climate scientists made big headway on this problem from 2001 to 2005 through a series of regional workshops. These meetings provided a

What counts as a heat wave?

Half the battle in assessing climate extremes is simply deciding what to measure. What defines an extreme spell of heat, aside from a sticky shirt or a wilting garden? Experts have tried out a variety of indices to capture climate extremes. In the realm of heat, these include the following. Caution: these may take a minute or two to digest.

▶ **Absolute thresholds** The number of days that exceed a given temperature. While this is nicely concrete, one threshold doesn't always fit all. A week of July afternoons at 35°C (95°F) might feel miserable in London but normal in air-conditioned Houston. Thus, each location might need a different threshold, which makes it hard to compare heat intensity among locations.

▶ **Monthly maximums and minimums** Changes over time in the highest and lowest single temperature observed during a given month of the year. These can provide useful, easy-to-grasp illustrations of a shifting climate, like the all-time highs recorded during the 2003 Euro heat wave. However, a single day or night of record warmth doesn't necessarily correspond to the kind of sustained multi-day heat that causes major problems.

▶ **Threshold departures** The number of days when temperatures climb above average by a fixed amount, such as 5°C (9°F). This gives a more location-appropriate sense of how unusual a hot stretch might be. However, it doesn't acknowledge that one city might normally have more variability than the next. For example, a 5°C jump in Denver's dry climate, where temperatures can gyrate wildly from day to night or across a few days, would be less obvious than the same leap in Miami's sultry summers, where temperatures often change little from day to day.

▶ **Percentile departures** The number of days that land among the hottest of all days in that month's long-term record, based on percentage (the hottest 10%, 5%, 1%, etc). This index provides a tailored-to-fit measure of a heat wave's intensity, based on each city's unique characteristics. No matter where you live, a day that's among the warmest 1% observed in the past decade or century means something (especially if such heat were to start occurring 5% or 10% of the time rather than 1%). Some researchers combine this with a measure of duration – for instance, the number of consecutive days on which the temperature reaches a given percentile ranking of heat.

venue in which countries could freely share the daily-scale data needed to understand changes in heat waves. Also, scientists agreed on a few key indices of heat, cold and precipitation that were used in preparing the 2007 IPCC report. Emerging from this work is the most detailed picture yet of how heat waves and other climate extremes are evolving across the world. Key findings include the following:

▶ On average, the level of heat that occurred only one day in ten during the 1950s is now observed about 15% more often. The strongest such trends are across Saharan Africa, Russia, central Europe, southern Australia, western Canada and Alaska.

▶ The average range in daily temperature shrank significantly across nearly half of the globe from the 1950s to the 1980s, especially in eastern Asia and the central United States. This is largely because the rise in low temperatures was stronger and more widespread (about 1°C or 1.8°F globally since the 1950s) than the rise in high temperatures (closer to 0.75°C). In other words, nights have warmed even more quickly than days. Interestingly, for reasons unknown, there's been little change in the daily temperature range since the 1980s.

How much of 2003's heat can be pinned on greenhouse gases? The ever-growing heat islands of modern European cities no doubt played a role, and these are a form of climate change in themselves (see p.175). Yet the continent-wide, rural-plus-urban scope of the heat points to something more going on. Many campaigners trumpeted the heat wave as a classic example of global warming at work, but it was a year later before scientists had marshalled a statistical case for this accusation. The resulting study by Peter Stott (University of Reading) and colleagues at the University of Oxford is a landmark – one of the first to lay odds on the likelihood that global warming was involved in a specific weather or climate event.

Using a climate model from the UK Hadley Centre, Stott's team simulated European summers since 1900 a number of times, sometimes including the century's ramp-up in greenhouse gases and sometimes omitting it. The idea was to see how likely it was that such heat might occur without human influence. The authors found that human-induced greenhouse emissions made it roughly four times more likely that a summer like 2003 might occur. In the US, a team led by NOAA's Martin Hoerling used similar techniques to find that human-produced emissions have led to a fifteen-fold jump in the risk of the type of record-setting heat the US saw in 2006.

Stott and colleagues also extended their Europe study through the year 2100 using the IPCC's high-emission A2 scenario. If the model pans out, the *typical* European summer of the 2040s would be even warmer than was the extreme summer of 2003. But even if this outlook turns out to be true, do warmer summers necessarily mean more, or longer, heat waves? Climate modellers are now confronting that question. In 2004, Gerald Meehl and Claudia Tebaldi of the US National Center for Atmospheric

Research examined the worst three-day stretches of heat by location. Their modelling shows these periods intensifying across all of the United States and Europe over the next century, especially across the US South and West and the Mediterranean. The model also showed typical heat waves in Chicago and Paris lengthening by a few days.

Other studies are adding fuel to the fire, as it were. A compilation of data from eight global models that fed into the 2007 IPCC report found the average length of heat waves (defined as a stretch of at least five days with highs 5°C above average) increasing by the year 2100 by at least two days across nearly all of the globe's interior land areas. If the models apply the IPCC's lower-emission B2 scenario, which assumes a good deal of progress towards energy alternatives, then heat waves still lengthen substantially until at least 2050. The IPCC concluded that heat waves are likely to become longer, more intense and more frequent across most continents, especially in central Europe, the western US and eastern Asia.

A massive European Union project called **PRUDENCE** has produced more specific projections of future European temperature. PRUDENCE employed nine regional climate models nested within two global models, producing 50 simulations for the IPCC's A2 scenario. Among the most harrowing findings in the model consensus:

- The number of days per year that reach 30°C (86°F) in the Paris region jumps from the current 6–9 days to 50 days by the end of this century. The region's longest consecutive streak of such days in a typical year leaps from 3 to 19, a value more typical of present-day Sicily or Spain.

- Unlike the global trend found in the past century, the hottest 10% of days across Europe warms even more than the coolest 10%, especially across the heart of Europe (from France to Hungary), where drier summers take hold.

Many other heat-prone regions of the world have yet to be studied in the detail shown above. However, Europe is an unusually instructive case in point. Long tempered by its proximity to water, Europe now appears to be losing some of that protection, as huge domes of hot, dry air build over the continent more frequently and gain more influence in summer.

A wildcard in the Atlantic

There's one character waiting in the wings that could change this picture dramatically: a slowdown or shutdown of the **Atlantic thermohaline**

circulation, which helps keep Europe unusually warm for its latitude. This is the scenario made famous in the 2004 film *The Day After Tomorrow* (see p.275). Although evidence shows that the thermohaline circulation has ground to a halt more than once in climate history, it's believed that this process takes at least a few years to play out, and sometimes many decades, rather than the few days portrayed in the movie.

In any case, a complete shutdown of the thermohaline circulation isn't expected anytime soon, but the IPCC deems a slowdown "very likely" over the next century. The best guess from the most sophisticated computer models is that the circulation might slow by 10% to 50% over the next century, assuming greenhouse gas emissions continue unabated, with the model average around 25%. If this happens, the expected climate warming might be nearly erased across the United Kingdom and diminished across many other parts of Europe. However, summers could still be warmer and more drought-prone across the UK and Europe than they are now. For a full discussion of the thermohaline shutdown hypothesis, see p.120.

Handling the heat

In the long run, even if the world tackles climate change wholeheartedly over the next few years, Europe will clearly need to adapt to the risk of heat waves like the one it endured in 2003. Better warning systems will help; more air conditioning, and the associated cost, seems inevitable. The latter may put a dent in Europe's goal of reducing greenhouse emissions, though it's possible that some of the energy spent to cool the continent will be counterbalanced by a drop in the need for wintertime heating fuel. However, nobody knows how many of the poorest and most vulnerable of Europeans will simply be left to suffer through future hot spells in un-air-conditioned misery.

Poverty is certainly a major co-factor in heat deaths across the **developing world**. The people of India are long accustomed to spells of intense heat during the late spring, just before the monsoon arrives, when temperatures can soar well above 40°C (104°F) across wide areas. As is the case elsewhere, it's the extremes on top of that already scorching norm that cause the most suffering. Several pre-monsoonal heat waves in recent years have each killed more than one thousand people, many of them landless workers forced by circumstance to toil in the elements.

As for the United States, much of its Sunbelt, including the vast majority of homes and businesses in places like Atlanta and Dallas, is already equipped with air conditioning. Even as temperatures soar further, these regions may prove fairly resilient – as long as the power stays on. Older cities in the Midwest and Northeast appear to be more at risk of occasional heat crises, even though their average summer readings fall short of those in the South, where air conditioning is the accepted standard. Indeed, fatalities appear to be more common in places where intense heat is only an occasional visitor.

Floods & droughts

Two sides of a catastrophic coin

Although it's natural to think of temperature first when we think of global warming, the impact of climate change on precipitation may be even more important in the long run for many places and many people. Too much rain at once can cause disastrous floods, while too little can make an area unproductive or even uninhabitable.

Anyone who was caught in the epic 24-hour rainfall of 944mm (37.2") in Mumbai, India, in July 2005, or who experienced Britain's destructive summer rains in 2007, might wonder if the planet's waterworks are starting to go into overdrive. Full-globe observations of rainfall are difficult to obtain, in no small part because 70% of the planet is covered by oceans (scientists are currently hard at work trying to improve the data). Land areas *are* now believed to be getting about 1% more precipitation than they did a hundred years ago – an increase of about 1cm (0.4") on average – but this small boost (which isn't statistically significant) hides much larger variations in space and time. For the United States, the increase is closer to 5%, and the range across most mid- and high latitudes of the Northern Hemisphere is anywhere from 5 to 20%, according to the 2007 IPCC assessment. Meanwhile, rainfall across many parts of the tropics and northern subtropics (from 10°S to 30°N) is on the decrease, although recent satellite studies hint that some tropical locations are getting up to 5% more rain than they did in the late 1970s.

Scientists expect these trends to continue over the next century, with a further increase in rain and snow poleward of about 50°N (including most of Canada, northern Europe and Russia) and in tropical areas that benefit from monsoons. Meanwhile, rainfall is projected to decrease across the subtropics, including much of Australia, southern Africa, the

Mediterranean and Caribbean, and the southwest US and Mexico. While confidence in this general picture has become more robust, there's still uncertainty about just where the transition zones between increasing and decreasing precipitation will lie.

When it comes to floods, though, it's the character of the precipitation that really counts. How often is rain concentrated in short, intense bursts that produce flash floods, or in multi-day torrents that can cause near-biblical floods across entire regions? And how often is it falling in more gentle, non-destructive ways? Here, the global-change signal is worrisome. Data show a clear ramp-up in **precipitation intensity** for the US, Europe and several other areas over the last century, especially since the 1970s. When it rains or snows in these places, it now tends to rain or snow harder, over periods ranging from a few hours to several days. Projections from computer models suggest that this trend will intensify in the decades to come, although the details will vary from place to place and not every location will follow the trend.

A woman looks out on flood waters in Mumbai, India, in August 2004
Punit Paranjpe/Reuters/Corbis

There's a cruel twist to this story. You might expect droughts to diminish on a global basis as rainfall goes up. But higher temperatures not only allow more rain-producing moisture to enter the atmosphere – they also suck more water out of the parched terrain where it hasn't been raining. Thus, in addition to triggering more rainfall, global warming could also increase the occurrence of drought, a seeming paradox that already appears to be taking shape. One study in 2005 by the US National Center for Atmospheric Research found that the percentage of Earth's land area undergoing serious drought has more than

A more potent monsoon for India?

There's no better example of seasonality in rainfall than the Asian monsoon, which delivers life-giving water together with deadly flooding to over a billion people from India to Southeast Asia. The monsoon normally kicks into gear as the Indian subcontinent cooks in the intense heat of late spring, which helps pull in moist air from the nearby ocean just as a hot day by the seashore will stimulate a sea breeze. While the basic process is the same from year to year, the strength of a given year's monsoon can vary greatly. For instance, heavy winter snows over the Tibetan plateau can slow the process by which land heats up in the spring and pulls in summer monsoon moisture. Also, El Niño events tend to dampen the Indian monsoon's strength. Seasonal forecasters have gained skill at incorporating these and other factors into outlooks for each year's monsoon, although most Indians still wait on tenterhooks until the first raindrops arrive.

Since the last ice age, the monsoon has apparently strengthened and weakened in tandem with Earth's temperature. About 10,000 years ago, summer sunlight across the Northern Hemisphere was a few percent stronger than today, due to a cyclic variation in Earth's orbit (see p.196). The extra sunlight not only helped melt the ice sheets but also intensified the Asian monsoon. After that effect started to wane about 6000 years ago, the trend has been towards weaker monsoons during cooler periods and stronger monsoons when Earth is relatively warm. Oddly enough, the link is especially strong with the North Atlantic: when those surface waters are warm, it's a sign that the global conveyor belt of ocean circulation is vigorous (see p.120), and this appears to help keep the Indian monsoon humming as well.

Most of the 21 global models analyzed in the 2007 IPCC report show an increase in summer monsoonal rains across south Asia over the next century, averaging 11% and ranging as high as 23%. What doesn't fit this picture is the 5–10% drop in average monsoon rainfall observed across India since the 1950s, even as global temperatures have continued to warm. This drying may be due to the rapid growth of sulphate pollution across parts of India, according to a 2006 study by Chul Eddy Chung and V. Ramanathan of the Scripps Institution of Oceanography. The soot and other pollutants blowing offshore appear to be blocking enough sunlight to shift heating patterns across the Indian Ocean and in turn weaken the circulations that drive India's monsoon. If the pollutants persist or increase, the monsoons may tend to stay relatively weak; if not, then the monsoons could gain strength as global warming proceeds, which could bring an increased chance of summers with excessive rains, such as Mumbai's 2005 inundation (which was more than 50% beyond the city's previous record). Further inland, the Thar Desert and other parts of western and northern India may see a drop in precipitation. These trends, if they're borne out, would lead to a sharper contrast between wet and dry lands. And in any given year, there's always the chance of monsoon failure, a much-feared event that in the past has led to massive famines.

doubled since the 1970s. Where and how the worst droughts manifest hinges on a number of factors, including a tangled web of influence through which the world's tropical oceans can help trigger multi-year drought in places as far-flung as the southwest United States and southern Africa. Even in a single location, both drought and flood can become more common if the timing of rainfall shifts in not-so-fortunate ways.

A wetter world (but not everywhere)

The stakes are high when it comes to assessing the future of floods and droughts, because both are notoriously deadly and destructive. According to the World Meteorological Organization, flooding affected 1.5 billion people between 1990 and 2001. China's Yangtze River flows over its banks so reliably during monsoon season (its floods and related crop failures have claimed more than a million lives since 1900) that it's prompted the world's largest flood-control project, the colossal and controversial Three Gorges Dam. Many other flood victims die on a smaller scale, unlucky enough to be caught solo or in a small group along a rampaging creek during a lightning-quick flash flood.

Though drought is a less dramatic killer than flooding, its toll is even more staggering. Monsoon failures across India reportedly killed more than a million people each in 1900 and 1965–67. More than thirty million Chinese died in the first half of the twentieth century as a result of drought and related famine, according to the US National Climatic Data Center. The World Meteorological Organization estimates that more than a million people died throughout Africa's Sahel during 25 years of poor rainfall that peaked in the devastating droughts of 1972–75 and 1984–85 (see box on p.66). Southern and East Africa have also been prone to recurring drought over the last several decades. Looking further back, the history of civilization is chequered with cultures believed to have met their downfall due to drought (see The Long View, p.193).

The roots of both flooding and drought lie in the physical process known as evaporation. As global warming heats the world's oceans, the water molecules near the sea surface become more energetic and tend to evaporate into the atmosphere more readily. Thus, the air gains water vapour (or, as it's often put, the air holds more water vapour). This is a nonlinear relationship, meaning the effect gets stronger for each additional degree of warming.

Because the connection between warmth and evaporation is so well established, scientists have considered it a fairly safe bet that a warmer

England and the rains of autumn

You'd have to have been a contemporary of King Charles I in order to have seen Yorkshire's River Ouse higher (in 1625) than it was in autumn 2000. A series of punishing rains and floods swept much of the United Kingdom during October and November 2000, flowing into over 10,000 homes and prompting more than 1400 flood warnings. The total cost was of the order of a billion pounds. It was the UK's most widespread flooding since March 1947, when many areas were inundated by the melting of a six-week snowpack. The floods in 2000 came during the wettest autumn in UK's 330-plus years of record-keeping. Moreover, the November damage could have been far worse, according to the consulting firm Risk Management Services: "It is clear that a flood event of even a slightly greater magnitude would have overtopped and/or breached many defences, causing a significantly greater loss."

By themselves, the floods of autumn 2000 – or summer 2007 (see opposite) – can't be laid at the doorstep of climate change, but the character of rainfall in the UK is clearly changing. So says Hayley Fowler of the University of Newcastle upon Tyne. As part of a European Commission-funded project, she and Chris Kilsby analysed rainfall extremes at 204 UK stations in nine regions for the period 1961 to 2000. She focused on "return periods" – the amount of time one would expect to go between experiencing extreme rainfall or flooding of a given magnitude.

The project produced a cloudburst of attention-grabbing statistics. Across eastern Scotland, for instance, a ten-day rainfall of 150mm (5.9") could be expected every fifty years during the mid-twentieth century. But since 1990, it's now moved closer to every eight years. Other parts of the UK have also seen substantial changes, with many return periods in the north less than half what they were before 1990. Such dramatic shifts have come unexpectedly early, says Fowler: "This pattern of change is the same as that projected by climate models under global warming for the end of the twenty-first century."

In the past, the northern and western UK would tend to experience its worst floods in autumn, while the southern UK would get them in wintertime. But now, the worst floods in the south – including London and environs – seem to be occurring in the September-to-November time period. That's likely a reflection of the recent shift in extreme rainfall events towards autumn, says Fowler. Not only does this confirm the perceptions of many weather-savvy Brits, it also bodes ill for the future. "The change in timing appears to be contributing to the recent increase in flooding across the south of the UK", says Fowler. "This has huge economic and social implications, especially if these trends continue under global warming, as climate models suggest."

If the trends do continue, they may dovetail with projections from the UK Climate Impacts Programme, whose twenty-first-century outlook predicts wetter winters and drier springs, summers and autumns across the UK. However, UKCIP also expects autumn-to-autumn variability in rainfall to increase markedly across southeast England. Thus, it's possible that tranquillity could alternate with torrents from one October to another.

atmosphere would carry more water vapour around the globe. That's been confirmed by satellite observations since 1988 over the ocean, where the amount of water vapour has been rising at just over 1% a decade. The air over some land areas has shown even more of a jump: roughly 5% per decade from 1973 to 1993, according to one study that examined the United States, the Caribbean and Hawaii.

Rain and snow tend to develop where air is converging and rising. If the air is warmer and has a bit more water vapour, it ought to rain or snow more intensely, all else being equal. Based on the physics involved, Kevin Trenberth of the National Center for Atmospheric Research estimates that, overall, precipitation intensity should rise by about 7% for every 1°C (1.8°F) of warming. However, rainfall and snowfall often vary greatly over small distances. Thus, a small increase in global precipitation could mask regional and local trends that are more dramatic, downwards as well as upwards. To complicate the picture further, rain or snow totals at some locations can rise or fall sharply for a year or more due to the climate cycles discussed below.

As with temperature, it's the extremes that matter most in rainfall and the lack of it. The UK's soggy summer of 2007 vaulted into the record books largely on the strength of two extremely wet days: June 25, which produced 103.1mm (4.06") of rain in Fylingdales, North Yorkshire, and July 20, when 120.8mm (4.76") fell at Pershore College in Worcestershire.

Searching for insurance papers in a flooded home near Doncaster, England, in June 2007
Corbis

In both cases, the amount of rain that fell over 24 hours was roughly half of what one would expect during an entire summer. The resulting floods put large sections of northern and western England under water. Hundreds of thousands of residents lost power and access to drinking water, and damages were estimated at more than £2 billion.

To help quantify the links between downpours and climate change, the scientists who came up with a set of criteria for measuring temperature extremes (see p.53) have done the same for precipitation. Those indices are a key part of research that fed into the IPCC's 2007 assessment. In one of the first related studies, Povl Frich and colleagues from the Danish Meteorological Institute confirmed that precipitation extremes seem to be increasing in general, especially across the planet's higher latitudes. Highlights from Frich's 2002 study, which looked at the second half of the twentieth century, included the following:

- **The number of days that see at least 10mm (0.4") of rain or melted snow** rose by 5–15% across large stretches of the United States, Europe, Russia, South Africa and Australia. These locations also tended to see upward trends in the peak five-day totals of rain or snow in a given year.

- **Much of North America and Europe** showed jumps in the fraction of rainfall and snowfall that fell on the soggiest 5% of all days with precipitation. In other words, the wettest days became wetter still.

As noted in the 2007 IPCC report, a number of studies zeroing in on regional trends have confirmed this general picture. They've shown intense rainfall events on the increase in a variety of places, including the Caribbean, the western United States, the United Kingdom and Italy. However, as Frich found, there's a lot of hidden variability: even a country where precipitation is on the increase can have a few locations that buck the overall trend.

Are floods increasing?

It's surprisingly hard to compile a global picture of whether floods are becoming more frequent or intense due to climate change. In part, that's because the chain of events that leads from an unusually heavy rain to a flood involves many factors other than immediate weather: for example, how wet the region's soils already are, how high rivers and reservoirs are running, what kind of flood-control devices are in place and – perhaps most critically – how much the landscape has been altered by development. One of the few attempts at a global flood census was published in the science journal *Nature* in 2002. Led by Christopher Milly of the US Geological Survey, the study examined 29 of the world's largest river basins. It found that 16 of their 21 biggest floods of the last century occurred in the century's second half (after 1953).

Land-use changes play a huge role in flooding potential. Deforestation appears to exacerbate the risk of flooding and landslides in most cases, since the water that falls can flow more quickly when it's unimpeded by trees and undergrowth. Water flows particularly easily across the acres of pavement that are laid down as cities expand into the countryside. Faster-flowing water is especially likely to feed into small-scale flash floods, which are even harder to monitor and analyse than larger river floods.

Timing is another important factor when looking at how global precipitation has evolved. For instance, are the heaviest one-hour deluges getting heavier? This is virtually impossible to calculate across the globe, because many national meteorological centres don't compile such fine-scale data or won't release it without collecting a hefty fee. Also, the heaviest of heavy downpours tend to be so localized that they can be measured only with the help of the dense observing networks found mainly in highly industrialized countries. This leaves us in the dark as to changes in flash floods across poorer nations.

Seasonal timing is critical when it comes to both flooding and drought. For example, much of northern and central Europe as well as the United Kingdom are expected to see a pair of coexisting trends in the coming century: heavier wintertime rains as well as drier, more drought-prone summers. The high-summer months of July and August have been getting drier across the UK since the 1970s, though this doesn't prohibit downpours (as the flooding rains of 2007 showed). There are also intriguing changes elsewhere in the year across the UK (see box on p.62).

Rising carbon dioxide levels may also influence both flooding and drought through their impact on plants. In a 2007 *Nature* paper, a team led by Richard Betts of the UK Met Office's Hadley Centre found that the average worldwide runoff from rain and snow increased a further six percent in a model with doubled carbon dioxide affecting plants, above and beyond the effects of carbon dioxide on global warming. The additional increase is due to plants' natural tendency to close their pores in conditions of enhanced CO_2, thus allowing more of a given round of precipitation to remain in the soil and drain into rivers instead of being taken up by the plant. All else being equal, this could make flooding more likely while ameliorating some aspects of drought.

Which way will the Sahel go?

Only a few years after the promising 1960s, when many African nations were shaking off their colonial past, weather stepped into the role of oppressor. A prolonged period of low rainfall – which hit hardest in 1972–75 and 1984–85 – caused drought-driven despair across much of the continent, with the north-central Sahel region at the epicentre. The region's name is derived from the Arabic *sahil,* or shore, and it's an apt label. Each summer, like the tide coming in, monsoon moisture sweeps north from the Gulf of Guinea. It brings the only substantial rain of the year to the semi-arid belt of land that extends from Senegal to Sudan across the great breadth of Africa. Stretching more than 7000km (4400 miles), the Sahel separates the wet tropics to the south from the parched Sahara to the north. There's a lot of climatic contrast packed into this thin ribbon, which spans only about 800km (500 miles) from north to south. Across the width of this ribbon from bottom to top, the average rainfall drops from about 750mm (30") – more than in London – to about 250mm (10"), less than in Los Angeles.

It doesn't take much of a shift in the monsoon to put the region in jeopardy. Major droughts struck the Sahel in the 1910s and 1940s, but the dry spell of the early 1970s was particularly intense. At a time when the media were covering Africa more thoroughly than in past decades, and environmentalism was just taking wing, the world soon saw how millions of Sahelians were suffering. Drawn faces and swollen bellies on the evening news helped launch a mass outpouring of relief. Another widespread drought struck in the mid-1980s, this one extending east to Ethiopia. This time the world's response – though unconscionably delayed until the BBC captured the misery on camera – was even more resounding. Bob Geldof launched his second career as humanitarian, organizing the supergroup charity single "Do They Know It's Christmas?", with Michael Jackson, Lionel Richie and friends following suit in the US with "We Are The World". The first Live Aid concert took place soon after.

Given all this, it may seem inconceivable to claim that the Sahel drought never happened. Yet a 2004 paper by Adrian Chappell (University of Salford) and Clive Agnew (University of Manchester) made that very claim – at least for the meteorological event, if not for the sociological one. Chappell and Agnew argued that changes in weather stations between 1931 and 1990 had produced a perceived drop in the region's rainfall. The total number of rain-measuring stations plummeted from nearly 200 at the onset of the long dry spell to less than 50 by the 1990s. Some of the remaining stations had been moved, jeopardizing their climatic consistency. Chappell and Agnew factored out the lost stations, attributed much of the presumed rainfall drop to those stations, and concluded – using the remaining stations – that no long-term drying had actually taken place. Though they didn't address why so many crop failures and deaths had occurred, the implication was that cultural rather than climatic factors were to blame.

The paper triggered a prompt rejoinder from an all-star cast of US and British climate scientists, led by Aiguo Dai of the US National Center for Atmospheric Research (NCAR). Titled "The Recent Sahel Drought is Real", the article was

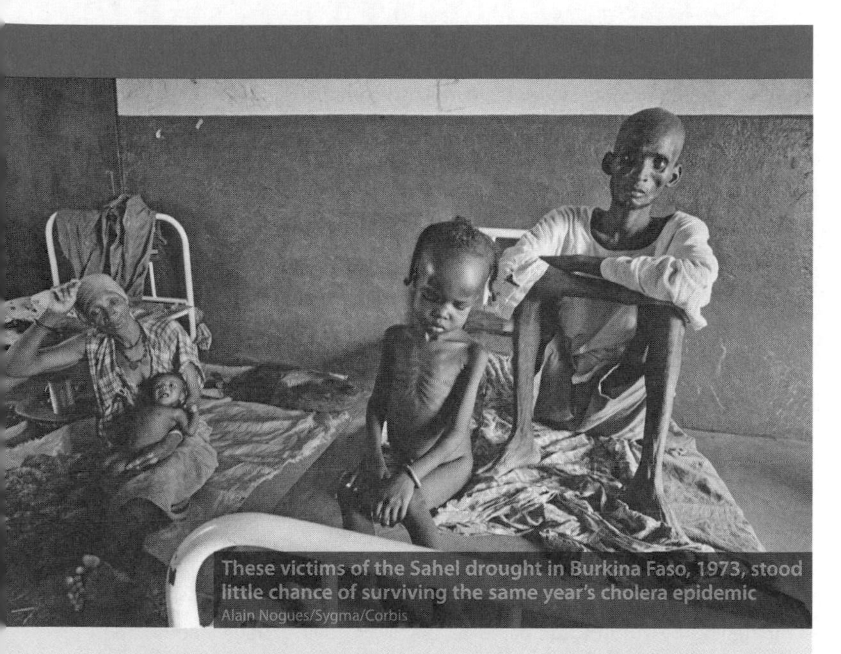

These victims of the Sahel drought in Burkina Faso, 1973, stood little chance of surviving the same year's cholera epidemic
Alain Nogues/Sygma/Corbis

peppered with unusually strong language. The authors called the Chappell-Agnew rainfall analysis "totally unrepresentative" and their conclusions about measuring drought in the Sahel and elsewhere "completely unfounded." Dai and colleagues found that the few Sahel weather stations that remained in place clearly showed the drying trend, and they noted that the loss of stations may itself be a telling sign of the drought's impact. In a fresh analysis spanning the years 1920 to 2003, Dai and his co-authors showed that rainfall during the worst of the drought, in 1984–85, fell nearly 30% short of the long-term average of about 500mm (19.7").

The Sahel's rainfall figures for recent years are both unsettling and encouraging. Though the region has moistened gradually, it's been a stop-and-go process. Some years (such as 1994, 1999 and 2003) have produced more rain than the pre-1970 average, but there have been notably dry years interspersed (occasionally triggered by strong El Niños, as in 1997–98). The outlook for Sahelian rain in the twenty-first century offers similarly mixed signals. In part, geography remains the Sahel's destiny: the region will stay poised between desert and wet tropics, and that location will inevitably produce whipsaws in year-to-year rainfall. If the variability goes up further, as is forecast for many parts of the world, then Sahelians could face even more challenges. The region's population is booming – it's projected to grow from fifty to a hundred million by 2050 – so the resulting stress on wood-based fuel sources and on food stocks will make the region even more vulnerable to climatic swings. The story of Lake Chad shows what can happen when climate and land use conspire. Once the sixth largest freshwater lake in ▶

the world, Lake Chad has shrivelled by 95% over the last forty years, a victim of the 1970s/80s droughts as well as intensive irrigation.

Despite these grim signs, there's some ground for optimism. Although many farmers and pastoralists have pushed into ever more marginal lands across the Sahel, there's also growing diversity in the ways land is used – a sign of resilience in a place where adaptation is essential – and more local involvement in development decisions. As for the climate outlook, when the 21 global models in the 2007 IPCC report are pooled, the average shows a 10–20% increase in Sahelian summer rains by 2080–2099 as compared to 1980–1999. However, there's a good deal of spread in the models, with everything from strong drying to a markedly wetter climate showing up in the mix. Some of the most encouraging recent projections come from James Hurrell (NCAR) and Martin Hoerling (US National Oceanic and Atmospheric Administration). In 2005 the two ran a series of eighty simulations using five different computer models. What emerged was a link between Sahel rains and the Atlantic. During the worst of the Sahel droughts, in the 1970s and 1980s, the North Atlantic surface waters ran cool relative to the South Atlantic's. That state of affairs reversed in the 1990s, as the North Atlantic began to warm more strongly. Hurrell and Hoerling believe the North Atlantic's warm-up helped pull monsoon moisture across the Sahel, bringing back the long-sought rains. Their simulations show further warming across the North Atlantic through at least 2050, with the Sahel moistening by 20–30% compared to the 1950–1999 average. The authors hasten to note that their results don't prove that greenhouse gases are directly driving the projected wet trend. In fact, paleoclimate data shows that even greater shifts in rainfall have occurred across this climatically precarious land in the distant past. From that angle, says Hurrell, "the recent African dryings appear to be neither unusual or extreme."

Defining drought

Anybody can see drought at work: wilting crops, bleached lawns, a receding lakeshore. Yet scholars have debated for many years how best to define this strange condition, the result not of the presence but of the absence of something.

Meteorological drought is the most obvious type: the rains stop, or at least diminish, for a given length of time. In places where it normally rains throughout the year, such as London or Fiji, you might see regular rain and still find yourself in a meteorological drought if the amounts don't measure up. One way to gauge a drought in such places is to add up the cumulative rainfall deficit over some period of time (eg 30, 60 or 90 days). Another would be to calculate how long it's been since a single day produced a given amount of rain. In semi-arid or desert locations, such as Melbourne or Phoenix, you could measure how many days it's been since it's rained at all. No matter where you are, it's the departure from average that makes the difference, since flora and fauna adapt to the long-term norms at a given locale. A year with 500mm (20") of rain would be disastrously dry for Tokyo but a deluge in Cairo.

Since drought affects us by drying up water supplies and damaging crops, experts now classify drought not only by rainfall deficits but also in terms of **hydrological** and **agricultural drought**. How low is the water level in the reservoir that your city relies on? Are the lands upstream getting enough rain or snow to help recharge the system? These factors, and others, help determine if a hydrological drought is under way. Especially when a dry spell first takes hold, it may take a while for rivers and lakes to show the effects. Likewise, long-parched ground may soak up enough of a drought-breaking rain to keep the hydrologic system from responding right away. Thus, hydrological drought often lags behind meteorological drought, a process that may unfold over a few weeks in a fairly moist climate or across months or even years in a semi-arid regime.

Crops can be affected in even more complex ways by drought, making a declaration of agricultural drought highly dependent on what's normally planted in a given location. If a dry spell is just starting, the topsoil may be parched but the deeper subsoil still moist enough to keep an established crop going. Conversely, a long-standing drought may break just in time to help germinating plants get started, even if the subsoil remains dry.

What do oceans have to do with drought?

Oceans might seem to be the last thing that would control the onset or departure of a drought. But the more scientists learn about the 70% of Earth's surface that's covered by water, the more they learn about how oceans can dictate the amount of water falling on the other 30%. And as greenhouse gases heat them up, the oceans may start issuing their droughty decrees in hard-to-predict ways.

Some parts of the world are dry by virtue of their location. Most of the planet's major deserts are clustered around 30°N and 30°S, where sinking air predominates, while rising air keeps most of the deep tropics between 15°N and 15°S quite rainy on a yearly average. The fuel for those showers and thunderstorms is the warm tropical water below. Should a large patch of ocean warm or cool substantially, it can take months or years for it to return to more typical temperatures. During that time, the anomalous water can reshuffle the climatic cards, bringing persistent wetness or dryness to various parts of the globe.

The drying of southern Australia

Water – and the lack of it – are key elements in the psyche of Australia. The unofficial national poem, Dorothea Mackellar's *My Country*, extols a land "of drought, and flooding rains." As a severe dry spell raged in 1888, Australian Henry Lawson bemoaned his fate in verse: "Beaten back in sad dejection, after years of weary toil / On that burning hot selection where the drought has gorged his spoil."

With much of the continent at the mercy of wild swings in precipitation, many Australians have viewed the southwestern coastal belt, roughly from Perth to Cape Leeuwin, as a corner of relative climatic sanity. One of the world's most biodiverse areas, it features a Mediterranean climate with hot, dry summers and mild, dependably damp winters. A 1920 book called *Australian Meteorology* noted the region's reliable moisture: "Here the rains rarely vary 10% from their average amount, and the lot of the farmer should be a happy one."

The farmers – and the urban water managers – haven't been so happy lately. Rainfall across far southwest and eastern Australia has declined notably over the last half-century. (It's also become wetter across the sparsely populated northwest, though the cause of that isn't entirely clear.) Since the mid-1970s, wet-season rainfall around Perth has consistently run some 10–15% lower than before, most noticeably in the late autumn and early winter. These days a wet year in Perth is one that merely reaches the long-term average of 869mm (34.2"). As of 2005, the city had gone 38 years without mustering a yearly total of 1000mm (39"), a mark once reached regularly. For a booming city with 1.5 million thirsty people, those are scary statistics.

The story is little better across the southeast, Australia's most populous corner. Places like Sydney – unlike Perth – do get a substantial amount of summer rain in the form of showers and thunderstorms. However, new analyses for the last fifty years show that even warm-season rains have slackened slightly across far southeast Australia. Intense drought has dogged the region for most of this decade, and relentless heat has helped deplete water for cities and agriculture even more than the rainfall deficit might suggest. In mid-2007, water supplies for both Melbourne and Sydney stood at less than half of capacity. The intensifying crisis helped push long-sceptical prime minister John Howard to publicly acknowledge the reality of climate change.

How much of Australia's drying is related to greenhouse gases? And will it continue? Long-term computer simulations hint that natural variability can push the region into and out of multi-decade dry spells. Like much of Australia, the southwest corner tends towards dryness during El Niño years, so the increase in El Niño's frequency and strength since the 1970s may be a factor. Furthermore, much of the countryside was deforested after European settlement, and some models show that the resulting landscape evaporates less moisture into the air, perhaps exacerbating drought. Even with all that in mind, however, the last few decades of drying do bear some of the fingerprints of global warming, especially for southwest Australia. Scientists have verified a poleward shift in the storm track that girdles the Southern Ocean, encircling Antarctica. The low-pressure centres

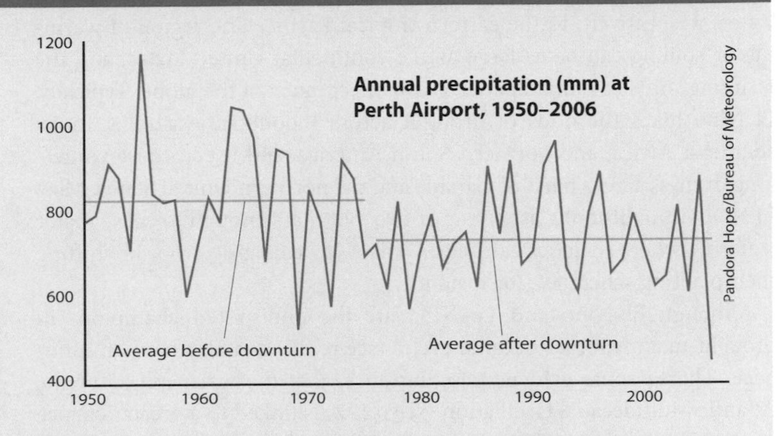

that race eastward along this track are the source of nearly all of Perth's rainfall, and they also provide important cool-season rains in Melbourne and Sydney.

Global computer models tend to agree that the storm track's winter position is likely to shift even further south as the century unfolds (see the Southern Annular Mode discussion, p.116). This could mean real trouble. Australia's Bureau of Meteorology and the nation's Commonwealth Scientific and Industrial Research Organization employed an ensemble of 23 computer models for a major set of climate projections released in 2007. The predictions showed an increased risk of further warming and drying across all but northernmost Australia in the coming century. Some models indicated winter and spring drying of up to 10–15% across southern Australia by 2030. Warming temperatures will only add to the water woes, helping evaporate what rain does fall before it can recharge water supplies.

In Perth, the dry shift has already wrought major impacts on water systems. Between 1997 and 2005, area stream-flows dropped 30% below those observed in the previous 23 years, which were themselves on the dry side. Faced with a deepening crisis, the city responded in 2005–06 by launching Australia's first major desalination plant in the industrial town of Kwinana, about 25km (16 miles) south of Perth. The largest such plant in the Southern Hemisphere, it uses reverse osmosis to render water from the Indian Ocean drinkable. The plant now supplies an estimated 17% of Perth's water needs; a second one is on order, with similar plants now in the offing for Sydney and Melbourne. Perth's plant is about as climate-friendly as they come. Powered by a nearby wind farm, it's being touted as the world's largest such facility to get its juice from renewable energy.

> **"We are trying to avoid the term 'drought' and saying this is the new reality."**
>
> **Ross Young, Water Services Association of Australia**

The supreme examples of such an ocean-driven shift are El Niño and La Niña (see p.118), which bring warmer- or cooler-than-normal surface waters, respectively, to the eastern tropical Pacific. The region of warming or cooling can be as large as the continental United States, and the resulting shifts in rainfall patterns can affect much of the globe. Typically, El Niño hikes the odds of drought across Indonesia, Australia, India, southeast Africa, and northern South America, and it can cause winter-long dryness across parts of Canada and the northern United States. Most El Niño droughts only last a year or two, but that's enough to cause major suffering where societies can't easily adapt (by changing crops or shifting their planting schedules, for instance).

Although El Niño and La Niña are the undisputed champions in drought-making, other oceanic cycles (see p.119) can also have an influence. There's some evidence, for instance, that the warm phase of the Atlantic Multidecadal Oscillation (see p.122) is linked to a greater chance of drought in the US Southwest. Similar cyclic warmings and coolings across the Pacific and Indian Oceans may play a role in sustained droughts that can parch regions such as the US Southwest and south Asia for years on end.

On top of these natural waxings and wanings, there's climate change. Warmer temperatures should help promote more evaporation from dry areas and thus create a tendency for droughts to be somewhat more severe where they do occur. Beyond that, the picture gets fuzzy, because it's exceedingly difficult for climate models to replicate the interwoven

Dry reservoir in Colorado
Carlye Calvin

network of ocean cycles and how they might evolve in a warmer climate. One of the most dramatic signals of the last few years has occurred over the Indian Ocean, where NASA satellite data show average surface water temperatures rose by around 0.25°C (0.45°F) from 1992 to 2000 – a substantial warming for this large a body of water. The NASA study shows that trade winds across the south Indian Ocean weakened enough to reduce up-welling and cooling currents by up to 70%. Should the Indian Ocean continue warming over the next century, as it does in some climate simulations, it may trigger persistent drought in some adjoining areas, particularly southern Africa. To the east, there's also some evidence that a quasi-permanent El Niño-like state could set up in the central Pacific, shifting the current zones of recurrent drought somewhat. All in all, it seems there's much yet to learn about how ocean-modulated droughts will unfold as the world warms.

The plough and its followers: farming and rainfall

Perhaps no other part of weather and climate humbles human ambition as does rainfall and the lack of it. For centuries we've flattered ourselves by thinking we can control the atmospheric tap, whether it be through rainmaking rituals or cloud seeding. Indeed, one of the first ways in which Western thinkers conceptualized human-induced climate change was the notion that "rains follow the plough" – an interesting spin-off from the manifest-destiny mindset that led to US and Australian expansion in the 1800s.

As Europeans moved into progressively drier parts of North America and Australia, they believed that cultivating the prairies and bushland – watering the crops and thus moistening the air above them – would help bring rainfall. Occasional setbacks, like Australia's vicious drought of 1888, didn't quash the belief at first. It wasn't until the 1930s, when unprecedented multi-year drought created the Dust Bowl across the US heartland, that farmers and land managers began to rethink their relationship to climate.

The flipside of this faith that humans could produce a rainy regime via agriculture was the belief that improper use could dry out the land permanently. The notion of **desertification** got its start in Africa's Sahel (see p.66), where it was postulated by colonial explorers as far back as the 1920s and long supported by actual events. During moist years, government policy in some African nations had encouraged farmers to cultivate

NOAA

northward towards the desert's edge. In turn, that pushed nomadic farmers and their herds into even more marginal territory further north. When the drought hit, both grazing and farming were too intensive for their land's now-fragile state. Wind and erosion took hold, and the zone of infertile land grew southward. At the time, it was accepted by many that the disaster was being produced, or at least perpetuated, by its victims.

Now it seems, researchers are coming around to the idea that neither drought nor rain may follow the plough in a predictable way. It stands to reason that both dry and wet conditions have a certain self-sustaining quality. All else being equal, when the ground is soaked, the air above it is more likely to produce further rain than when the ground is dusty. But the big question is, how powerful is this process against the vast energy reserves of the oceans and the natural variations of weather?

Thus far, there's no consensus on exactly how much of a difference land use can make. In any event, drought-prone areas can take simple, useful steps to reduce their vulnerability. Some of the techniques that proved helpful in the US Great Plains, such as wind-breaking belts of trees, were successfully adapted to the Sahel. Prudent land management can also help people get the most benefit out of the rain that does fall. At the same time, a bit of modesty is in order: oceanic shifts and other natural processes appear to be able to produce drought in some cases no matter how much we do to keep it away.

The big melt

Climate change in overdrive

If there's any place where global warming leaps from the abstract to the instantly tangible, it's in the frigid vastness of the Arctic. Here, the ice is melting, the ground is heaving, plants and animals are moving, and people are finding themselves bewildered by the changes unfolding year by year. Due to a set of mutually reinforcing processes, climate change appears to be progressing in the Arctic more quickly than in any other region on Earth.

Not that some Arctic residents haven't had to adapt to climate change before. Signs of Stone Age life dating back nearly 40,000 years – well before the last Ice Age ended – have been found near the Arctic Circle in northwest Russia. What appears to be different this time is the pace of the change and the sheer warmth now manifesting. Long-time residents of western Canada and Alaska have seen wintertime temperatures climb as much as 4°C (7°F) since the 1950s. That's several times beyond the average global pace. Across parts of the Arctic, recent summers have been the balmiest in at least 400 years.

Some of the permafrost that's undergirded Alaska for centuries appears to be thawing. Sea levels are slowly rising across the Arctic, as they are elsewhere around the globe, and longer ice-free stretches are increasing the risk of damaging, coast-pounding waves. And experts believe the twenty-first century will likely bring a summer when the Arctic Ocean is entirely free of ice for the first time in nearly a million years. As we'll see, these changes threaten to reorder Arctic ecologies in dangerous and unpredictable ways. On the other side of the globe, there are signs that Antarctica's enormous ice sheet is building slowly in the interior but eroding dramatically along some of its fringes.

"The defining feature of the present Arctic system – permanent ice – is almost gone."

Jonathan Overpeck, University of Arizona

The big melt isn't limited to the poles. Many glaciers in the highest mountains of the tropics and mid-latitudes are receding at a startling pace. It's tempting to chalk this up to a warmer atmosphere, but in fact drying may be a bigger culprit than warming in the erosion of glaciers on some tropical peaks, including the one best known to many: Africa's Kilimanjaro (see p.96).

On thin ice

The North Pole sits at the heart of an ocean ringed by the north fringes of eight countries: **Canada**, **Denmark** (via **Greenland**), **Finland**, **Iceland**, **Norway**, **Russia**, **Sweden** and the **United States**. All this adjoining land provides plenty of toeholds for **sea ice**, which overspreads the entire Arctic Ocean and portions of nearby waters each winter. This ice takes on many roles: a home base for polar bears, a roadbed for people, and a shield for coastal towns and underwater creatures.

In the summer, as warm continental air sweeps poleward from North America and Eurasia, the ice releases its grip on land and contracts into the Arctic Ocean. Much of it then disintegrates into a messy patchwork that allows icebreaker cruise ships to penetrate all the way to the North Pole (see box). Eventually, winter returns, and by February the Arctic Ocean is once again encased in ice.

Nobody expects the Arctic to completely lose its wintertime ice cover, but summer is another matter. As of 2007, the coverage in September (when the ice is usually at its minimum) is less than 80% of what it was in the 1970s, according to the US National Snow and Ice Data Center. Even the wintertime ice extent is shrinking by close to 4% per decade, with the depletion becoming especially rapid since 2003. Satellite images paint a

Minimum Arctic sea ice in 1979...

...and 2005

NASA

Where did that North Pole go?

A group of Arctic tourists got less than they bargained for in the summer of 2000 when they headed for the North Pole aboard an icebreaker ship. Planning to set foot on the ice at 90°N, they made it to the pole – and found only a few scattered floes strewn among the cold water. The group's unnerving experience soon made it onto the front page of *The New York Times*. A paleontologist and a zoologist aboard the ship were quoted beneath the headline "Ages-Old Icecap at North Pole is Now Liquid, Scientists Find." Climate experts hastened to point out that winds and summer warmth can easily push open sections of ice as far north as the pole in a typical summer. But the image of an unnaturally soggy North Pole persisted, a symbol of the very real Arctic melting that's been gaining momentum over the last few years. Extreme swimmer Lewis Gordon Pugh called on the archetype in 2007 when he swam a kilometre amid chunks of ice at the pole to draw attention to climate change.

stark portrait of these trends, which have reached truly dramatic proportions (see graphic). The summer of 2007 smashed all prior records, with even long-experienced ice watchers surprised by the pace of melt. By that September, the ice extent was around just over four million square kilometres, more than a million square kilometres less than its previous low, set in 2005. Only two-thirds of the usual late-summer pack remained, and ice-free waters stretched from Alaska and Siberia nearly to the North Pole. For the first time in memory, the fabled Northwest Passage above Canada was clear enough of ice so that a typical ocean-going vessel could have sailed directly from Europe to Asia unimpeded.

The Arctic's sea ice isn't just contracting horizontally; it's getting shallower as well. A batch of data on ice thickness was collected by US nuclear submarines at the height of the Cold War and declassified in the late 1990s. A comparison to more recent data revealed that sea ice had thinned by up to 40% over the intervening forty years. Where the ice in some spots had averaged 3m (10ft) thick, it now extended down to only about 2m (6.6ft). The US submarines traversed only a few select routes below the ice, and it's quite difficult to track the thickness more widely on an ongoing basis, but the average appears to be down by as much as 15% since the 1960s. Moreover, the widespread summer melt is eliminating much of the thick multiyear ice, making for a thinner winter coat that's easier to erode when the sun returns.

"The rules are starting to change and what's changing the rules is the input of greenhouse gases."

Mark Serreze, US National Snow and Ice Data Center

What makes the Arctic so vulnerable?

Trouble seems to beget trouble when it comes to far-northern climate. As a team of scientists noted in the journal *Eos* in 2005, "The Arctic system balances on the freezing point of water. Each summer, the system swings toward the liquid phase; each winter, it returns to the solid phase." While a tropical city might warm a degree or two without obvious effects, the same amount of warming could transform parts of the Arctic, were it to bring the system from just below to just above the freezing point. If the tropics are the lumbering long-haul trucks of the climate world, responding slowly but powerfully to greenhouse gases, the Arctic is a finely tuned sports car, reacting sensitively to even small changes.

Much of this responsiveness is because of **positive feedbacks**: self-reinforcing processes that tend to amplify change. In its definitive 2005 study, the Arctic Climate Impacts Assessment (ACIA) noted several positive feedbacks that allow the Arctic to warm more quickly than other parts of the globe in response to a given increase in greenhouse gases. These include:

▶ **A darkening surface** Several kinds of change in the Arctic all help to produce a darker surface that absorbs more sunlight and, in turn, warms the air above it more strongly. Open ocean typically absorbs more than 90% of the solar energy reaching it, while snow and ice absorb as little as 10%, reflecting the rest of the sunlight off their bright white surface. Thus, a patch of ocean that remains unfrozen can soak up as much as nine times more solar energy. Likewise, when snow cover over land disappears for longer periods, the exposed tundra absorbs more of the sunlight than the snow did. Shrubs and forests, in turn, absorb more solar energy than the lighter tundra does, so as these larger plants migrate northwards (see p.85), they act to further warm the region. (This is in contrast to the tropics, where trees tend to cool the climate as they catch rainfall and evaporate it back into the air.) And pollution wafting into the Arctic from more populated areas helps darken some of the snow and ice, increasing its absorptivity. About a third of the soot in the Arctic atmosphere arrives from South Asia, according to a 2005 study by James Hansen (NASA), and a 2007 study led by Charles Zender (University of California, Irvine) found that the effects of Arctic soot may explain a third or more of the region's warming since preindustrial times.

▶ **A more direct route to warming** As the Sun passes over the tropics, much of its energy goes into evaporating water, leading to the sultry air we associate with the Caribbean or Indonesia. But the evaporation rate in the frigid Arctic is far less. Thus, of the sunlight that reaches the Arctic, a higher percentage goes directly into warming the air.

▶ **A thinner atmosphere** The Arctic's troposphere, or "weather layer", typically extends about 8–10km (5–6 miles) up, compared to 16–18km (10–11 miles) in the tropics. In the shallow troposphere of the poles, it takes less energy to produce a given amount of warming. There's another wrinkle as well. Especially during clear, calm cold spells, Arctic landscapes often experience

Katy Williamson/DK images

a thin, ground-hugging layer of air (an inversion) that's much colder than the air just a few hundred metres higher up. When the inversion breaks up, the surface air can warm dramatically. Thus, anything that punctures inversions or makes them form less often – such as a change in atmospheric circulation, or a loss of snow cover – could produce major ground-level warming.

Potential changes in **heat-trapping clouds** aren't a feedback effect per se, but they do represent a huge uncertainty in polar regions – and elsewhere across the globe (see p.185). US satellites show that, between 1982 and 1999, Arctic cloud cover increased in spring and summer by 5–10% and decreased in winter by about 15%. Overall, these changes point to an increased cooling effect over time, because the low clouds and fog of Arctic summertime screen out the midnight Sun, whereas clear skies in winter allow heat to easily escape to space. If these trends continue, they might help counter the positive feedbacks above. But for now, the Arctic's headlong warm-up seems to be crashing through any cloud-erected barriers.

With all this in mind, we might hope to find a few **negative feedbacks** – processes that tend to dampen change over time. One possibility is the vast flow of heat from the tropics towards the poles. If the Arctic continues to warm at a faster pace than lower latitudes, as expected, then the pole-to-equator contrast would be weakened and the flow of heat ought to slacken. However, those weakened winds will carry more moisture than before – and that's a warming influence, all else being equal.

All of these feedbacks, plus the tendency of glacial melting to accelerate (see p.89), point towards a potential cascade of effects – a so-called "tipping point", as it's been characterized by some scientists and journalists. In fact, there may be a number of tipping points, each with its own setting. If so, each point we pass makes it that much harder to avoid the others. Conversely, the sooner that greenhouse warming is stabilized, the less likely the Arctic will enter a period of runaway change.

How long can the Arctic Ocean hang onto its summertime ice? Until the last few years, the best consensus among top climate modellers was that the Arctic would retain at least some summer ice until at least 2080, perhaps into the next century. But as new data and model results come in, that consensus is fragmenting even more quickly than the ice itself. The breakneck pace of recent summertime loss has raised concern that the iceless-summer benchmark could be upon the Arctic waters far more quickly than scientists once thought. Some experts are now projecting the milestone to arrive by the 2030s or even sooner. This is no minor development: as far as anyone can tell, the Arctic hasn't been free of major ice cover at any point in the last 800,000 years. Of course, it's possible that the intense melting of the last few years is a short-term variation superimposed on a slower long-term trend. However, there's no obvious natural cause for the recent speedup.

People, animals and ice

So what's wrong with a few extra weeks of open ocean in a place that's usually locked tight in ice? Indeed, the prospect of ice-free summers has many industries and entrepreneurs salivating, especially those hoping to cash in on trans-Arctic shipping (see p.276). Along the Arctic coast, however, the increasingly tenuous state of the ice during autumn and spring complicates life for indigenous residents. Subsistence whalers and seal hunters use the ice sheets to venture beyond the coastline during the transition seasons. The timing of those seasons, and the quality of the ice, has now become less dependable. Also, the near-shore ice normally helps keep wind-driven waves well offshore during the violent weather of autumn and spring. But storms now smash into coastal towns unimpeded by ice well into autumn. The resulting erosion threatens the very survival of some Arctic settlements (see p.83).

The loss of sea ice has especially grave implications for **polar bears**. These white giants retreat to land-based dens by winter and then bring their hungry young across the thinning springtime ice to find seals. After the ice is gone, the bears spend their summers fasting on land, which has become a more stressful affair for both mothers and gestating cubs now that ice-free spells are lengthening. At Hudson's Bay, near the southern end of the polar bears' range, the summer fast is now a month longer than it was decades ago, and the average birthweight of cubs dropped 15% from 1981 to 1998. The WWF estimates that, if this trend continues, females may become too thin to reproduce by the year 2012.

> **"You don't have to be a polar scientist to see that if you take away all the sea ice, you don't have polar bears any more."**
>
> **Andrew Derocher, University of Alberta**

Greater extremes in precipitation (see p.58) are also taking their toll. Heavier and more frequent spring rains have collapsed some of the bears' dens, and fires that rage through drought-stricken boreal forests destroy permafrost and the dens lodged within it. In the end, though, it comes down to sea ice. Beyond a certain point, there simply won't be enough of it over a large enough area to sustain polar bears as they now live. In a 2007 study, the US Geological Survey estimated that two-thirds of Earth's polar bears will be gone by 2050. Unless they can quickly adapt to land-based life like their brown-bear cousins and predecessors – or unless Arctic warming can be slowed down – polar bears appear to be headed for extinction in the wild before the century is out.

Several species of **seal** are also at serious risk from ice loss. **Ringed seals**, the Arctic's most common type, are the main food source for many Inuit communities as well as for polar bears. In spring, the seals nurse their pups in lairs atop the snow-coated ice sheet, so the stability of springtime ice is critical for them. A recent survey of ringed seal populations near the northwest coast of Hudson's Bay showed a seesawing pattern, with the numbers dropping in the 1970s and 1990s but rising in the 1980s, in

Polar bear scrounging at Barrow
Gary Braasch

Move or drown: one town's tough choice

Poised at a geographic crossroads, Shishmaref, Alaska, also lies in the crosshairs of a warming climate, earning the village of six hundred people a modest measure of fame. Shishmaref is perched on a slender island the length of New York's Central Park but barely half as wide. It lies off the north coast of the Seward Peninsula, not far from Russia (in fact, Shishmaref is closer to Moscow than to Washington DC). Inupiat peoples, with close ethnic ties to their counterparts across the Bering Strait, have lived on the island for centuries. Today, like many indigenous communities, their town is a hybrid of the modern (a school, contemporary homes) and the frontier (no television, no running water).

Each winter the Chukchi Sea surrounds Shishmaref with a thick layer of ice that's integral to the subsistence-based, hunting-and-fishing culture. The ice also keeps out seawater that would otherwise smash into the island's cliffs during violent winter storms. Residents have long counted on the freeze-up to occur in October, but lately there have been patches of open water near shore as late as December, leaving the town at the mercy of autumn elements. Recent October storms have flung winds of 144kph (90mph) and waves of nearly 4m (13ft) at the tiny island, whose top elevation is only 6.7m (22ft).

"It's not our fault that the permafrost is melting, or that there's global warming that's causing us to go farther away from our home in Shishmaref. But we'll survive."

Luci Eningowuk, Shishmaref resident

In an earlier time, the people of Shishmaref might have simply packed up and moved on, but abandoning a modern township with ancient roots isn't so easy. Though it goes against their cultural grain, locals pleaded for state and federal officials to recognize their plight. An Erosion and Relocation Commission was formed in 2001, and in 2002 the townspeople voted by a count of 161 to 20 to move. By 2007 it still wasn't certain where Shishmaref would relocate, although the commission was pushing for Tin Creek, a mainland site about 8km (5 miles) away – close enough that many homes could be moved there and townspeople could maintain connections to the land and sea. The US Congress had considered moving the residents to the larger towns of Kotzebue or Nome, but the idea of their community's identity being swallowed up didn't sit well with many residents.

Wherever they end up, the move will be a costly process for Shishmaref, with the tab estimated at $180 million. The commission established a website in 2007 (shishmarefrelocation.com) to publicize its plight and seek supporters. Other settlements along the Arctic coastlines of Alaska, Canada and Russia may also be forced to move, a prospect that serves as eloquent – and expensive – testimony to a climate in flux.

sync with the snowfall that the seals use to build their winter dens. As the ice retreats, ringed seals may get a short-term boost from new access to Arctic cod and other prey that become easier to snatch. However, pups will find themselves exposed to the cold ocean – and to predators – at earlier points in the spring. There's no evidence that ringed seals will take to land, which goes for several other seal species as well, so in the long run the loss of summertime ice may prove devastating. In contrast, grey seals and other temperate species already accustomed to land have a better shot at dealing with reduced Arctic ice successfully.

Birds – made for mobility – appear to have a head start in adapting to changes across the Arctic. The task won't be easy for all species, though. White gulls are scavengers that feed mainly on the carcasses of polar bears and ringed seals, so a decline in those species will at least force the gulls to find other meal tickets. Fish and other marine creatures may also find their diets shaken up, as schools shift with the warming oceans and as industrial fishing moves northward into formerly ice-encased areas.

It's hard enough for some forms of Arctic wildlife to deal with climate change, but **pollution** may be making matters worse. A witch's brew of contaminants, including mercury and other heavy metals, has drifted into the Arctic from industrial nations over the years. Many of these toxins are currently locked in snow and ice. A warming climate could melt this snow and ice, releasing decades' worth of pollutants into rivers, ponds and oceans. Some of these contaminants can work their way up the food chain, and they've already been found in the fat reserves of some creatures near the top of the chain – notably polar bears.

A softening landscape

As the Arctic Ocean slowly sheds its mantle of year-round ice, terrestrial life nearby is taking on new forms. Perhaps the most emblematic is the "drunken forest" – larch and spruce trees that lean at crazy, random angles as the soil beneath them thaws out.

Permafrost – land that's been frozen for at least two years – holds the key to this and many other transitions across the Arctic landscape. Permafrost covers about a quarter of the Northern Hemisphere's land area, including half of Canada, most of Alaska and much of northern Russia, as well as a small part of Scandinavia and several mountainous regions closer to the Equator. On the fringes of the Arctic, the permafrost is **discontinuous**, forming in a mosaic of favoured locations. In the coldest areas, **continuous permafrost** extends throughout the landscape.

Building damaged by melting permafrost in Chersky, Russia
Vladimir Romanovsky

Permafrost is topped by an **active layer** that typically melts and refreezes each year. The permafrost itself – which starts anywhere from a few centimetres to a few tens of metres below ground – remains below freezing year-round. It can extend hundreds of metres further down, until the soil temperature again rises above freezing en route to Earth's molten centre. The thicknesses of the active layer and the permafrost depend on both recent and past climate, making them convenient diaries of global change.

It's the deepening of the active layer and the thawing of adjacent permafrost that's causing trouble in certain parts of the Arctic. Tucked within the permafrost is water, some of it in crystals bonded to the soil structure and some in much larger ice beds or wedges. If a thaw descends below the active layer, it can cause pockets of ice in the topmost permafrost to liquify and create an underground lake (a **thermokarst**). Should this lake then drain off, the resulting cavity may fill with surface soil slumping inward. Such is the process that has put many trees and buildings in a slow-motion fall, leaning at ever-sharper angles over time, especially across the fast-warming lower Arctic between about 55°N and 65°N.

Because of the variability of year-to-year warming and the random locations of ice pockets, it's almost impossible to predict which house or tree in a melt zone will start leaning next. In **Yakutsk,** the major city of eastern Siberia, many buildings are on stilts a few metres deep. Even so, over 300 structures and an airport runway have been compromised by unstable permafrost (perhaps coupled with less-than-optimal building practices). Siberia's permafrost has warmed more than 1°C (1.8°F) on average since the 1960s, and computer models tag vast stretches of Siberia at high risk for more structural problems as the century unfolds.

Fairbanks, Alaska, is a poster child for permafrost trouble. The city of 30,000 lies atop a fairly thin active layer – only a few metres deep in many locations. Record warmth in the past twenty years has percolated downward, heating permafrost across Alaska by 1–2°C (1.8–3.6°F) to its warmest levels since the last ice age in some spots. Around Fairbanks, some sections of permafrost have warmed by more than 3°C (5.4°F), bringing its temperature above –1°C (30°F) in some locations. The results are startling: drunken spruce trees, heaving bike trails and sinkholes pockmarking the landscape. Thus far the sinking has been patchy, focused where digging and construction have opened up layers of landscape to warming and melting. More widespread problems could occur if and when the bulk of Fairbanks' permafrost gets to the dangerous side of the freezing point.

Trees aren't just toppling across the Arctic; they're also burning. In 2004 and 2005, **fires** swept across more than 10% of the tundra and forest in Alaska's interior, ravaging nearly 49,000 square km (19,000 square miles). Across Siberia, which holds half of the world's evergreen forest, the fire losses have reportedly risen tenfold over the last few decades. In 2004 alone, fire consumed an area almost the size of Great Britain – 220,000 square km (85,000 square miles). Sharply warmer summers in recent years have helped dry out the landscape much sooner than usual, lengthening the fire season and making the fires that start that much more intense. Insects are playing a part, too. Epic infestations of pine and spruce bark beetles have swarmed across western North America over the last decade. The bugs have killed off millions of trees and increased the forests' vulnerability to flame. These infestations have been driven in part by winters too warm to kill off the insects (see p.155).

To be sure, a milder Arctic paves the way for the northward migration of many flora and fauna. The problem is that many parts of the ecosystem can't change with the fluidity of the climate itself. For instance, the weather in thirty or forty years may be conducive to boreal forests across some regions now covered with nutrient-poor tundra. However, it might take decades longer for decaying grasses to prime the soil so it can support tree growth. Still, projections for the next century suggest a massive increase in high-latitude forest, especially across western Siberia and north-central Canada, over the next century. This would pull an increasing amount of carbon dioxide out of the atmosphere, though probably not enough to counteract the warming effect caused by the forest's propensity to absorb heat (see p.78).

Methane lurking in the muck

Among all the byproducts of the melting Arctic, one stands out in its sheer horror-movie potential. Trapped within permafrost are billions of tons of **methane hydrates** (also known as **methane clathrates**). These are molecules in which water and the potent greenhouse gas methane are bonded under high pressure and/or low temperature. Besides their presence in permafrost, methane hydrates are even more extensive in seafloor sediments around the margins of continents across the globe. In their supercompacted form, methane hydrates are more than 150 times more concentrated than gaseous methane. Though it's not clear what will happen to methane hydrates as the permafrost continues melting, scientists point to the possibility that vast quantities of methane could be released – perhaps enough to dwarf the greenhouse gases emitted by human activity. Should that happen, global warming could go well beyond current projections. It's believed that methane hydrates may have been involved in some of the most intense warming episodes in Earth's history (see p.211). Nevertheless, governments from the US, China and Japan are reportedly investigating the potential of methane hydrates as a fuel source: there may be twice as much of it as there is coal, oil and natural gas elsewhere on Earth.

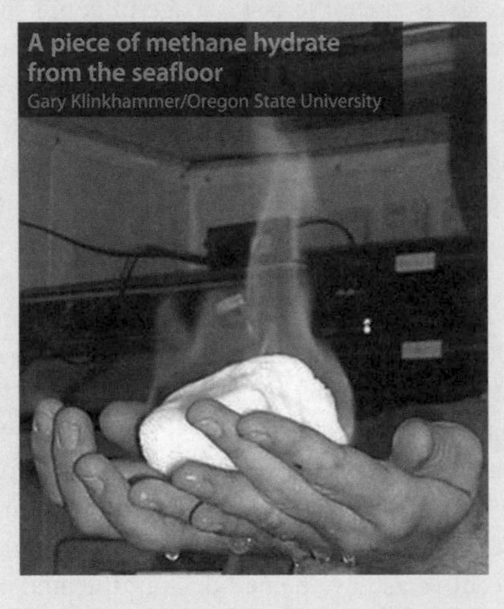
A piece of methane hydrate from the seafloor
Gary Klinkhammer/Oregon State University

The Arctic's signature land beasts – **caribou**, **reindeer** and their cousins – face their own challenges. Heavier spring runoff and rains across the North American Arctic could widen rivers enough to jeopardize the seasonal trek of the continent's vast herds of caribou, some of which hold more than 100,000 animals. In Lapland, autumn and early winter are already proving treacherous. Reindeer normally browse below the surface of early-season snow to reach the tender lichen that spread at ground level. But in the past few years, bouts of freezing rain have become more common as warmer-than-usual air fed by the Gulf Stream flows atop the Arctic's frigid surface air and rains into it. When ice encases the snow and the underlying lichen, the reindeer lose their access to food. No major die-offs have been reported, but the trend has raised concerns among ecologists and local herders.

The Arctic's patchwork of **freshwater lakes** and **rivers** is also being rewoven as the climate warms. Across a Texas-sized patch of Siberia, researchers used satellite photos to identify more than a hundred small lakes that disappeared between the 1970s and 1990s. These were victims of thermokarsts that allowed the waters to drain out, while new lakes opened to the north.

How will Greenland's fate affect ours?

The biggest chunk of land in the far north, **Greenland** has languished in frozen obscurity for centuries. Two areas on its chilly southwest coast were populated by Erik the Red and a few hundred other Danes and Icelanders around 1000 AD. (To this day, Greenland remains a self-governing part of Denmark.) The Western Settlement disappeared around 1350 during a marked cool-down, while the Eastern Settlement hung on for some 200 years more before giving way. Since then, the island's brutal cold has kept human occupation limited to a scattering of Inuit towns and, in the last few decades, a sprinkling of seasonal research camps. Several of these are parked on top of the colossal **ice sheet** that covers about 85% of the island.

Greenland's vast coating of ice could end up shaping where millions of people live a few generations from now. Actually, "sheet" may not be the best word for this icy monolith. It extends over 1.3 million square kilometres (500,000 square miles) – the size of France and Spain combined – and the bulk of it is more than 2km (1.2 miles) high. That makes the ice itself a topographic feature on par with the US Appalachian Mountains. Although some of the sheet probably melted in between ice ages, scientists believe it's been at least three million years since Greenland has been ice-free. Under the sheer weight of all its ice, Greenland's centre has sunk roughly 300m (1,000ft) below sea level.

Some models indicate that a 2°C (3.6°F) rise in average global temperature, if sustained, would ensure the ice sheet's eventual demise. Right now we're on a trajectory that could bring global temperature well above that 2°C point this century, assuming that greenhouse emissions continue to climb at their current pace. For Greenland's ice, the point of no return wouldn't be obvious at first. Once under way, though, the melting will become virtually unstoppable through a set of positive feedbacks. For example, as meltwater runs off from the highest elevations, the sheet's newly lowered

Courtesy Russell Huff and Konrad Steffen/University of Colorado CIRES

summit would encounter warmer air temperatures. This would allow for still more melting, and so on, putting the ice in a cascade towards oblivion.

Scientists are debating whether the amount of water locked up in Greenland ice has increased or decreased over the last several decades. Most studies point to a loss, but the total is hard to assess given the sheet's vast scope. If and when the entire Greenland ice sheet were to melt into the sea, it would trigger a truly catastrophic sea-level rise of more than 7m (23ft). Such an inundation would swamp coastal cities around the world and, in some low-lying areas, close the book on millennia of human history.

Thankfully, most glaciologists believe it would take centuries for the entire sheet to melt. However, there's still cause for concern in the shorter term, because Greenland's peripheral glaciers appear to be losing ice at a faster clip than either scientists or their computer models had reckoned. Some of the ice processes now at work aren't represented well – or at all – in computer models because of their complicated nature. Yet even a small fraction of Greenland's ice entering the ocean over the next few decades could push sea-level rise well beyond the current consensus estimates.

What does seem clear is that there's an increasing amount of water cycling through the system. Since 1992, European Space Agency satellites have used laser-based altimeters to measure the ice sheet's height from above. The ESA data indicate that the sheet is gaining about 5cm (2") of snow a year. If so, this wouldn't be a surprise. As overall warming brings the frigid air flowing atop the sheet closer to 0°C (32°F), the air holds more moisture and snowfall is enhanced.

Greenland's ice sheet should be losing much more mass on its sides than it gains on top before long, if it isn't already. In 2005, a week-long thaw spread across all of southern Greenland. Even atop South Dome (elevation 2900m, or 9600ft), scientists reported three days of melting. The changes have been equally dramatic along Greenland's margins. This is apparently due to increased warm-weather runoff as well as an important process called **dynamic thinning**. When meltwater seeps through a flowing glacier, it can lubricate the base and hasten the glacier's seaward flow. This dynamic thinning, together with the enhanced runoff from melting, led to a 35% increase in the annual rate of ice loss in 2003 compared to 1993–99, according to a multinational study that combined NASA altimeter data with computer models.

As scientists watch glaciers more closely, they're discovering how quickly these rivers of ice can speed up – and, at times, slow down. A flurry of well-publicized reports in 2006 revealed that several of Greenland's largest glaciers were speeding towards the sea at a pace far greater than expected. Despite the acceleration, so much melt is occurring at the forward edge of these glaciers that the net effect can be a retreat of that edge. (Think of dipping a chunk of ice very slowly into a glass of hot water: even as the whole chunk goes downward, the bottom edge of the ice is melting its way upward.)

One of the world's fastest-moving glaciers is **Kangerdlugssuaq**, through which 4% of Greenland's ice flows slowly towards the Atlantic. The glacier advanced seaward at an annual pace of 5km (3 miles) in 2001, but almost three times more quickly – 14km (9 miles) – in 2005. The glacier's forward edge pulled back a full 5km (3 miles) in 2005, by far its largest retreat on record. **Jakobshavn Isbrae**, the largest of Greenland's ice-draining glaciers and another fast mover, slowed slightly from 1985 to 1992 but then more than doubled its seaward velocity in the following decade. Yet in the summer of 2006, the thinning rates of Kangerdlugssuaq and another large east-coast glacier, Helheim, dropped to near zero. It's now appearing that these highly dynamic systems can move seaward in pulses that strengthen and weaken from year to year, but atop longer-term depletion apparently induced by climate change.

Not all of Greenland's glaciers are ideally positioned for a speed-up, though. It appears that dynamic thinning works best on glaciers that lie atop a layer of bedrock that slopes smoothly towards the sea. Scientists from NASA and elsewhere are now working to map the topography beneath key glaciers to help predict which ones are flowing on favourable slopes.

Taken as a whole, the recent accelerations in Greenland hint at a faster-than-expected response to warming. One study finds that the meltwater from Jakobshavn Isbrae, all by itself, has pushed up the rate of global sea-level rise by 4%. "We are witnessing enormous changes", Eric Rignot of the California Institute of Technology told the Associated Press. It's fair to say that Greenland's ice is one of the biggest unknowns looming over our planet's coastlines.

What about Antarctica?

Strange as it may seem, Antarctica has stayed mainly on the sidelines in the saga of global warming – at least until recently. This vast, lonely land took centre stage in the mid-1980s when the infamous ozone hole was discovered a few kilometres up. The Montreal Protocol and resulting actions by world governments are expected to eliminate the hole by later in this century (see p.29).

Almost two thirds of the planet's fresh water is locked in Antarctica's ice sheet, ready to raise sea levels by spectacular amounts if the ice were to melt. Fortunately, nobody expects the entire sheet to start melting anytime soon. Temperatures in the interior of Antarctica are shockingly cold: Vostok Station, for example, has never recorded a temperature above freezing.

It's hard to know exactly what's happening to temperatures across the great bulk of Antarctica, which has only a few reporting stations. There are signs of a cooldown at the surface, a finding often seized upon by climate-change sceptics. At the South Pole itself, surface readings have been cooling by about 0.2°C (0.36°F) per decade. Satellite measurements of the ice sheet's temperature also show cooling over much of the interior. The reason appears to be twofold. First, overall global warming has led to a tightening of the ring of upper-level winds that encircles the continent. This helps to keep the coldest air focused over the continent and to minimize its dilution by cold fronts that sweep off the ice and into the Southern Ocean. At the same

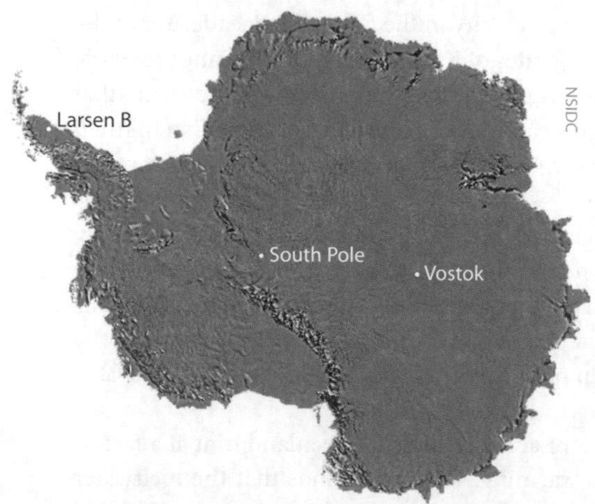

time, ozone depletion in the Antarctic stratosphere over the last several decades has produced a marked cooldown in late winter and spring. This has also helped to strengthen the Antarctic vortex.

Once the ozone hole begins healing in earnest later this century, it's expected that a slight warming will proceed across most of Antarctica, growing over time. Across the interior, that warm-up may do little at first other than slowly build the ice sheet, in much the same way as Greenland's appears to be growing on top. Slightly warmer air (that is, air that's less bitterly cold) should bring more moisture with it, resulting in a boost to the continent's tiny snowfall amounts. That's already been the case across East Antarctica, according to a 2005 report in the journal *Science* by a team of US and British scientists. They found the ice sheet across the East Antarctic interior thickening by about 2.2cm (0.9") per year since the mid-1990s, somewhat more than most global computer models had predicted.

While it's getting a bit cooler at the surface across parts of interior Antarctica, it's getting a lot warmer above ground and across the continent's far western reaches. The upper-level warming was announced in a 2006 paper in *Nature* by John Turner of the British Antarctic Survey, after long-term data from weather balloons became available for analysis. Looking at South Pole Station plus eight coastal stations, Turner and colleagues found that the air at a height of about 5.5km (3.4 miles) above sea level had warmed by 1.5–2.1°C (2.7–3.8°F) since the 1970s. Frigid surface air often decouples from the air immediately above it, which may explain how surface cooling and upper-air warming could coexist in the heart of Antarctica.

As for surface warming, the most dramatic example is the Antarctic Peninsula, which juts northward towards South America. Annual average temperatures there – one of the few parts of the continent that sees major thawing and refreezing each year – have soared by as much as 2.5°C (4.5°F) since the 1950s. That's the most rapid rise in annual average observed anywhere on Earth over that time span.

It's at this western edge where the Antarctic ice sheet appears to be fraying most rapidly. The Larsen B shelf made headlines in early 2002 when a Delaware-sized chunk – perhaps 12,000 years old – broke off (see photo overleaf). Fortunately for us, such dramatic calving doesn't affect sea level directly, since most of the ice was already in the water. More concerning is whether such breakups might open the door to a faster seaward flow of ice from glaciers upstream. The hydrostatic force of the ocean pushes back on large ice shelves that are grounded on offshore bedrock; this process

Inset shows linear melt ponds atop the decaying Larsen B ice shelf less than two years before its collapse

helps stanch the flow of ice from upstream. When the shelves break up, the adjacent glacier flows more easily into the sea. A NASA-led team found that glaciers flowing into the Larsen B shelf increased their speed eightfold after the shelf's 2002 breakup.

Another group led by NASA's Robert Thomas found that glaciers in the Amundsen Sea area of West Antarctica were thinning much faster in 2002–03 than in the 1990s, as far as 300km (190 miles) inland. If the floating shelves in this area were to break up, *à la* Larsen B, they could allow for the eventual drainage of glaciers that together hold enough ice to raise sea levels by 130cm (51"). "Although these glaciers are the fastest in Antarctica, they are likely to flow considerably faster once the ice shelves are removed", wrote Thomas and his colleagues in *Science*.

It's a good thing that climate models project the vast bulk of Antarctica's ice to stay onshore for at least the next century and probably much longer than that. Were it all to melt, global seas would rise by a cataclysmic 63m (208ft) – a scenario ripe for Hollywood, if nothing else.

Tropics and mid-latitudes: goodbye glaciers?

Far above the lush forests and warm seas of the tropics, a few icy sentries keep watch on our planet's climate from a unique vantage point. These are Earth's tropical glaciers, perched atop the Andes and other peaks that extend from 4600 to more than 6900 metres (15,100–22,600 feet) above sea level. Further north, in the mid-latitudes, thousands of small glaciers and a few larger ice caps are sprinkled across parts of the Rockies and Sierras, the Himalayas and other Central Asia ranges, and the Alps.

These scattered outposts of ice aren't very good at hiding the strain they're under. Because low-latitude glaciers tend to be much smaller than their polar cousins, they often respond more quickly to climate change. A melting trend that kicked in during the mid-1800s has accelerated in recent decades across many of the tropical and mid-latitude glaciers

Lonnie Thompson: putting tropical ice on the map

He may not be able to save them, but **Lonnie Thompson** has done as much as anyone to document and publicize the plight of tropical glaciers. Based at Ohio State University, Thompson paid his first field visit to Peru's Quelccaya in 1974 as a graduate student. He has since spent more than 840 days above 5486m (18,000ft) as well as 53 consecutive days at around 6096m (20,000ft). Early on in his career, Thompson toiled for years – often working on a shoestring with few collaborators – to collect data from the Andes and other regions of vulnerable ice. Few colleagues thought the tropics had much to offer glaciologists compared to the much vaster ice fields at the poles. Thompson proved them wrong in 1983 by drilling 164m (538ft) into the Quelccaya ice cap and retrieving a unique core that documents 1500 years of climate history near the Equator. Thompson's subsequent expeditions have dug even deeper into climate's past. An ice core he retrieved in 1987 at Sajama, Bolivia, spans 25,000 years, while the Guliya cores from China's Western Kunlun range extend back more than 700,000 years.

After more than three decades of work, Thompson feels a growing urgency to capture the climate clues offered in tropical ice before it's gone. When he re-cored Quelccaya in 2003, meltwater seeping downwards had already compromised some of the ice layers from earlier years. His return to Kilimanjaro in 2006 after field work there in 2000 showed the mountain's ice continuing to dwindle. "The accelerating retreat of most tropical glaciers is very telling", says Thompson, "for it is occurring in a part of the world known for its temperature stability." Because glaciers are "great integrators of climate", as Thompson puts it, "the fact that most glaciers, from the poles to the tropics, are speaking in one voice and telling us the planet is warming should be a concern to us all."

monitored by scientists. By later in this century, some of these remnants of the last ice age could be little more than history themselves.

Like rivers in slow motion, glaciers are fluid entities, with fresh water flowing through them. Snow accumulates on top and gets pushed downward and outward by the snowfall of succeeding years. For a healthy glacier, the change in its total ice mass per year – its **mass balance** – is either zero or positive. Today, however, most of the world's tropical and mid-latitude glaciers have a mass balance seriously in the red. In short, they're shrinking.

It isn't just melting that spells doom for ice. Many glaciers erode largely through a process called **sublimation**. When dry air flows over ice, the moisture in the ice can pass directly from its frozen state to its vapour state without any intervening liquid form. Sublimation can occur even when the air is below freezing, and it's the dominant process of glacier loss in some areas. Overall, however, melting is the main culprit. It takes eight times less energy for a molecule of ice to melt as it does for it to sublimate. Under the right weather conditions, direct sunshine can melt glacial ice at temperatures as cold as -10°C (14°F). And, of course, a reduction in snowfall can take its toll on a glacier even if average temperatures stay constant.

It's the unique mix of conditions at each location that shapes the behaviour of tropical and mid-latitude glaciers. Those who study glaciers fear that this complex picture sometimes gets lost in the scramble to interpret glacial loss as a simple product of warmer temperatures. Still, they agree on the main point: that climate change is eroding many glaciers at a dramatic pace.

Glimming the glaciers

People in Scotland once used "glim" the way we now use "glimpse" or "glance". The acronym-happy world of atmospheric science has folded this archaic word into a name for the biggest effort to date to trace the world's receding glaciers from above. GLIMS (Global Land Ice Measurements from Space) is employing imagery from two NASA satellites to compile an annual progress (or regress) report on each of the world's glaciers. The Americans behind GLIMS, including scientists at the US Geological Survey and the National Snow and Ice Data Center, are linking their pictorial census with software tools to help analyse the ice retreat. They've built a public-friendly website and a global network of over sixty institutions dedicated to monitoring the health of glaciers.

Trouble towards the Equator

Thanks to a century of visits by adventurers, photographers and scientists, the ice cap atop Africa's **Mount Kilimanjaro** is the most widely publicized of the tropics' disappearing glaciers (see box overleaf). But others tell a similar tale of woe. Not far away from Kilimanjaro, **Mount Kenya** lost seven of its eighteen glaciers over the last century; those that remain cover a paltry area less than a square kilometre (0.4 square miles). On the Indonesian island of Irian Jaya, the glacial area atop Mount Jaya shrank by 7.5% from 2000 to 2002. Its ice now covers only about 20% of the area it did in 1950.

All of these are mere slivers next to the ice atop the **Andes** of South America. With their string of peaks and high valleys, many extending above 5000m (16,000ft), the Andes hold more than 99% of the world's tropical ice. In Peru, north of Lima, more than 700 glaciers dot Peru's **Cordillera Blanca**, or White Mountains (a name deriving mainly from the range's granodiorite rock, rather than the ice it sports). Further south along the Andes, not far from Bolivia, sits the vast, high **Quelccaya** ice cap, sprawling across a city-sized area of 54 square km (21 square miles).

Peru's tropical glaciers are far more than playgrounds for climbers and backdrops for photos: they carry both spiritual and practical significance. The ice melt provides water and hydroelectric power that's vital during the dry season for much of the country, including up to 80% of the water supply for the perennially parched Pacific coast. **Lima** spends months shrouded in clouds and drizzle, but it averages less than 20mm (0.8") of rain a year, which makes it the world's second-largest desert city. Its eight million residents now depend on glacial ice that is measurably receding. So grave is Lima's risk of losing power that the city has begun expanding its portfolio of "alternative" energy sources – that is, power plants that burn natural gas. As for the potential water shortage, there's no solution in sight, though engineers have suggested a tunnel through the Andes.

What might be called The Case of the Disappearing Tropical Glaciers is the latest chapter of an ancient tale. Tropical ice has been on the wane ever since the end of the last ice age. The meltdown has unfolded in massive fits and starts, sometimes shaping the destiny of whole civilizations. The glaciers' most recent decline began in the mid-1800s, towards the end of the Little Ice Age but before human-produced greenhouse gases started to rise substantially. It's likely that a blend of forces, possibly including a slight rise in solar output, got the trend started. However, in recent decades greenhouse gases appear to have taken the reins, with assistance

The fast-disappearing snows of Kilimanjaro

… and then they were out and Compie turned his head and grinned and pointed and there, ahead, all he could see, as wide as all the world, great, high, and unbelievably white in the sun, was the square top of Kilimanjaro.
The Snows of Kilimanjaro, 1938

It's a safe bet that Ernest Hemingway wasn't thinking of the greenhouse effect when he set his tale of a man dying of gangrene near a mountain whose ice cap is itself dying. Yet the title of Hemingway's classic short story *The Snows of Kilimanjaro* is now a standard piece of climate-change iconography. Africa's tallest mountain is indeed losing its ice, and quickly, in a process that started decades before Hemingway penned his story. Of the three peaks that make up this giant, extinct volcano, only one (Kibo, the tallest, at 5893m/19,334ft) has anything left of the vast ice cap that apparently covered all three peaks long ago. Kibo's ice shrank by 80% in the twentieth century, and some reports have its thickness going down by roughly 50cm (20") per year since the 1960s. Although the ice can hold its own in a good snow year, the retreat has been steady over the long term, leading to a fairly precise forecast of the ice cap's disappearance between 2015 and 2020.

But Kilimanjaro is anything but a textbook case of "warmth equals melting". In fact, it's become something of a flash point, a topic of spirited debate among glaciologists as well as a favourite of climate-change doubters who extrapolate from a single mountain in an attempt to disprove a world full of receding glaciers.

What makes the story of this ice so slippery? For one thing, the average air temperature over Kibo's ice cap runs well below freezing, hovering close to -7°C (21°F) year round. That in itself doesn't rule out climate change as a factor, but it means something more than sheer warming is probably at work. The ice cap's shape offers more clues. Instead of being rounded, it features steep, sharp sides about 20m (66ft) high. Georg Kaser of the University of Innsbruck argues that if air temperature alone were shifting the balance towards melting, the ice cap would quickly revert to a smoother-edged contour. This suggests the Sun rather than the air is the prime melting agent.

More evidence for the Sun's key role is the east-west orientation of Kilimanjaro's ice cap. Since the mountain sits just 370km (230 miles) south of the Equator, the north- and south-facing walls each face the midday Sun at different times of year. These happen to coincide with the start of the peak's two dry seasons (roughly December to February and June to September), which means few clouds sit between the Sun and the ice. Moreover, the local winds generally stay light enough so that the Sun-melted surface doesn't dry out and refreeze in the chilly air.

While all of this helps to explain the shape of Kilimanjaro's ice, it doesn't tell us what caused it to start disappearing. In a 2003 paper, Kaser and colleagues offer one possible solution to the mystery. Their chief suspect is a regional drying (confirmed by a drop in the level of nearby Lake Victoria) that began around 1880 and may have triggered the melting that continues today. What's unknown is how the ice cap behaved in other prolonged dry spells before 1880. Ice cores

Mount Kilimanjaro on February 17, 1993 and February 21, 2000. Short-term conditions can affect such comparisons, but the melting trend is clear.

NASA

analysed by glaciologist Lonnie Thompson indicate that some parts of the cap have survived for more than 11,000 years. However, Kaser believes the bulk of the cap may have built up as recently as a few hundred years ago. He doesn't rule out global warming as a trigger for changes in air circulation – perhaps driven by the nearby Indian Ocean – that could be helping to dry out the peak. Three new weather stations installed on and near the ice since 2000 should help clarify how much warming and drying is going on. Meanwhile, new computer simulations are beginning to link the small-scale weather around Kilimanjaro to the regional and global atmosphere.

Of course, the interaction of people with Kilimanjaro adds more twists to the tale. Innsbruck's Thomas Mölg suggests that, by chopping down trees that catch rain and re-evaporate the moisture on the mountainside below, farmers may have played a role in the recent local drying. It's a notion supported by work elsewhere but unproven at Kilimanjaro. In another form of weather modification – this one intentional – geologist Euan Nisbet of the University of London briefly suggested that giant white tarps be thrown, *à la* the artist Christo, over the ice cap's edges to deflect sunlight and halt the melting.

Apart from its symbolism, the loss of Kilimanjaro's snow may not harm the region all that much. Tourism could suffer, though the mountain itself should certainly remain a draw. And it seems unlikely that locals will experience major water shortages if the glacier disappears. Measurements show that 90% of the ice that escapes Kilimanjaro goes directly into the atmosphere. Much of the remaining runoff appears to evaporate before it reaches springs and wells downstream, which get moisture from wet-season rains.

Although the fate of the ice cap appears sealed, Kaser believes a few of the smaller fringe glaciers – which lack Sun-facing cliffs – could hang on to life. In any event, some scientists would prefer that the spotlight be turned away from this celebrity victim and towards the thousands of other glaciers at risk from climate change. As Raymond Pierrehumbert (University of Chicago) noted on the RealClimate blog, "If Hemingway had written *The Snows of Chacaltaya,* life would be much simpler."

in the Andes from El Niño, the periodic warming of the eastern tropical Pacific. When an El Niño is in progress, temperatures generally rise across the Andean glaciers, and summer rainfall and snowfall decrease. Since the 1970s, El Niños have been unusually frequent.

Today, even the most diehard sceptics are hard-pressed to deny that tropical glaciers are vanishing before our eyes. The Andes offer especially vivid examples of low-latitude ice loss, particularly in some of the most accessible parts of the ice (which often happen to be the lowest-lying and the most tenuous to begin with).

Qori Kalis Glacier in 1978 and 2000

Lonnie Thompson, Byrd Polar Research Center, Ohio State University

▸ **Qori Kalis** The largest glacier flowing out of the Quelccaya ice cap, Qori Kalis was retreating at about 4.7m (15.4ft) per year in the 1960s and 1970s. Since 2000, the front of this thin ice tongue has disintegrated rapidly, producing a retreat of more than 1000m (3300ft). All told, more than a fifth of the ice in Qori Kalis has disappeared since 1978 (see photos), and in 2007 glaciologist Lonnie Thompson (p.93) predicted that the glacier may vanish by 2012.

▸ **Chacaltaya** For half a century, a ski lift took adventurers onto Bolivia's Chacaltaya glacier, billed as the world's highest developed ski area (the summit is at 5345m/17,530ft). But the lift hasn't been used since 1998, and by 2007 Chacaltaya had lost some 80% of its surface area, with local experts predicting its demise within a year or two. Franz Gutiérrez, a 50-year member of the local ski club, told *The New York Times*, "This is a tragedy I can hardly bear to witness."

Mid-latitude melt

The signs of glacier decline are no less dramatic as you move poleward. Patagonia is home to South America's largest icefields, as well as glaciers that cascade all the way to sea level. As with Antarctica and Greenland, the calving of glaciers that spill into the sea appears to be uncorking the flow of ice from higher up. NASA teamed up with Chile's Centro de Estudios Científicos to analyse the Patagonian glaciers in 2003. Combining space-shuttle imagery with topographic data, they found that the rate of glacial thinning had more than doubled between 1995 and 2000, making it one of the fastest thinnings on the planet. Only about half of the ice loss can be explained directly by warmer and drier conditions, according to the researchers. The rest may be due to more-unstable glacier behaviour, such as dynamic thinning (see p.89) and enhanced calving.

The Northern Hemisphere has a much fuller palette of mid-latitude ice than the tropics, but even here things are changing quickly. Glacier decline is obvious to millions in and near the **Alps**, whose ice has lost 30–40% of its surface area and about half its overall mass since 1850. (A few individual glaciers have bucked the trend due to local variations in climate.) At heights of 1800m (5900ft), parts of the French Alps have been experiencing winters up to 3°C (5°F) warmer than those of the 1960s. Some of the Alps' smaller glaciers lost as much as 10% of their ice mass in the warm summer of 2003 alone. A linear extrapolation from recent trends indicates that some 75% of Switzerland's glaciers could be gone by the year 2050. In reality, of course, the trend could accelerate or decelerate, though there's no particular reason to expect the latter.

Robert Shepard

Signs mark the retreating ice in Glacier National Park, Montana

For ice climbers and ordinary hikers, the alpine warming carries real physical risk. Parts of the Alps saw the active layer – the zone above permafrost that melts and refreezes each year – double in depth to around 9m (30ft) in 2003. Massive ice and rock slides, such as that year's avalanche on the Matterhorn that triggered the rescue of some seventy climbers, may be increasing due to the melting of permafrost and the resulting instability of rock and ice on steep slopes.

In Montana, **Glacier National Park**, one of the jewels in the US park system, has lost nearly 80% of the 150 or so glaciers it had in 1850. Most of the remainder are likely to vanish by 2030, according to a model developed by the US Geological Survey. One observer has already suggested changing the park's name to Glacier Moraine National Park, a title less likely to become obsolete. And a group of twelve nongovernmental organizations in 2006 petitioned the United Nations to redesignate the US park and its Canadian neighbour, **Waterton Lakes National Park** – together named a World Heritage Site in 1995 – as a World Heritage Site in Danger.

The loss of both ice and permafrost are major issues for the Central Asian mountains – the **Himalayas**, the vast **Tibetan Plateau**, and a number of nearby ranges. Together, these are home to the largest non-polar set of ice caps and glaciers on Earth: more than 100,000 in all, covering an area the size of Pennsylvania. Average temperatures across parts of Nepal have climbed 1°C (1.8°F) since the 1970s. The extent of snow and ice cover across the eastern Himalayas has dropped by around 30% in that same period.

Meanwhile, thanks to very moist monsoon flows, the southeast flank of the Himalayas is getting more than 20% *more* snowfall than it did in 1970, enough to keep a few glaciers steady or growing even as temperatures rise. The possibility of increased run-off from these growing glaciers poses

Will rocks help this railroad run?

Its political implications may trouble those who favour Tibet's independence, but a new railway connecting the Tibetan plateau to China's Qinghai province has faced another issue – climate change – head on. The Qinghai-Xizang railway, extending from Lhasa across eastern Tibet to Golmud, China, is the world's highest and longest, stretching 1118km (695 miles) on a route that's mostly at altitudes above 4000m (13,000ft). Its pressurized cars traverse land that's about half permafrost, which raises the spectre of buckling track should recent warming continue as expected. To keep trouble at bay, engineers called on a surprisingly low-tech tool: crushed rock. A bed of stones up to 1m (3ft) deep along the route helps shield the ground from summer sunlight but allows winter cold to seep through. In a year-long test under a railroad embankment, the permafrost ended up with a net cooling. However, at least one Chinese permafrost expert, Wu Ziwang, has warned that thawing soil could destabilize parts of the new railway in as little as a decade.

problems for flood and water management downhill. A 1999 study by the United Nations Environment Programme (UNEP) found more than forty glacial lakes in **Bhutan** and **Nepal** primed for overflow in the next few years. (One such lake has grown sixfold since the 1950s.) While lowlands may be able to benefit from stepped-up hydroelectric power, the WWF warns of growing flood risks in the next few decades.

It's conceivable that the glaciers might someday retreat enough to jeopardize streamflows and related agriculture across south and east Asia – perhaps even threatening the water supplies on which hundreds of millions of Asians depend – but that prospect isn't yet firmed up in climate simulations. Certainly, though, the timing of streamflows might change as melt seasons lengthen and the monsoon (presumably) strengthens. Even without such complicating factors, the sheer growth in population across this part of Asia looks set to strain water resources.

The picture is a bit more nuanced further north, in the sub-Arctic lands of North America and Europe. Certainly, **Alaska's** southern glaciers are eroding quickly, especially those that extend into the Pacific Ocean. The front edge of the vast **Columbia Glacier**, at the end of Prince William Sound, has moved about 15km (9 miles) upstream since 1980. But the sub-Arctic also holds some of the most publicized exceptions to global glacier retreat. Several of **Norway**'s coastal glaciers grew through the 1990s

Glacial melt lakes in the Bhutanese Himalayas
Jeffrey Kargel, USGS/NASA JPL/AGU

– the leading edge of **Briksdal** advanced by more than 300m (1000ft). However, that interval was bracketed by an earlier sixty-year decline in coastal glacier extent, as well as another retreat that began around 2000. Further inland, virtually all of the glaciers in interior Norway have been steadily shrinking.

Many climate-change sceptics have pointed to the growth of specific glaciers as evidence against global change. However, since global warming isn't a uniform process – some regions saw little temperature rise in the twentieth century, while others saw a great deal (see p.4) – it's easy to see why a few individual glaciers scattered across the globe have grown or held their own in recent decades, especially in places where warming hasn't yet pushed increased precipitation from snow towards rain.

A rocky future for skiers

Thanks to a warming climate, skiing conditions are heading downhill at some of the most popular resorts around the world. Skiing depends critically on each winter's snowfall, so there will remain both good and bad years. Overall, though, the good years are becoming less frequent and the bad ones getting worse. In 2006–07, many slopes in the Alps and the northeast US remained snow-free from autumn into January, ruining holiday ski plans and scotching more than a half-dozen World Cup events.

The future looks bleakest at lower latitudes and/or lower elevations, where winter temperatures often run close to the freezing point. A 2003 report from the United Nations Environment Programme (UNEP) estimates that the Alps' average snow line will rise 200–300m (660–990ft) by the 2040s and predicts that "many mountain villages, above all in the central and eastern parts of Austria, will lose their winter industry because of climate change." Half of Italy's ski resorts lie below 1300m (4300ft), and many German resorts also sit at relatively low altitudes. There's lots of income at stake: 8% of Switzerland's workforce, and 5% of Austria's gross national product, are ski-based. Though the highest resorts in the Alps and Colorado may hang on, a 2007 report produced for Halifax Travel Insurance foresees quota systems as demand skyrockets for the few resorts where snow remains reliable. The firm ATMOSresearch projects that ski seasons across many resorts in the western US could be 10–20 days shorter on average by 2025.

Resorts are already being forced to adapt. Switzerland's Tortin glacier, which supports the **Verbier** ski area, has receded so much that in 2005 the resort placed an insulating sheet the size of a city block atop the glacier's edge to slow summer melting. Other ski areas are thinking of moving uphill. Austrian promoters want to develop some 30km (45 miles) of slopes in the wilderness area of Piz Val Gronda, hundreds of metres above the popular **Silvretta** resort. However, the Alpine nations have signed a treaty designed to limit high-altitude development, and Austria's Tyrol government has banned all new ski areas.

More than melting?

It's impossible to quantify just how much of the blame for the overall retreat in low- and mid-latitude glaciers can be pinned on human activity. Temperature change takes time to filter through the dense world of water: glacier behaviour today reflects in part the conditions present years ago.

That said, tropical ice appears to be one of the most reliable of all witnesses to human-induced change. For starters, the wild temperature swings often found at higher latitudes are muted near the Equator, especially as you go up in height. There's also little variation as you go from west to east around the globe. As a result, the average readings vary only about 1.5°C (2.7°F) across the entire equatorial belt at glacial heights of

Snowpacks across the **US West** are projected to become smaller and more variable over the coming century. Many ski resorts in the **US Northeast** – already vulnerable to winter warm spells – could become marginal. In response to all this, seventy of the nation's ski areas have joined forces with the Natural Resources Defense Council to promote energy efficiency and renewables and push for federal emission controls. Their motto: "Keep Winter Cool".

Competitors in the 2003 Biathlon World Cup ski on artificial snow trails after a complete lack of natural snow
Stefan Matzke/NewSport/Corbis

Skiers who flock to the modest mountains of southeast Australia savour every flake of the white stuff. But at least one part of the **Snowy Mountains** has seen a 40% drop in the depth of spring snow since the 1960s, according to Australia's Bureau of Meteorology. The future looks even gloomier for snow-lovers, according to a 2003 report prepared by the Commonwealth Scientific and Industrial Research Organization. CSIRO predicts that the area covered by snow for at least fifty days each year could shrink by 18–60% by 2020 and by 38–96% by 2050, depending on global emissions levels.

about 6km (3.7 miles). This relative evenness of temperature means that even a small trend should stand out more noticeably than it would in mid-latitudes. It also means that any warming trend should be reflected throughout the tropics, which happens to jibe with the near-universal shrinkage of low-latitude glaciers.

On the other hand, the tropics aren't the best-sampled part of the world, meteorologically speaking. A few weather stations are cropping up at the highest tropical altitudes, but observations remain sparse, both on the ground and aloft. As a result, scientists are debating exactly how much temperatures across the higher altitudes of the tropics have risen. Thus far, the jury seems to be settling on a warming somewhat less than the

Picturing glaciers as they go

It may be the most ambitious round of time-lapse photography ever attempted – and certainly the most sobering. From 2007 into 2009, a set of 26 cameras stationed at glaciers across the Western Hemisphere – from Greenland and Iceland to Alaska, the Rockies and the Alps – will take daylight photos once per hour. The anticipated result: more than 300,000 images of a disappearing world. The Extreme Ice Survey (www.extremeicesurvey.org) is the brainchild of US photographer James Balog, a trained geologist and regular contributor to *National Geographic*. Balog teamed with glaciologists, computer scientists and experts in still and video imagery to assemble the project's network of resilient cameras, non-polluting power sources and uplink facilities. It's envisioned primarily as a way to convey the immediacy and scope of Earth's ice loss to a broad audience through videos, Internet-based outreach and a 2009 book.

global surface average, but still enough to tip glaciers towards melting. Across eastern Africa, the scanty data show little if any warming trend, which means that drying and other factors may be teaming up to produce Kilimanjaro's rapid ice loss (see p.96).

Aside from temperature, there's a range of other processes that will help shape the future of tropical and mid-latitude glaciers. These can vary from region to region and even on the scale of individual mountains, which leads to a patchwork of local advances and retreats within the larger fabric of overall glacier loss.

▶ **Sunlight** not only raises the air temperature over glaciers, but it can also directly melt glacial ice even when the air remains well below freezing (as noted previously). Local changes in cloud cover could either enhance or detract from Sun-induced ice loss.

▶ **Moisture** in the air can also help or hinder glacier growth, depending on location. In the tropics, where glaciers typically sit at heights well below freezing year round, drying can allow for more snow and ice to slowly leave glaciers via sublimation (see p.94). The melting process is different in the mid-latitudes, where temperatures often warm above freezing in summer. If those balmy temperatures came with extra atmospheric moisture, they could help keep the temperature at the glacier surface above freezing and thus hasten its melting.

As glaciers continue to shrink, their situation gets increasingly difficult, because the ratio of their surface area to their ice mass goes up. This gives the atmosphere more area to work on, causing them to change state more quickly – just as a pound of frozen peas defrosts more quickly than a pound of frozen chicken breasts.

Perhaps the only way for glaciers outside the poles to avoid death by warming would be through a boost in snowfall. That doesn't appear to be happening on a large scale, despite an general increase in atmospheric water vapour worldwide. The snowfall on glaciers atop mid-latitude mountains is fed by strong, moist, upper-level winds, such as those near the west coasts of North America or Norway. Glaciers in these areas may stand the best shot at survival, since the winds reaching them could bear more moisture in years to come. But even that might be a long shot, according to at least one study of glacier balance. It estimates that, to keep a glacier alive, a 1°C rise in average temperature would need to be counteracted by a 25% increase in average snowfall – a massive amount when it comes to long-term climatology.

Oceans

A problem on the rise

Glaciers calving like there's no tomorrow, sea ice dwindling to patches – it's easy to find drama in the warming of our planet's coldest waters. But only a small part of Earth's oceans ever produce ice. What about tropical and mid-latitude waters, the ones that together span more than half the planet? If appearances can deceive, then these oceans are master tricksters. They might not look much different to the naked eye than they did a century ago, but they've gone through quite a transformation – warming up and growing more acidic, among other changes. It appears they've hidden the true impact of our energy choices quite skilfully, absorbing some of the heat and CO_2 that would otherwise have warmed the atmosphere.

Many scientists suspected as much for years, but the first solid numbers came from a team led by Sydney Levitus of the US National Oceanic and Atmospheric Administration. After assembling a variety of deep-ocean measurements for the period 1948–98, Levitus announced that every major ocean exhibited warming down to at least 1000m (3300ft), and that the top 300m (1000ft) had warmed globally by about 0.3°C (0.54°F). You wouldn't be able to tell the difference in temperature by taking a quick dip, but the total heat added to the oceans represents at least 10% of the energy trapped by human-produced greenhouse gases in those fifty years. Water expands as it warms, and the temperature increase translates to a rise in sea level of about 25mm (1"). Again, that may not sound like much, but – as we'll see – it's on top of the sea-level rise from all other sources.

The undersea storage of vast amounts of heat has serious implications for humanity's future. Although the topmost layer of the ocean stays in balance with the atmosphere over timescales of a month or longer, the much deeper layer below the **thermocline** is more insulated. This means the heat it slowly absorbs from above will take a long time to work its way back out. Even if greenhouse gases magically returned to their pre-industrial levels tomorrow, it would take many decades for the heat tucked away in the deep oceans to work its way out of the climate system. This idea of "climate commitment" is a key theme in research leading up to the 2007 IPCC report. Some of these studies estimate that if global greenhouse gas emissions had levelled off in 2000 (which they didn't), an additional warming of at least 0.5°C (0.8°F) beyond the year-2000 global average would be guaranteed from twentieth-century emissions alone.

Of course, as climate change unfolds, the oceans will do far more than absorb and release heat. Ocean-driven storms, primarily tropical cyclones, are showing signs of strengthening as their fuel source heats up (see p.130). Sea levels are rising, too. Much of this is due to the expansion of ocean water as it warms, as noted above, but over time the rise will be increasingly enhanced by glacial melting. Furthermore, the uneven pattern of ocean warming may influence how, where and when a variety of ocean-atmosphere cycles unfold. This could spell the difference between a life of plenty or a life of poverty for millions of people living close to the land in drought- or flood-prone areas (see p.58).

All of this has made oceans one of the favourite trump cards of climate activists. The spectre of islands disappearing under the sea does make for a potent image. And yet the image isn't as crystal-clear as one might think. We still have limited data on the three-dimensional reality of oceans: how warm they are, where and how fast their currents are flowing, and so forth. There are big questions yet to be answered, including what "sea level" actually is.

From sticks to satellites: measuring sea level

Mean sea level (MSL), the average height of the ocean surface at a given point, sounds as if it should be an easy enough thing to calculate. Subtract the astronomical tides, factor out the storms that come and go, and *voilà* – you've got your average. In truth, however, MSL is enormously difficult to nail down. For a start, the tides depend upon the locations of the Sun and Moon within a set of cycles that takes 18.6 years to fully repeat itself. Even when you factor this out, the underlying MSL isn't the same everywhere, but varies across different sections of the open ocean by as much as 50cm (20"), depending on the surrounding topography and the warmth of the underlying waters. The effects of global warming complicate things further.

The simplest and oldest method of measuring MSL relative to some point along the coast is through basic **tidal gauges** – essentially, sticks

A tidal gauge in use, 1926
NOAA

in the sand. A technological leap came in the 1800s with **wells** that used floating gauges to factor out the effects of waves. Similar techniques are still used at many coastal stations, though they've been joined lately by far more precise methods. Modern gauges measure the time it takes sound waves to bounce off the water's surface and back, converting that into distance and then to sea-level height.

Radar altimeters aboard satellites work in a similar way, typically sending microwave energy from space to the sea surface – the open ocean as well as coastal areas. One key advantage of satellites, when they're coupled with Global Positioning System (GPS) data, is that they provide a measure of MSL that's independent of a land-based feature. The **TOPEX/Poseidon** satellite, a collaboration between the US and French space agencies, began gauging ocean heights across much of the world in 1992, ending its long run in 2006. Its successor, **Jason-1**, went into service in 2001. Jason maps 95% of the ocean surface every ten days with a resolution as fine as 2cm (0.8"). Even sharper-eyed is **GRACE (Gravity Recovery and Climate Experiment)**. This European satellite system, launched in 2002, tracks MSL with an error of less than 1mm (0.04"). Because it measures subtle changes in the gravitational tug exerted by the oceans as well as Earth itself, GRACE is able to help scientists sort through the small-scale continental adjustments that may influence sea level.

Those geological tics can have a surprisingly big influence on sea level. Coastlines rise and fall with the clashing of tectonic plates. In some places, such as parts of Alaska and Japan, the motions are so pronounced that tidal gauges can't be used as a reliable index of long-term sea-level change. Other coastal locations, such as Venice and New Orleans, are plagued by

Scandinavia's bounceback

The Gulf of Bothnia, bordering Sweden and Finland, offers a classic example of how the after-effects of the last ice age can disguise global warming at work. An international team used GPS data and a computer model in 2001 to track the vertical change in both land and sea. They found that the region's mean sea level was rising at about 2mm (0.08") per year as the ocean warmed. But that rise in MSL was obscured across much of the Gulf by a much larger rise in the coastline itself. Like a piece of dough springing back slowly from the push of a baker's thumb, the coast is still rising after the release of thousands of years of pressure from glacial ice. The rebound is as much as 10mm (0.4") per year, which is enough to more than cancel out sea-level rise and produce a net drop in sea level from the perspective of towns on the coast.

gradual subsidence. Still other regions are steadily rising, on the rebound after being squashed beneath kilometre-thick ice sheets for thousands of years. All of these ups and downs have to be factored out of sea-level data to find out how the sea itself is changing.

Balancing the sea-level budget

With all these challenges in mind, the Intergovernmental Panel on Climate Change has done its best to pool the data and estimate global sea-level rise. The second IPCC assessment, in 1996, pegged the twentieth-century MSL rise in the range of **10–25cm (4–10")**. The 2007 IPCC narrowed the range only slightly, to 12–22cm (5–9") over the century or 1.2–2.2mm per year. That may seem like a fairly wide margin of error for measuring something that's already occurred. Yet it disguises an even more stark discrepancy: something called the **sea level enigma** that's puzzled experts for the last few years. This lively scientific controversy has played out largely below the media radar, but it's worth knowing about.

Mark Meier of the US Institute for Arctic and Alpine Research put the enigma succinctly in the title of a 2002 paper, "Sea level is rising: Do we know why?" In short, the factors that raise sea level don't quite add up to the gauge-measured twentieth-century rise of 1.5–2.0mm per year.

The 2000 paper by Levitus, noted above, proved that the oceans had warmed deeply and dramatically, but showed an MSL rise of a mere 0.5mm per year from ocean warming. That's only about 25–33% of the gauge-observed rise. Where did the rest come from?

The melting of ice-caps and glaciers is the other big source of sea-level rise. Here, the question marks in each region make the total amount of

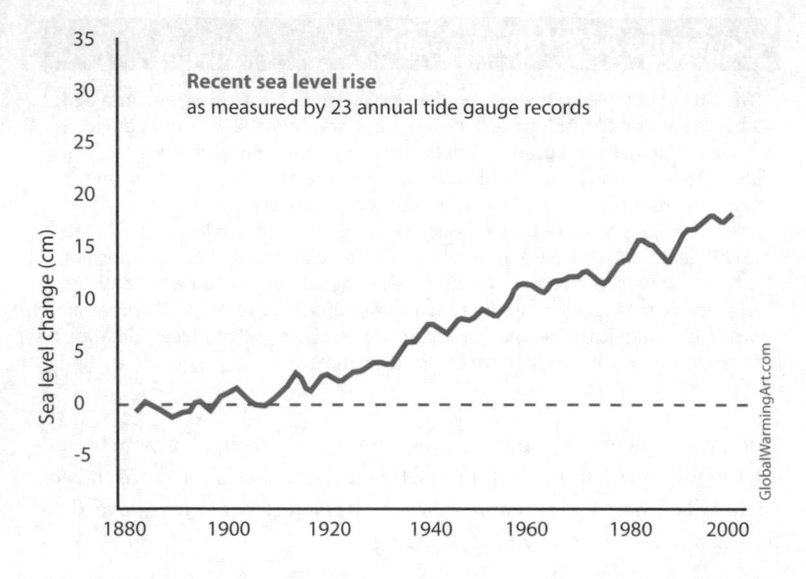

melt highly uncertain. Until recently, for example, Antarctica was expected to work against sea-level rise this century as snowfall rates increase across its interior (see p.91). However, recent reports indicate that the erosion of large glaciers along the west fringe of Antarctica may be enough to make the continent a net contributor to sea-level rise. Greenland's ice sheet may also be a net source of sea-level rise by this point. It was in a roughly steady state through most of the last century until melting began to accelerate in the 1990s. Smaller glaciers elsewhere around the globe have been melting even more rapidly. All told, the 2001 IPCC report put the estimated ocean rise from land-based ice melt over the last century at roughly 0.5mm per year, or about 25–33% of the gauge-observed rise.

Together, then, ocean expansion and ice melt appear to account for only about half to three-quarters of the observed century-scale rise in global MSL. Clearly, something's fishy: either the oceans are warming more, the ice is melting more, extra water is heading seaward from land, or sea level isn't rising as fast as we think. Whatever the truth, the rise in MSL appears to be accelerating. Data from TOPEX and Jason show that the rate quickened to roughly 2.8mm per year in the late 1990s, with signs of a further increase after 2000.

In a flurry of papers since 2001, researchers have floated several ideas on where the enigmatic remainder of sea-level rise may have come from. These include:

▶ **Incomplete data** The Levitus study lacked data for parts of the Southern Hemisphere, as well as for water below about 3000m (10,000ft). If warming is greater than believed in these hard-to-sample areas, it could help solve the puzzle.

▶ **Glacier melt** could be the culprit, and evidence for this case has been mounting. A 2006 analysis led by NASA's Eric Rignot, based on radar altimetry, shows that the Greenland ice sheet was pouring nearly twice as much water into the sea in 2005 as it was in 1996. Other glaciers, from Alaska to Patagonia, may also be releasing more meltwater than we once thought.

With the recent numbers for glacial melt at hand, the 2007 IPCC report came closer to balancing the books on sea level rise. It estimated that the Greenland and Antarctic ice sheets poured meltwater into the ocean at roughly four times the pace in 1993–2003 as compared to the longer-term average. Taking into account the increased melt from these sources and glaciers elsewhere, the IPCC was able to explain all but about 10% of the 1993–2003 MSL rise, though a full accounting remains elusive.

How high the sea?

Assuming humans continue to burn fossil fuels, it's a safe bet that sea level will continue to rise. But by how much? And what does this mean for people on or near the coast?

The 2001 IPCC report projected a rise in sea level by the year 2100 of 90–880mm (3.5–35.0"), with a midpoint of around 480mm (19"). However, in its 2007 report, the IPCC surprised many onlookers by pulling the range towards the lower end while narrowing it somewhat. The panel is now projecting a 21st-century MSL rise of **180–590mm (7.0–23.0")**, Part of the drop is a matter of framing: the earlier outlook was for the year 2100, while the more recent projection is for 2080–2099. But the new range also intentionally omits some factors, such as a potential speed-up in dynamic thinning of glaciers (see p.89), because they weren't considered well enough understood to include in the projections. Instead, the IPCC noted that these factors could add another 100–200mm to the top of the official range and that "larger values cannot be excluded." The earlier report, in contrast, did take a stab at incorporating uncertainties in Greenland melt in its MSL range. Thus, the new, lower range shouldn't be taken as a source of comfort – it simply emphasizes the better-known factors while leaving out some of the gnawing (and growing) uncertainties.

Are the Maldives and Tuvalu doomed?

They sit thousands of kilometres apart, but the Maldives and Tuvalu have much in common. As two of the smallest, least populous nations on Earth, they're perched atop low-lying coral atolls, from which they watch the sea and wonder how long they can keep it from swallowing their countries whole. The world is watching too. These two tiny, once-obscure countries have been at the forefront of climate change coverage since the late 1980s. In part that's due to the eloquence of their leaders, who have been driven onto the world stage by palpable fear of what the future may hold and anger at those deemed responsible. Maldives president Maumoon Abdul Gayoom made this plea in 1990 before the United Nations: "It is our earnest hope that the world community heed our voice – that of low-lying states – and save us from the ignominy of becoming environmental refugees." In 2002, Tuvalu threatened the United States and Australia with a lawsuit to be brought before the International Court of Justice, charging the two nations with reckless greenhouse-gas emissions (the lawsuit never materialized, though). And in 2003, Tuvalu's prime minister, Saufatu Sapo'aga, told the UN that the world's carbon-guzzling nations were committing "a slow and insidious form of terrorism against us."

There's no doubt that water is lapping at the shores of both countries, and the IPCC projections of sea level rise offer no reassurance. The average elevation is around 1m (3ft) in the Maldives and around 2m (6ft) in Tuvalu, while the highest terrain in each nation is around 3m (10ft) and 5m (16ft) respectively. The 2007 IPCC high-end projection for global sea-level rise of 590mm (23") by the 2080s and 2090s – with the risk of even larger rises from dynamic thinning of glaciers – would bring major impacts to both countries. Even now, storms are capable of flooding large parts of each island chain.

The two hundred islands of the Maldives sit due south of India. They're so close to 0° latitude that they're at little risk from tropical cyclones, which need some distance from the Equator in order to develop. But even far-off storms can send swells to worrisome heights, and flooding can also occur when intense monsoon rains team up with tides. The Maldives' capital, Malé, is now fortified by a seawall 3m (10ft) high, built in the late 1980s and 1990s with the help of over $60 million provided by Japan. However, on another island, Kandholhudoo, over half of the residents had already committed themselves to leaving before the catastrophic Asian tsunami of 2004 struck. The waters, topping around 1.5m (6ft), ruined almost every house and sped up plans to move residents to a nearby island, which will be bolstered by landfill.

Small as they are, the Maldives resemble an empire compared to Tuvalu, which is located well east of New Guinea and far to the north of New Zealand. Tuvalu's nine atolls only encompass around 23 square km (9 square miles) in land area. Stretching from 6° to 10°S, Tuvalu is far enough south to get battered by the occasional tropical cyclone. As with the Maldives, rogue waves and swells are an ever-present threat, and for both island chains the impact of such one-off events will only get worse as sea level gradually rises.

On the face of it, the residents of the two countries appear to have little choice but to either abandon their homelands or engineer their way to safety with expensive landfills and seawalls. But not everyone is convinced the islands are doomed. That's because atolls like the Maldives and Tuvalu form on the edge of coral reefs that sweep in narrow arcs around huge lagoons. The islands themselves are made up of the pulverized remnants of corals and other reef organisms. When the climate isn't changing and the coral are healthy, the islands appear to grow slowly over time, although there's a lot of uncertainty about how quickly this occurs. Without humans intervening, big waves occasionally overwash the islands, scouring sediment from one side of the island and depositing it on the other. In theory, the island doesn't drown so much as it shifts backward. As sea levels rise, the islands should shift position on their reefs, generally moving towards their lagoons and growing taller.

This scenario is championed by geographer Paul Kench of the University of Auckland. In studying a variety of small landforms, such as tropical atolls as well as the barrier islands common along the US Gulf and Atlantic coasts, Kench has come to believe that these lands are more resilient than they're given credit for. In a 2005 report for the journal *Geology*, Kench says that the Maldives formed atop their reef roughly five thousand years ago at a time of higher sea level. Kench also reports that the Maldives held up well in the 2004 tsunami. Although some of the eastern Maldives lost up to 9% of their land area in the waves, much of the sediment was redeposited on the far side of each island, as one might expect.

However, Kench doesn't foresee an easy path ahead for the Maldives, Tuvalu and other reef-based populations. The problem is that human settlement interrupts the natural growth process of atolls described above. Climate change may stunt the process further. In a study partially supported by the World Bank, Kench and colleagues surveyed atoll and reef islands and adapted a model originally designed for sandy barrier islands in order to explain the life cycle of the coral-based atolls. Normally, the growing reefs provide extra sediment that helps build up the islands. As sea level rises and the ocean warms, Kench believes the heat-stressed coral (see p.125) will devote most of their energy to self-preservation, providing little or no sediment for the island. Thus, it seems, the residents will have to make do with the sediment already beneath their feet. For this reason, says Kench, "island nations should make conservation of island and nearshore sand resources a high priority".

Tuvaluan kids watch as a high tide floods their town
Gary Braasch

Years of haphazard land use by indigenous residents as well as non-residents (such as the US forces that occupied Tuvalu in World War II) have triggered chronic erosion and left a difficult legacy for islanders to overcome. As for the future, if Kench's model is correct, it may spell survival for some atoll islands – at least for the next century – but constant flooding and woe for their residents. Tuvalu is exploring options for its citizens to resettle in Australia or New Zealand, where over 4000 Tuvaluans already live. In 2006 a retired Tuvaluan scientist, Don Kennedy, began lobbying to move the nation's 9000 remaining residents to the Fijian island of Kioa. There, ethnic ties with Tuvalu are strong – and the land is higher.

It's also important to note that the IPCC ranges reflect many scenarios and possibilities, some of which are discussed above. For starters, we don't know exactly how much greenhouse gas will be emitted, how much global temperatures will warm and how quickly Earth's oceans and ice will respond to that warming. As higher temperatures allow more moisture to filter into Antarctica, increased snowfall could help reduce the net MSL rise by anywhere from 20 to 120mm (0.8–4.7"), according to a mid-range emissions scenario in the IPCC's 2007 assessment. Global ocean expansion due to heating could add anywhere from 130 to 320mm (5.1–12.6"). The other big question marks include the Greenland ice sheet, other glaciers, and rain and melted snow running off from land.

The seemingly modest sea-level projections in the 2007 IPCC report piqued the interest of the world's press and of many scientists. Some took issue with the IPCC's assumption that dynamic thinning of Greenland and Antarctic glaciers (see p.89) would proceed at the same pace throughout the 21st century as it did in the 1990s, rather than accelerating further. During the same week it went public, the IPCC outlook was complemented by a *Science* paper from Stefan Rahmstorf of the Potsdam Institute for Climate Impact Research. He found that if sea-level rise in the 21st century stays proportional to the increase in global air temperature

Parts of Florida susceptible to a sea-level rise of 6m (20ft). The paler areas show high-population zones.
Jeremy L. Weiss and Jonathan T. Overpeck, The University of Arizona

– a relation that held fairly firm in the 20th century – then an MSL boost of 500–1400mm (20–55") above 1990 levels could be in the cards by the century's end.

A sea-level rise on the order of 500mm (20"), near the high end of the IPCC range and the low end of Rahmstorf's range, wouldn't be enough in itself to inundate the world's major coastal cities. But it could certainly lead to an array of serious problems, especially for populous, low-lying regions. Some tropical islands may become uninhabitable (see p.112) and, significantly, higher sea levels also mean a higher base on which to pile storm surges, tsunamis and all other ocean disturbances. These are likely to be the real troublemakers as sea level rises over the next several decades. A region at particular risk is the Ganges Delta of eastern **India and Bangladesh**. While most of Bangladesh is expected to remain above water even if the ocean rises several metres, such a rise could still displace millions of people in this extremely low-lying country. Millions of other Bangladeshi people who remain near the coast will be vulnerable to catastrophic flooding. (See p.135 for more on surges from tropical cyclones and coastal storms.) In addition, the IPCC's 2007 report stressed that MSL rises won't be uniform across the world's oceans. Some areas – including European coastlines – may see a few centimetres more than others, due to the geologic and oceanographic factors noted above.

In the longer range, much will depend on how long it takes global society to cut back on greenhouse emissions. As odd as it seems to contemplate, it's not out of the question that most of the world's coastal cities could be largely or completely under water in a few hundred years. While the risk

Swimmers and sea level

If the water level in a bathtub goes up when you get in, couldn't people in the ocean be pushing up sea level? That's the semi-serious question posed in the late 1990s by Gregory Pasternack of the University of California, Davis. While teaching short courses for high-school teachers in Maryland, Pasternack was inspired by a state politician who claimed that an excess of boats and boaters, rather than climate change, was behind the rise in global sea level. As part of a classroom exercise he developed, Pasternack encouraged students to measure how much water they displace in a bathtub, then to extrapolate that figure to a "world of swimmers". If everyone on Earth took a dip in the sea at the same time, they might occupy a volume on the order of a third of a cubic kilometre. But spread out over the vast area occupied by oceans, that translates to a sea-level rise of a mere 0.0009mm (0.000035") – around a hundredth of the width of a human hair.

is seldom put in such stark terms, it's a logical outcome of the amount of warming expected to occur over the next several centuries unless major emission reductions take place. According to some estimates, sea levels ran up to 6m (20ft) above current MSL during the last interglacial period more than 100,000 years ago, enough to inundate all or most of nearly every low-lying coastal city. The warming back then was produced by changes in Earth's orbit around the Sun (see p.196), but greenhouse gases could bring as much or more warming in the next century and beyond.

James Hansen (NASA) is one of a handful of scientists cautioning that MSL rises of the order of metres are possible not only in the distant

NAM, SAM and other climate cycles

After ENSO, the next-biggest cycles of climate variability are two circulations that play out over the North and South Poles, affecting ocean as well as land. The **Northern Annular Mode** and **Southern Annular Mode** (or **NAM** and **SAM)** refer to oscillations in the strength and structure of the upper-level winds that encircle the Arctic Ocean and Antarctica, respectively. Think of bracelets of strong wind, centered a few kilometres high, that loop around the North and South Poles respectively.

The SAM is the more straightforward of the two. Unblocked by land, its westerly winds howl above the Southern Ocean, circulating around a cold vortex locked over the South Pole. The circulation alternates between a tighter, faster-flowing ring (the positive mode) and a looser, more variable ring (the negative mode) that allows cold air to spill out from the pole across the Southern Ocean more easily. Since the 1960s, the SAM has trended more towards its tighter positive mode, which is likely related to ozone depletion (see p.28). This trend is projected by climate models to continue even after the ozone hole heals later this century. Among the implications for climate are a potential drop in winter rainfall for parts of Australia (see p.70).

On the other side of the globe, the NAM (also called the **Arctic Oscillation**) is a more complicated beast. The NAM encounters three continents, the thick ice sheets of Greenland and the warm currents of the North Atlantic – all of which interfere with the uniformity of the vortex and help to produce a more variable circulation. Still, most climate models are consistent in projecting a tighter, more positive NAM as the century unfolds.

In and near the North Atlantic, the NAM manifests itself as the **North Atlantic Oscillation** (NAO), defined by pressure variations between Iceland and the Azores. It has a strong effect on European weather, especially in winter. The NAM and NAO tend to vary in tandem, shifting positive or negative for periods lasting from a week or so to a month or more. Their strongest effects on climate occur in the winter. When the NAO turns negative, the polar jet stream tends to future but within this century. Hansen speculates that a 3°C (5°F) warming – near the IPCC's mid-range estimate of global temperature rise over the next hundred years – would eventually melt enough of the Greenland and Antarctic ice sheets to produce a spectacular sea-level rise of 25m (80ft). On top of that, the simple warming and expansion of ocean water could eventually add several more metres to sea level. In a 2007 paper for *Environmental Research Letters*, Hansen notes that, according to data from the GRACE satellite system, Greenland and Antarctica are losing around 150 cubic kilometres (36 cubic miles) of ice annually, a rate that's doubled over the last few years. He feels the IPCC and other groups need

buckle northwards across the Atlantic and then back south. This often throws northern Europe, eastern Canada and the northeast United States into extended cold spells, while milder, slow-moving storms drench the Mediterranean. A positive NAO, on the other hand, keeps Arctic air locked to the north and sends the pole-encircling winds more directly from the Atlantic across northern Europe, increasing the odds for mild and wet weather there as the Mediterranean stays dry. Thus, the projected shift of the NAM and NAO towards positive modes under global warming is one of the strongest clues to the type of climate Europeans can expect as greenhouse gases build up.

It's hard to assess ocean cycles that span more than a few years from peak to peak, because sea-surface temperature data before the advent of satellites in the 1970s is notoriously spotty. But many scientists are intrigued by the **Pacific Decadal Oscillation** (PDO), a seesaw of rising and falling temperatures across the northwest Pacific, and the **Atlantic Multidecadal Oscillation** (AMO), in which temperatures across much of the North Atlantic alternately warm and cool. Both cycles go for about 20–30 years before switching to their opposite phase. In both cases, the physical driver isn't yet known, though there may be a link to the global oceanic conveyor belt (see p.120). There's evidence for a link between the PDO and ENSO: when the PDO is in its positive phase, El Niños are more likely. And in the Atlantic, the AMO appears to boost the risk of Atlantic hurricanes, though it also complicates assessment of a well-publicized and much-feared risk to Europe's climate (see p.119).

Additional ocean cycles continue to pop up. The **Indian Ocean Dipole** (IOD) got its first widespread exposure in 1999 via the journal *Nature*. Like a smaller sibling of El Niño, the IOD involves abnormally warm and cool patches of ocean that build and decay along the Equator. A positive mode of the IOD (ie warm surface waters focused in the western Indian Ocean) helps boost monsoon rainfall across India. Given that some ocean cycles like the IOD are only now being analysed in depth, it'll take time to understand how climate change will affect each of them and how various cycles might interact in a warmer climate.

to emphasize the risks of higher-end MSL rise, in spite of the inherent uncertainties. "In a case such as ice sheet instability and sea level rise", says Hansen, "there is a danger in excessive caution. We may rue reticence, if it serves to lock in future disasters."

Most researchers don't appear worried that sea-level rises of this magnitude will occur over the next few decades. It's later in the century, and especially beyond 2100, that the risk of such impacts will clearly rise more steeply. The catch is that the greenhouse gases we're now adding to the atmosphere at unprecedented rates will continue to warm the climate for at least a century. Thus, the greenhouse ship is becoming progressively harder to turn around, and the seas are rising around it.

Climate change and El Niño

The oceans play a vast role in shaping the vagaries of weather and climate, and much of that influence comes through a set of **ocean-atmosphere cycles**. Linked to arrangements of high and low pressure centres over various parts of the world, these cycles each alternate between two "modes", producing recognizable, repetitive weather patterns – drought, excessive rainfall, unusual warmth or cold, and so forth. These can unfold half a world away from their oceanic triggers, and they're natural parts of climate. It's possible that global warming will tamper with them, but right now, scientists have more questions than answers about whether and how this will happen.

Globally, the most important ocean-atmosphere cycle is **ENSO** (El Niño/Southern Oscillation), whose two modes are known as **El Niño** and **La Niña**. ENSO is based in the tropical Pacific Ocean, which spans a third of the globe from Ecuador to Indonesia. Trade winds blow more or less continuously across this huge area from east to west, pushing warm water towards Indonesia, where the resulting balmy seas help generate persistent showers and thunderstorms. The cold, upwelled water off Ecuador and Peru, meanwhile, stabilizes the air there and produces the region's legendary aridity (Peru's capital, Lima, gets about the same amount of rain each year as Cairo). About every two to seven years, the trade winds weaken or reverse, the surface layer of warm water deepens and expands into the eastern tropical Pacific, and an El Niño sets in, typically lasting one or two years. The flip side, La Niña, occurs when trade winds are stronger than average, pushing cooler-than-usual water westward into the central tropical Pacific. About half of the time, neither El Niño or La Niña is in progress and the Pacific is neutral.

El Niño increases the odds of drought across Indonesia, Australia, India, southeast Africa and northern South America. It tends to produce mild, dry winters in Canada and the northern US and cool, moist winters in the US South and Southwest. It also raises the chances of hurricanes in the Atlantic, and lowers them for parts of the North Pacific. La Niña, in general, has the opposite effects.

Currently, even the best computer models struggle to depict the tropical Pacific's arrangement of ocean currents and rainfall patterns during neutral periods, so it's not yet clear how global warming will affect the mix. Some models suggest an intensified ENSO, while others point to a weaker one. Some show a sustained La Niña-like scenario. As a whole, though, the model ensembles in the 2007 IPCC report point toward a future climate that could resemble a semi-permanent El Niño, with individual El Niños and La Niñas still occurring on top of that new base state.

If there's been any trend in recent decades, it's in favour of El Niño. The period 1977–2007 saw only two moderately strong La Niñas, compared to four potent El Niños, including the two strongest ones of the last century. This trend could be part of a "natural" multi-decade pattern (more on this below), or it could reflect a warming caused by climate change. Towards Indonesia, the sea appears to be almost as warm as it can get, averaging above 29°C (84°F). Any further heating might trigger enough storminess to cool that patch of ocean back down. But, according to Gerald Meehl of the US National Center for Atmospheric Research, it's possible that the warm pool could build eastward into the central Pacific, moving the region's typical state closer to that of El Niño (as suggested by the IPCC analysis mentioned above).

Will the Atlantic turn cold on Britain?

"Britain faces big chill as ocean current slows down", warned the London *Times*. "Gulf Stream 'engine' weakening", cautioned *The Independent*. These headlines from 2005 depict a chilly alternative to the standard portrait of a warmer climate in the United Kingdom of the future. Even as global warming appears to be proceeding full steam ahead, there are hints that the UK and parts of the European continent could be in for a countervailing force – a cooling influence that might inhibit human-induced warming in some areas.

The storyline will be familiar to those who caught 2004's blockbuster *The Day After Tomorrow* (see p.275), which took the idea to cartoonish extremes. The nugget of truth in the film is that the northward flow of

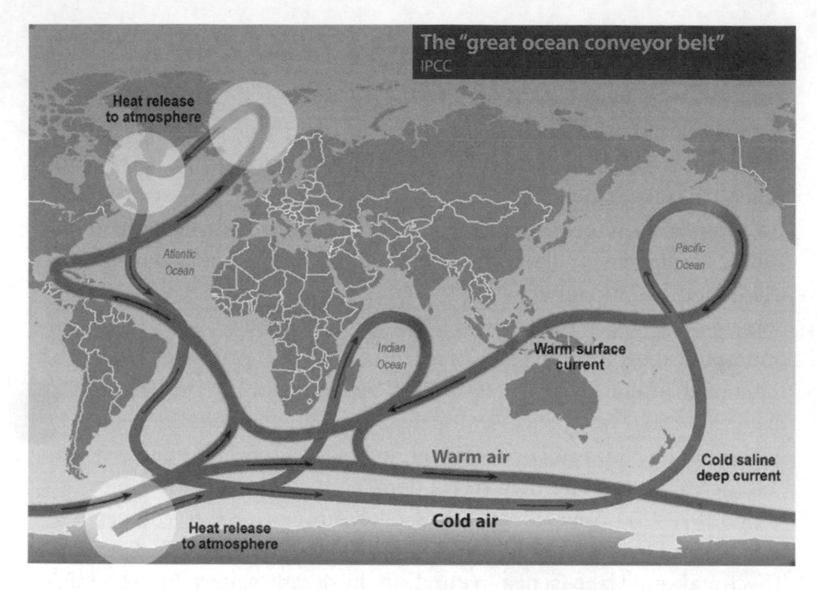

warm water into the far North Atlantic could slow or even shut down entirely, which would plunge much of northwest Europe into a climate more akin to that of Canada, Russia or other countries at the same latitude. In Hollywood's view, this occurs in a matter of days, flash-freezing almost everyone from Washington to Scotland. In real life, any cooling should be far less dramatic and would unfold over at least a few years, more likely decades to centuries. Even so, the possibility is worth taking seriously.

The Atlantic flow is part of the global loop of ocean circulation (see graphic above). It's a system whose mechanics are well mapped but not fully understood. One of the most important branches is often referred to as the **Atlantic thermohaline circulation,** because it's related to both temperature ("thermo") and saltiness ("haline"). Many scientists now use a different name: the Atlantic **meridional overturning circulation** (MOC). "Meridional" refers to its north-south orientation, and "overturning" to the somersault it performs in the far north. Before it gets to that point, the MOC is a vast conveyor belt of warm water that flows around South Africa and then northward. In the North Atlantic, prevailing winds sculpt the sea-surface flow into a clockwise gyre, sending warm water north along the US coast in the **Gulf Stream** and bringing cool water south from Spain in the Canary Current. An extension of the Gulf Stream called the **North Atlantic Drift Current** pushes warm surface water even further northeast towards the British Isles. (Journalists often use the term "Gulf Stream" to refer to the North Atlantic Drift Current, but strictly speaking that isn't correct.)

The flip side of the conveyor occurs when the warm surface water approaches Iceland and splits around it, heading for the Labrador and Norwegian Seas. Along the way, it gradually chills and sinks. Then, at several depths ranging from about 2 to 4 km (6600–13,100 feet), the frigid deep water begins the return trip to the South Atlantic, thus balancing the warm northward conveyor and completing the MOC.

By any measure, the MOC is a serious workhorse. As it passes between the Bahamas and the Canaries, around 25°N, the circulation brings an average of roughly a petawatt of heat energy northward, day in and day out. That's enough energy to run ten trillion 100-watt light bulbs, or more than fifty times the energy being used by all the homes, offices, and factories on Earth at any moment.

Scientists have speculated for more than a century that this giant natural turbine might have more than one speed setting. Today, based on paleoclimate evidence, many ocean specialists see the MOC as having three distinct modes: essentially "faster", "slower" and "off". During the last ice age, the MOC appears to have shifted among these three modes. Since Earth began to warm up about 12,000 years ago, the MOC has stayed fairly robust, except for a couple of dramatic weakenings at around 11,500 and 8,200 years ago, as glacial meltwater poured into the North Atlantic and suppressed the circulation. These events were enough to cool the winter climate in and around the present-day UK by as much as 10°C (18°F) below present-day readings. In other words, London in a typical January would have felt more like St. Petersburg does today.

The main worry for this century is the increasing flow of freshwater into the far North Atlantic. This comes from increased rainfall and snowfall across the Arctic and adjacent land areas, as well as from melting glaciers. Because fresh water is less dense than salty water at the same temperature, the fresh water has more trouble sinking and forming the Atlantic's cold, deep return flow. Analyses led by the UK's Centre for Environment, Fisheries and Aquaculture Science show that the upper 1000m (3300ft) of the Nordic seas – the main source of sinking water for the Atlantic – have become steadily fresher since the 1960s.

Computer models agree that the flow of fresh water into the far North Atlantic ought to continue increasing. That could put a substantial brake on the MOC of the twenty-first century. In response to this concern, the UK's Natural Environment Research Council is funding a £20 million study called RAPID, launched in 2001 and due to run until 2008. The project has already produced more realistic computer models of how the MOC might evolve this century. At the time of RAPID's launch, the UK Hadley Centre had produced simulations in which the MOC was halted in its tracks. The results showed a net cooldown of around 1–3°C (1.8–5.4°F) by 2050 across most of the UK, Scandinavia, Greenland and the adjacent North Atlantic, with warming elsewhere around the globe. However, as the models have become more sophisticated due to RAPID and other studies, they've produced a more modest slowdown of the MOC. A 2005 comparison of eleven different models coordinated by Jonathan Gregory of the University of Reading showed the MOC weakening by anywhere from 10% to 50%, even with a quadrupling of greenhouse gases. The 2007 IPCC report concurs with this general picture. In all of these more recent models, a net warming is shown for the British Isles and nearby countries, with greenhouse gases more than compensating for the MOC slowdown.

On top of these big changes, there are signs of a more subtle speed-up/slow-down cycle in the MOC, with each phase peaking about every sixty or seventy years. Measurements in the deep Atlantic are too sparse to show this trend, and there's no solid physical explanation for it. Nevertheless, sea-surface temperatures across the North Atlantic show several distinct rises and falls since the late 1800s – a pattern referred to as the **Atlantic Multidecadal Oscillation** – and some scientists take this as evidence of the MOC waxing and waning. Interestingly, global air temperatures have risen and fallen over a roughly similar timeframe, which leads some scientists to wonder how the two processes might be intertwined.

If the sixty- to seventy-year cycle is robust, then the MOC ought to undergo a slight natural slowdown starting around the 2010s and lasting into the 2040s or thereabouts. The challenge will be to untangle this from the potentially larger human-induced slowdown. Improved measurements should help. In 2004, the RAPID project installed a set of moored instruments (maintained in collaboration with the US) that monitors the MOC continuously along 26.5°N. These data have already countered a dramatic and much-publicized 2005 study published in the journal *Nature*. In that paper a team led by Harry Bryden of the UK National Oceanography Centre found that the MOC slowed by about 30% overall between 1957 and 2004. But by 2007, data from the RAPID mooring showed that the MOC's strength can vary by a factor of nine over a year's time, which suggests that natural ups and downs might account for the apparent long-term drop found by the Bryden team.

#300 14-12-2017 11:36AM
Item(s) checked out to p10575893.

TITLE: The rough guide to climate change
DUE DATE: 11-01-18

Thanks for using Merritt Library.
TEL: 250 378-4737 WEB: www.tnrdlib.ca

As the data from RAPID continue to amass, we'll get an even clearer sense of the MOC's recent behaviour and how it may change with the climate. In the meantime, Britons might be best off planning for a warmer climate while learning to live with a bit of uncertainty about exactly how warm it might be.

Living in a changing sea

The creatures that make their home in the world's oceans might seem to have much more flexibility to deal with climate change than do their land-based cousins. But do they? Marine biologists are still sorting through the many interconnections among ocean species and trying to map out how things might change as the sea warms up.

As with life on land, there will be winners and losers. Migratory fish should in principle be able to shift their ranges relatively easily. But not all components of the oceanic food chain will change in unison. A small rise in ocean temperature can affect the mix that exists among the tiniest marine plants and animals (**phytoplankton** and **zooplankton**). That, in turn, can produce mass die-offs of seabirds and fish further up the chain. A 2006 study using NASA satellite data showed a global-scale link between warmer oceans and a drop in phytoplankton abundance and growth rates.

Scientists can zero in on processes like this by observing the sharp regional warmings produced by El Niño. Off the coast of Peru, the warm water of El Niño cuts off the upwelling of colder, nutrient-rich waters that support a variety of marine life. Back in 1972, an El Niño intensified the effects of overfishing and decimated Peru's lucrative anchovy industry. Many birds that fed on the anchovies died off, and in turn the local supply of *guano* – bird droppings used as fertilizer – plummeted. Such chains of events aren't limited to the tropics. In 1997, unusually warm waters developed in the eastern Bering Sea (which lies between Alaska and Russia), killing off the type of phytoplankton that normally dominate during the summer. The disturbance worked its way up to devastate local zooplankton, seabirds and salmon.

As global warming unfolds, it's possible that some of the most dramatic impacts on ocean life will occur through intense seasonal and regional spikes in water temperature driven by El Niño and other ocean cycles. It's somewhat analogous to the way climate change makes its presence most obvious and discomforting to humans during extreme heat waves.

There's more to consider than warming alone. Some ocean life is more dependent on seasonal changes in light than on temperature, especially at higher latitudes. Other species are more purely temperature-driven. Since cycles of sunlight aren't changing, the most light-dependent creatures haven't shifted their timing, whereas many species that are more temperature-dependent are starting their yearly growth cycle at earlier points in the warm season. A 2004 study of the North Sea by Martin Edwards and Anthony Richardson of the Sir Alister Hardy Foundation for Ocean Science found that the differences between these groups is leading to a large-scale mismatch, threatening the seasonal synchrony in growth and feeding times that keeps much of the sea's food chain humming.

Ocean chemistry is changing, too, as the waters soak up enormous amounts of carbon dioxide. Each year a net influx of roughly seven gigatonnes of CO_2 – including close to 25% of all the carbon dioxide produced by human activity – goes into the sea. In the well-mixed upper part of the ocean, where the absorption happens, the extra CO_2 is gradually making the water more acidic – or, to be more precise, the water is becoming less alkaline. Average pH values in the top 100m (330ft) of the ocean have decreased from around 8.15 in pre-industrial times to about 8.05 today. That change may seem small, but because pH is measured on a logarithmic scale, every change of one point is equal to a tenfold increase or decrease. A 0.1-point drop in pH translates to a 30% increase in hydrogen ions throughout the upper ocean. A 2004 report on ocean acidification by the UK's Royal Society estimates that the average pH will drop to 7.65 by the year 2100. A pH value of 7 is neutral, so the oceans will still be slightly alkaline, but less so than they've been for hundreds of thousands of years.

Scientists are only beginning to probe the impacts of acidification on marine life. One group that's likely to be affected is corals and algae that secrete tiny shells of calcium carbonate. As acidification proceeds, the carbonate ions they rely on will become more scarce. On the plus side, it appears higher ocean temperatures tend to speed up the calcification process. For some species, this positive effect might outweigh the negative effects from the growing scarcity of carbonate. However, these interactions aren't well mapped out yet.

Even if humans can turn the tide on greenhouse-gas emissions, it will take tens of thousands of years more before the ocean returns to its pre-industrial pH levels. Some engineering fixes have been proposed – for instance, adding mammoth amounts of limestone to the ocean to reduce its acidity – but the Royal Society warns that these are likely to succeed only on a local level, if even then. Any such technical fix is laden with ecological question marks: for instance, could an attempt to de-acidify the oceans cause something even more harmful to occur?

Coral reefs at risk

The most famous oceanic victims of climate change may be the magnificent **coral reefs** that lace the edges of the world's subtropical and tropical waters. More than a quarter of the planet's coral reefs have already been destroyed or extensively damaged by various types of human activity and by spikes in water temperature that may be intensified by overall ocean warming. Nobody expects climate change to completely eliminate the world's reefs, but many individual coral species may face extinction.

It's amazing that coral reefs manage to produce the colour, texture and overall splendour that they do. After all, they grow in the oceanic equivalent of a desert: the nutrient-poor waters within 30m (100ft) of the surface. The reefs' success in this harsh environment is because of their unique status as animal, vegetable and mineral. Coral reefs consist of **polyps** – tentacled animals that resemble sea anemones – connected atop the reef by a layer of living tissue. Inside the tissue are microscopic **algae**

Coral reefs such as this one in the Red Sea are at risk in a warming world
DiMaggio/Kalish/CORBIS

that live in symbiosis with the polyps. The algae produce the reefs' vivid colours and feed the polyps through photosynthesis, while the polyps provide the algae with nutrients in their waste. The polyps also secrete **limestone** to build the skeletons that make up the reef itself.

Coral reefs are durable life forms – as a group, they've existed for at least forty million years – but they're also quite selective about where they choose to live. The microalgae need access to light in order to photosynthesize, and the skeleton needs a foundation, so coral reefs form in shallow, clear water atop a base that's at least moderately firm, if not bedrock. Finally, the waters around coral reefs need to be warm, but not too warm. Generally, the average water temperature should be above 27°C (81°F), with annual lows no cooler than 18°C (62°F). If the waters spend much time at temperatures more than 1–2°C (1.8–3.6°F) above their typical summertime readings, the algae can begin to photosynthesize at a rate too fast for the coral to handle. To protect its own tissue, the coral may expel the microalgae, allowing the white skeletons to show through the translucent polyps. When this occurs on a widespread basis, it's known as **mass bleaching** (although chlorine has nothing to do with it).

The record-setting El Niño of 1997–98 led to unusually warm water in much of the tropical Pacific, triggering a bleaching disaster unlike anything ever recorded. Some 16% of the world's coral reefs were damaged in that event alone. A weaker El Niño in 2002 bleached up to 95% of individual reefs across parts of the western Pacific. Some coral species appear to be bouncing back from the 1997–98 bleaching, although full recovery could take up to twenty years. Other species – especially those with lavishly branched forms and thin tissues – don't appear to be so lucky. Experiments hint that some coral may be able to adapt to bleaching over time by teaming up with different microalgae more suited to warmer temperatures. The upshot over the long term may be a thinning of the herd, as it were, leaving behind a hardier but less diverse and less spectacular set of coral reefs.

To complicate the picture, global warming isn't the only threat the reefs face:

- **Agricultural fertilizer** that washes offshore can deposit nitrogen and phosphorus that encourage the formation of seaweed or phytoplankton at the expense of polyps.

- **Overfishing** can also promote seaweed growth atop reefs by removing many of the plant-eating fish that would otherwise keep the growth in line.

▶ **Sediment** from rivers can block needed sunlight, interfere with the polyps' feeding, or even bury a reef whole. Logging and other land-based activities have produced enough sediment downstream to put more than 20% of the coral reefs off Southeast Asia at risk, according to the World Resources Institute.

The ultimate impact of climate change on reefs depends in part on how extensive these other stressors become. By itself, the influence of global warming is a mixed bag. Sea-level rise is a potential positive. Many older coral reefs have built upwards to the maximum height that temperatures and other conditions allow, so a higher sea surface would give these reefs more vertical room to grow. Meanwhile, the lowest-lying coral reefs could probably grow upward quickly enough to keep up with modest sea level rises – a typical vertical growth rate is about 1m (3ft) in a thousand years. All else being equal, warmer temperatures appear to speed up the calcification process that builds reefs. And the global spread of warmer waters will add a sliver of territory around the current margins of coral-reef viability, though that extra real estate is limited by other requirements such as a firm sea surface.

Together, these pluses may not be enough to counter the negatives produced by the environmental stresses noted above, not to mention global warming itself. Along with causing bleaching, warmer waters may nourish the enemies of reefs as much or more than they help the reef itself. One study by Drew Harvell of Cornell University found that some pathogens tend to thrive at temperatures 0.6°C (1.0°F) or more higher than the optimal readings for coral themselves. Another risk is the ocean acidification discussed above, which can both limit the amount of carbonate available for reef building and perhaps erode the reef skeletons directly, in both cases slowing the rate of reef growth. And tropical cyclones may become more intense with time (see p.133), although it appears that a healthy reef can recover from hurricane or typhoon damage fairly quickly.

In short, coral reefs should still populate the world of our grandchildren, but they may not be in the same locations or glow with the same brilliance that snorkelers and scuba divers now savour. A 2004 report by the Pew Center for Global Climate Change summarized the situation: "Coral reefs of the future will be fewer and probably very different in community composition than those that presently exist."

Hurricanes & other storms

Rough waters

Spinning their way across the warmer parts of the globe, tropical cyclones throw fear into millions of coastal dwellers each year. Depending on which ocean they're in, these huge weather systems are also known as hurricanes, cyclones or typhoons. Whatever you call them, they've been on a rampage lately. Five hurricanes struck or side-swiped the United States in 2004, with four of those slapping woebegone Florida. Meanwhile, a record ten typhoons raked Japan.

Things got even worse in 2005, at least for North America. Based on measurements of their lowest pressure, the year brought three of the six most powerful Atlantic hurricanes ever observed: **Wilma** (the strongest), **Rita** (fourth strongest) and **Katrina** (sixth strongest). All three made devastating strikes on coastal areas. Wilma slammed Cancún and south Florida, Rita pounded the Texas/Louisiana border, and Katrina swept through Miami before bringing mind-boggling misery and more than 1800 deaths to the New Orleans area and the Mississippi coast. Not to be outdone, Australia was hammered in early 2006 by a trio of cyclones, each packing winds above 250kph (155mph) either offshore or on the coast. It's the first time the region has recorded three storms of that calibre in a single season.

Did global warming have anything to do with all this havoc? That question couldn't be dodged in the wake of 2005's violent Atlantic season. As if on order, several major studies made a compelling argument that tropical cyclones worldwide are indeed getting stronger. Meanwhile, a number of hurricane experts (many of them forecasters) insisted that global warming played only a minor role at best in the recent rash of activity, and that

Rescue workers in New Orleans after hurricane Katrina
Jocelyn Augustino/FEMA

it will be far outweighed by other factors in the future. All this has made tropical cyclones one of the liveliest arenas of climate-change debate.

To the north and south of the tropics, people tend to be more concerned with **coastal storms**. These wintry blasts of rain, snow and wind sweep through places such as New England, British Columbia and Western Europe, packing less punch than hurricanes but often affecting much larger areas. Coastal storms appear to be shifting gradually polewards over time. Though the trends aren't highly dramatic, they could have major importance for the densely populated coastlines of North America and Eurasia.

Well away from the coasts, the most intense bouts of wind and rain are likely to occur with severe **thunderstorms**, which occasionally pack the tiny but powerful windstorms known as **tornadoes**. (Many hurricanes also produce tornadoes as well.) Although tornadoes are being reported more often in many parts of the world, it looks as if that can be pinned on greater awareness of them. Thankfully, there's no reason to believe that the most violent tornadoes and thunderstorms have become more intense or frequent due to global warming, though some hints of future change are now showing up (see p.143).

A hurricane rolls down towards Rio

People in Brazil aren't used to hurricanes. In fact, no hurricane had ever been reported there until March 28, 2004, when a mysterious system packing winds up to 137kph (85mph) swept onto the coast of Santa Catarina province. Forecasters from Brazil and from the US National Hurricane Center had been tracking the storm by satellite. Although no tropical cyclones had ever been officially recorded in the South Atlantic, the swirl of clouds had the clear eye and other telltale features of a bona fide hurricane. Despite the absence of a precedent, Brazil's meteorological service issued warnings and got people out of harm's way. The storm ended up destroying many hundreds of coastal homes, but only one death was reported.

Christened Hurricane Catarina for its landfall location, the storm continued to stir things up even after it died. Some meteorologists questioned whether it was truly a hurricane, and a few climate-change activists pointed to Catarina as an illustration of a planet gone awry. In fact, it turns out that Catarina had some of the classic earmarks of a hurricane, but not all of them. For instance, the showers and thunderstorms circling its centre were more shallow than usual. And Catarina can't be easily pinned on climate change, because the waters over which it formed were actually slightly cooler than average for that time of year. The storm was clearly unprecedented in its landfall impacts, but analysts poring over satellite records found at least one similar (if weaker) system in 1994 that stayed well out to sea. In short, Catarina remains a cipher – a bizarre weather event that pushed the buttons of a world increasingly jittery about climate change.

A taste of things to come?

Two main questions loom over any discussion of future tropical cyclones: will there be more of them in a warmer world, and will they be more intense? In its 2001 report, the **IPCC** noted that no trend had yet been identified in tropical cyclone intensities, as measured by winds and pressures, adding that "past and future changes in tropical cyclone location and frequency are uncertain". Things had changed dramatically by the time of the 2007 IPCC report, which reflects a growing body of evidence that the number of intense hurricanes and typhoons has jumped in recent decades.

The notion of stronger hurricanes and typhoons in a warmer world makes physical sense. **Warm ocean waters** help give birth to tropical cyclones and provide the fuel they need to grow. Generally, the cyclones can't survive for long over sea surfaces that are cooler than about 26°C (79°F). The larger the area of ocean with waters above that threshold, and the longer those waters stay warm in a given year, the more opportunity there is for a tropical cyclone to blossom if all other conditions allow for it.

Not all areas are conducive: within about 5° of the Equator, the winds can't easily gather into a rotating pattern. And the South Atlantic doesn't seem to have the right combination of upper-level winds and warm surface water to produce hurricanes, although 2004 brought a major surprise (see box opposite). That leaves the prime breeding waters for tropical cyclones between about 10° and 30° across the North Atlantic, North Pacific, South Pacific and Indian Oceans (see map on p.134).

The terrible 2005 trio of Katrina, Rita and Wilma show how important warm water is for hurricanes. Katrina and Rita both surged to their peak intensity – Category 5 (see p.136) – just as they passed over a patch of the central Gulf of Mexico heated by an infusion of deep, warm water from the Caribbean. This channel, called the **Loop Current** for its arcing path into and back out of the Gulf, is often a hurricane booster. In 2005 it featured some of the warmest waters ever observed in the open Gulf, topping 32°C (90°F) in places. Similarly toasty waters had also overspread much of the Caribbean, where Wilma intensified in an astonishing twelve hours from Category 1 status to attain the lowest barometric pressure ever observed at sea level in the Western Hemisphere: 882 hectopascals (26.05 inches of mercury).

To keep a truly intense hurricane at top strength, the ocean's warmth needs to extend deep – ideally at least 100m (330ft). Otherwise, the powerful winds and waves can churn up enough cold water from below to dilute the fuel source and diminish a slow-moving cyclone in a matter of hours. Both Rita and Katrina began weakening after they left the Loop Current; they began to pull up cooler subsurface waters, while drier air and stronger upper-level winds from the United States started to infiltrate. By the time they struck the **Gulf Coast**, both storms were down to Category 3 strength. That didn't matter much, though, because the storms had already piled up fearsome amounts of water that slammed into the coast in the form of punishing storm surges. Katrina's surge was a US record: at least 8.2m (27ft) above sea level at its worst, with waves atop that, destroying the homes of thousands who thought they lived at safe elevations. The surge also helped lead to the failure of several levees in and around **New Orleans**, which in turn produced the cataclysmic flood that left most of the city submerged and paralysed (see box overleaf).

The wreckage of the 2005 Atlantic hurricane season extended far beyond US shores. Ocean temperatures were markedly above average

Katrina and climate change

The damage and misery wrought by 2005's Hurricane Katrina topped anything produced by a single US storm for many decades. More than 1800 people died, mostly in the New Orleans area, where public evacuation options outside of the wretched Superdome and convention centre were nearly non-existent. The storm's toll in property damage soared well into the tens of billions of US dollars, and that doesn't include the cost of restoring the New Orleans levee system to its previous condition – much less the billions more needed to make the system able to withstand a direct hit from a Category 5 hurricane.

Some observers linked the Katrina debacle to climate change from the outset. In a *Boston Globe* editorial, journalist Ross Gelbspan declared, "The hurricane that struck Louisiana yesterday was nicknamed Katrina by the National Weather Service. Its real name is global warming." Others denied any connection between Katrina and climate change, a viewpoint that came to dominate US media and legislative discussion before long. It's true that no weather event can be blamed solely on climate change, and certainly a storm like Katrina doesn't require global warming in order to flex its muscle. Though they're quite rare, hurricanes on par with Katrina have developed in the Atlantic since records began. However, the gradual warming of tropical waters over the last several decades has made it easier for storms like Katrina to intensify when other conditions are right. Perhaps the most telling hint of a shift in climate isn't Katrina itself so much as Emily, Katrina, Rita and Wilma combined. It's the first time four, or even three, Atlantic hurricanes have reached Category 5 strength in a single year in records that date back to 1851.

NASA-GSFC, data from NOAA GOES

As for the landfall in New Orleans, that was simply the luck (albeit the bad luck) of the meteorological draw. Several near-misses had grazed the city since 1965, when Hurricane Betsy produced flooding that killed dozens. Betsy prompted the US Corps of Engineers to build the city-girdling levee system that was in place when Katrina struck. That system was built to withstand only a Category 3 storm, a fateful move sure to be analysed by critics for years to come – especially since Katrina itself had weakened to Category 3 strength by the time it made landfall.

across vast stretches of the North Atlantic tropics and subtropics. All that long-lasting warmth, coupled with nearly ideal upper-level winds, led to 27 named Atlantic storms in 2005, which smashed the record of 21 set in 1933. For the first time, the US National Hurricane Center exhausted its alphabetic Atlantic list of 21 names and moved into the Greek alphabet. Thus we saw **Alpha**, **Beta**, **Gamma**, **Delta**, **Epsilon** and finally – on December 30 – **Zeta**, a fitting postscript. Still another system was upgraded to tropical-storm strength posthumously, bringing the season's final total to 28. This swarm of storms was noteworthy not just for its sheer size but for its unusual geographic spread as well. The remnants of Tropical Storm Delta brought rare hurricane-force wind gusts to the Canary Islands, where seven people died and more than 200,000 residents lost power. Flooding and mudslides produced in part by **Hurricane Stan** killed over 1000 people in Guatemala, El Salvador and Mexico. Oddly, even Spain got its first-ever tropical storm, **Vince**, which made landfall on the southwest coast near Huelva.

Keeping count: will there be more cyclones in the future?

The sheer number of Atlantic hurricanes in 2005 led many to ask whether tropical cyclone frequency might already be on the rise due to climate change. One might think that a larger, more long-lasting zone of warm-enough water would be enough to spawn more and more tropical cyclones. And indeed, 2005 saw Atlantic waters that were unusually warm over vast areas for long periods. Yet the question of frequency is a complex one, for several reasons.

▶ **Other factors** Even over sufficiently warm water, only a small fraction of the weak circulations that form in a given year spin up to cyclone status. A number of factors can lead to their early demise: unfavourable upper winds, an influx of dry air, or competition from nearby systems, to name a few. Warm water is critical, but it isn't enough to guarantee that a hurricane will form.

▶ **Variable records** It's hard to know exactly how many tropical cyclones roamed Earth before **satellite monitoring** became routine in the 1970s. Although ships did a good job of reporting stronger tropical cyclones in the pre-satellite era – at least across major shipping lanes – it's likely that some of the weaker, smaller systems over the open ocean were missed then but are being caught now.

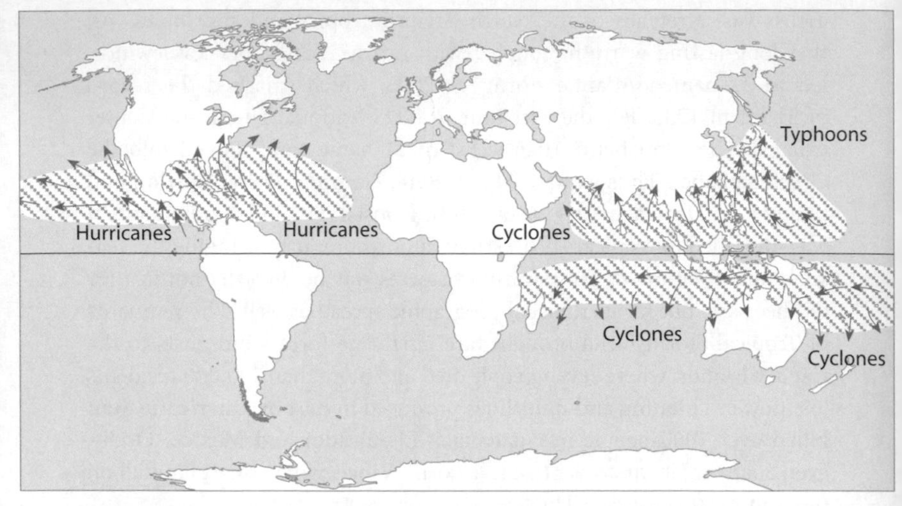

Regions prone to hurricanes, typhoons and cyclones (different names for the same phenomenon)

▶ **Complex global interrelations** For reasons not fully understood, the planet does a good job of conserving the total number of tropical cyclones. About 40–60 per year attain hurricane strength, and there's been no significant trend up or down during the few decades of reliable data. The frequency stays fairly constant because the planet's ocean basins appear to trade off in their periods of peak production. Typically, when the Atlantic is producing hurricanes galore, parts of the Pacific are relatively quiet, and vice versa. In 2005, while the Atlantic was setting hurricane-frequency records, the northeast Pacific saw only seven hurricanes, compared to a yearly average of nine. Some of this seesaw effect is due to **El Niño** and **La Niña**, the cyclic warmings and coolings of the eastern tropical Pacific (see p.118). The large-scale air circulations driven by La Niña encourage Atlantic hurricanes but suppress them in some parts of the Pacific; the opposite is true of El Niño.

At present, **computers** aren't of much help in assessing whether we'll see more tropical cyclones in a warmer world. Most hurricanes and typhoons are small enough to fall in between the grid points of the comprehensive global models that map climate out to a century or more. It's also tough for a model to pick up on the meteorological subtleties that allow one system to become a hurricane and another not to – the same unknowns that plague tropical weather forecasters on an everyday basis. An early sign of progress came in 2006, when a group of Japanese scientists led by Kazuyoshi Oouchi published results from a globe-spanning model with 20km (12-mile) resolution, enough to follow individual tropical cyclones in a broad-brush fashion. The results suggested that in a mid-range emissions scenario, the late-21st-century Earth could see fewer tropical cyclones but a higher fraction of intense ones. The North Atlantic – where ocean temperatures and other conditions are often marginal for hurricane formation – is the one place where some models indicate an increase in tropical cyclones over time. Within a few years, more models should have enough resolution to be able to depict tropical cyclones more precisely and shed more light on the frequency problem both globally and regionally.

The question of whether or not we'll see more tropical cyclones is problematic for another reason: most of those cyclones never reach land. Even if a given ocean began to spawn more cyclones, it doesn't necessarily mean that more of them will hit the coastline. Indeed, even a slow season can bring destructive landfalls. In 1992, the Atlantic was far less busy than average, yet it still produced Florida's catastrophic **Hurricane Andrew**. Conversely, it's possible for most of a busy season's systems to churn harmlessly out to sea.

Packing more power

What about strength? Here, the tune is different, and more ominous. Though the IPCC saw little evidence of change in its 2001 report, subsequent studies have found a jump in the power of tropical cyclones on a global basis, as noted in the 2007 report. An increase in hurricane strength is both plausible on a physical basis and apparent in some of the most recent modelling.

Two key studies pushed this topic into the headlines in 2005, where it shared space with the year's bumper crop of Atlantic super-storms. Kerry Emanuel of the Massachusetts Institute of Technology took a new look at tropical cyclones since the 1970s, when satellite monitoring became widespread. He looked not just at the raw wind speeds but at the destructive force they packed over their lifespan. This force,

"While we don't know precisely how global warming will change hurricanes, that's not really the point. What matters is that today, we know enough to be worried."

Chris Mooney, *Storm World*

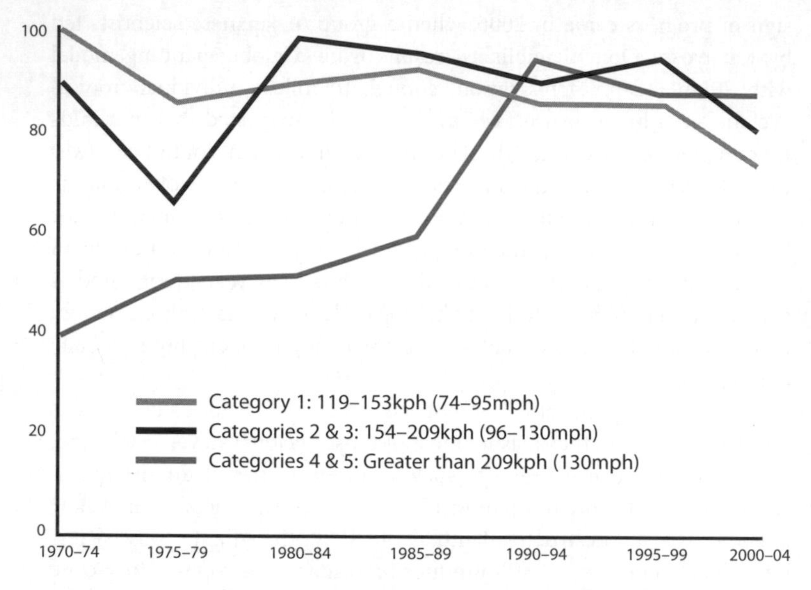

Tropical cyclone totals for each five-year period, 1970–2004
While the total number of tropical cyclones around the globe ranked Category 1 to 5 hasn't changed much in the last thirty years, the number of Category 4 and 5 cyclones – those that produce the worst damage – has nearly doubled.

Adapted from P. J. Webster et al, *Science* 309: 1844–1846 (September 16, 2005). Reprinted with permission from AAAS

the **dissipation of power**, is proportional to the wind speed cubed. Thus, a wind of 100 kilometres per hour would be twice as damaging as a wind of 80kph. Small increases in wind speed, then, can produce major ramp-ups in destructive force. From this vantage point, Emanuel discovered that tropical cyclones in the Atlantic and northwest Pacific are now packing nearly twice the power they did in the 1970s – the result of the strongest storms getting even stronger as well as lasting longer.

A team led by Peter Webster of the Georgia Institute of Technology found similar results when they ranked tropical cyclones since the 1970s by **intensity category** (see graph). As expected, they found little change in the total number of tropical cyclones in all categories – a sign of the global balance noted above. The eye-opener was their discovery that the strongest cyclones, those at Category 4 or 5, were almost 50% more frequent globally in 1990–2004 than they were in 1975–1989. Put another way, Category 4 and 5 storms went from comprising about 25% of all tropical cyclones globally to comprising about 35% of them.

Predictably, these studies provoked a storm of controversy. They also exposed a gulf between climate-change researchers, few of whom had done detailed studies of tropical cyclone behaviour, and hurricane specialists, who knew the cyclones and the data by heart but weren't always experts on the larger-scale processes related to global warming. In his 2007 book *Storm World*, science writer Chris Mooney cast the battle as only the most recent example of a tiff between observationalists and theoreticians with roots extending back to the 1800s.

One of the most vocal critics of the new and worrisome findings was William Gray (of Colorado State University), who founded the field of seasonal hurricane prediction. A long-time sceptic of climate change, Gray assailed the studies on various technical grounds. His scepticism got substantial play in US media and gave him an audience among sympathetic members of Congress, where he testified in 2005.

Also dismissive of a climate-change link was the **US National Oceanic and Atmospheric Administration** (NOAA), whose **National Hurricane Center** (NHC) forecasts Atlantic storms. The initial wrap-up of the 2005 hurricane season on NOAA's website made no mention of global warming, a reflection of the agency's top-down efforts at the time to put forth a unified stance on issues related to global warming. Individually, some NOAA scientists acknowledged that human-induced climate change has the potential for "slightly increasing the intensity of tropical cyclones", in the words of Christopher Landsea, who oversees research at NHC. However, Landsea and colleagues suggested that the apparent surge in top-strength hurricanes could be at least in part due to inconsistencies in the official record. For instance, lower-resolution satellites used in the 1970s may not have sensed the tiny areas of high, cold cloud tops associated with the strongest cyclones. Other observational tools, such as the instrument packages dropped from aircraft into hurricanes, have also evolved over the years, as have the analysis techniques that classify the storms. Landsea and colleagues are now in the midst of a multi-year effort to reanalyse several centuries of Atlantic hurricanes, which will likely result in some hurricanes being upgraded and others downgraded.

Another question mark is the **Atlantic Multidecadal Oscillation**. It's been known for years that, over the past century, sea-surface temperatures in the North Atlantic have alternated between warming and cooling for periods of about 30–40 years each. Many have tied this waxing and waning to presumed cycles in the **Atlantic thermohaline circulation**. This circulation is only now being measured with some precision (see p.56), and its dynamics are poorly understood. Nevertheless, Gray showed that the frequency and average intensity of Atlantic hurricanes has risen and fallen every few decades in sync with the AMO – lower from the 1890s to the 1920s, higher from that point through the 1950s, then lower again from the 1960s to the 1990s. Based on this, Gray predicted an increase in Atlantic hurricane action in the mid-1990s that would continue for several decades. Thus far, his vision has proven remarkably accurate.

But even if there is a strong physical basis for the AMO, it doesn't mean climate change can be ignored. The AMO only affects hurricane formation in the Atlantic, whereas the studies mentioned above show a strengthening trend across several other ocean regions as well. Moreover, the recent Atlantic up-tick is a combined product of the AMO *and* general ocean warming. A 2006 study by Kevin Trenberth (of the US National Center for Atmospheric Research) found that global warming had more than twice the influence of the AMO in producing 2005's extra hurricane-fuelling warmth in the waters of the tropical North Atlantic. Thus, if and when the AMO cycle finally turns towards a quieter Atlantic – perhaps around 2015–2025 – things may not calm down as much as promised.

Despite their limited ability to project tropical cyclones into the future, several global computer models agree that overall warming ought to help induce a modest increase in strength. One of the most thorough studies, led by Thomas Knutson of NOAA in 2004, used nine different models and three regions (the Atlantic and the northeast and northwest Pacific). The consensus was for an increase of a few percent in peak tropical cyclone winds by the late 21st century. This translates to the typical hurricane gaining about a half-category in strength. (If anything, current events might be outrunning this projection.) In more general terms, the 2007 IPCC report deemed "likely" an increase in both peak wind speeds and rainfall intensities within tropical cyclones across most of the globe.

Surges and downpours

Extreme **storm surges** will continue to loom as a catastrophic possibility from tropical cyclones in the century to come, but the shape of that threat will shift as sea levels rise and the coast itself evolves. **Subsidence** and the loss of protective **wetlands** are already making many areas more vulnerable, including the Texas and Louisiana coastlines hammered in 2005. According to scientists at the UK's Hadley Centre and the University of Middlesex, sea-level rise due to climate change by the year 2050 could put twenty million people around the world at risk of serious flooding who wouldn't otherwise have been affected. This assumes "business as usual" emissions and a mid-range estimate for sea-level rise this century of 27cm (10.6"). The stretch of coastline from India to Vietnam, plus Indonesia, would see the most people threatened.

More than any place on Earth, the **Bay of Bengal** exemplifies the deadly risks of tropical-cyclone surge. Fully half of the world's deadliest surge tragedies have taken place in eastern India and present-day Bangladesh. Half a million Bangladeshi people died as a result of a catastrophic 1970 landfall (the event that prompted a famous relief concert by ex-Beatle George Harrison). Rising sea level only adds to the Bay of Bengal's flood risk, while soaring coastal populations make even more people vulnerable to a given flood. And the intricate delta-style coastline would be extremely difficult to protect through engineering even if the nations could afford such a fix. Using a regional model for the Bay of Bengal, Jason Lowe at the Hadley Centre found that extreme storm surges by 2050 could affect the India/Bangladesh coast in complex ways. Some areas become more vulnerable than others in the model, as storm behaviour shifts and as the higher sea level causes changes in tidal and surge behaviour.

Landfalling tropical cyclones also produce a great deal of rain, on coastlines as well as far inland. That may get worse as moisture increases in the atmosphere and precipitation increases on a global scale. **Inland flooding** from heavy rain is already a serious threat in many hurricane-prone areas. With Katrina as the great exception, river flooding has been the leading cause of US hurricane deaths in recent decades. It's also a major problem for agriculture.

With all this mind, despite the holes in our knowledge, it would seem sensible to prepare for the worst. One major problem for planners to take on board is the explosive population growth and development taking place on many cyclone-prone coasts. Roger Pielke Jr of the University of Colorado argues that these factors could easily boost the human and financial costs of hurricane damage this century far more than any intensification related to global warming would – a reminder that demographics have everything to do with the ultimate impact of climate change. Researchers from both sides of the hurricane/climate-change debate laid down their arms in 2006 to craft a joint statement stressing the human element of the problem: "We are optimistic that continued research will eventually resolve much of the current controversy over the effect of climate change on hurricanes. But the more urgent problem of our lemming-like march to the sea requires immediate and sustained attention."

Coastal concerns beyond the tropics

Hurricanes and typhoons may cause horrific damage, but coastal storms threaten the wellbeing of millions of people who live well away from the tropics. Unlike tropical cyclones, which need a cocoon of warm, humid air in order to grow, coastal storms – part of the larger group known as **extra-tropical storms** – thrive on the contrasts between cold continental air and warm marine air. Moving with the polar jet stream, in which they're typically embedded, coastal storms strike many of the major coastlines at mid- and high latitudes, mostly from autumn through spring. In the US, they're often dubbed **nor'easters**, a reference to the strong onshore winds they bring. The lack of a mountain barrier in Western Europe allows coastal storms to plough well inland with heavy rain and high wind. That was the case over the last week of 1999, when an especially vicious pair of post-Christmas storms dubbed **Lothar** and **Martin** pounded much of the region with winds as strong as 219kph (136mph). The storms killed more than 140 people and ruined centuries-old treescapes from France's Versailles to Germany's Black Forest.

As with hurricanes, the challenge in projecting coastal storms is two-fold: will there be more of them, and will they become more intense? A few studies show that extratropical cyclones in general have become fewer in number but stronger on average, although the intensity change appears to be minor at best. These cyclones have also shifted polewards in the Northern Hemisphere, a sign of the jet stream shifting north in response to a warming atmosphere. Across Europe, this jibes with the observed tendency since the 1970s for milder, stormier winters occurring to the north and drier-than-average winters near the Mediterranean – the calling cards of a positive **North Atlantic Oscillation**, or NAO (see p.116).

What about the future? Since extratropical cyclones tend to be several times larger than hurricanes – often spanning more than 1000km (600 miles) – they're more easily simulated by global computer models. Although early modelling work didn't always agree on how the storms might evolve, the picture is now sharpening a bit. A study by Jeff Yin (of the National Center for Atmospheric Research), carried out in support of the 2007 IPCC report, looked at the results from fifteen advanced global models. The consensus picture could be labelled "more of the same". It shows extratropical cyclones continuing to shift polewards, just as they've been doing in recent decades. The total kinetic energy within these storms increases, too, and other studies indicate that this stronger punch could play out as it has in recent years, with storms that are less frequent but stronger on average. Should all this pan out, lower-latitude coastal regions such as southern California and the Mediterranean may find themselves experiencing more prolonged and/or intense dry spells in winter, as windy, rainy, fast-moving storms track further north over time.

Any such regional projection is framed by uncertainty at this point. Smaller-scale effects that aren't well modelled or understood could overpower a global trend in particular locations. And although climate-change models tend to make the NAO increasingly positive in the decades to come, that's hardly a done deal. There's plenty of natural variation in the NAO, both on a yearly basis (the index can slide back and forth several times in a single winter) and from year to year. In fact, the NAO began trending away from its positive phase after the year 2000, and in 2005–06 Europeans saw several spells of the bitter, snowy, negative-NAO weather that had become scarce over the preceding few years.

Coastal-storm flooding: a deepening problem

Even if coastal storms don't change drastically, the extreme high-water levels they generate should increase. As with tropical cyclones, this is an outgrowth of the overall rise in global sea level. One region where high

Venice is sinking as the water rises, and its new defence system may not be enough to protect it from severe storms
Michael S. Yamashita/Corbis

population meets up with a high risk for surge from coastal storms is the **North Sea**. The Netherlands have spent centuries defending themselves from the sea with a massive series of dykes. Much of coastal England, which sits slightly higher on average, gets along with a variety of dams and other flood-control devices. Both countries found themselves unprepared for the history-making North Sea storm of 1 February 1953, which sent huge surges into the British, Dutch and Belgian coasts. Over 1800 people died in the Netherlands, where a powerful surge overpowered dykes and levees in the lower Rhine Valley. England saw more than 300 deaths on its east coast, including 58 on Canvey Island, from where more than 10,000 residents were safely evacuated.

The 1953 storm triggered a new age of massive efforts to tackle coastal flooding. The Netherlands spent over $5 billion on its **Delta Project**, completed over four decades. Its centrepiece is a pair of doors, each about 20m (66ft) high, that can close off the Rhine if a major coastal storm looms. Along the same lines, the 1953 storm inspired Britain to enhance seawalls along its east coast and to spend over half a billion pounds on the **Thames**

Trouble in the Thames

We don't know exactly how people dealt with the Thames flood of 9 AD – the earliest one for which records exist – but we do know that people living near this great British river have faced flood after flood over the succeeding two millennia. Many Brits hoped that the great Thames Barrier would be the end-all of flood control for the region. But in fact that project was only designed to keep London and other Thames cities safe until the year 2030. Thanks to a combination of subsidence and rising sea levels, the water level in the Thames estuary is reportedly increasing by as much as 3cm (1.2") per decade. The area at risk of flooding includes 1.25 million people (more than the entire city of Birmingham), and a frenzy of new construction is luring hundreds of thousands of new residents to the Thames Gateway area.

With this in mind, the UK's Environment Agency has extended its usual thirty-year window of concern and begun the planning process for TE2100, an acronym that denotes flood control for the Thames estuary up to the year 2100. The plan is expected to include a suite of new devices to be built between 2015 and 2040 with climate change in mind. A massive set of studies is taking the region's climate, environment, land use and social structure into account.

If regional climate models are anywhere near correct, the TE2100 project looks like a prudent move. In a 2002 study, the UK Climate Impacts Programme showed that the storm surge one might expect once every fifty years across southeast England could be as much as 1.4m (4.7ft) higher by the 2080s, assuming a high-end estimate for global emissions of greenhouse gases.

Barrier, which was built between 1972 and 1982. Spanning the Thames in eastern London at a width of 520m (1700ft), it includes ten massive gates, six of which rise from concrete sills in the riverbed and swing to form a single wall.

The Thames Barrier was employed only eleven times in its first decade of service. In 1999, the operating rules were changed, allowing for more frequent closures in order to keep tidal waters from blocking river runoff produced by heavy rain. This change makes it hard to assess how much climate change may have affected the closure rate. For whatever reason, the gates are swinging shut much more frequently than before – eighteen times in the year 2003 alone. It may close thirty times in a typical year by the 2030s. The region now has embarked on a hundred-year plan to protect the Thames Valley from the increased surge risk related to climate change (see box opposite). One of the many options being considered is a far wider barrier across the Thames estuary that would span 16km (10 miles).

In another well-publicized initiative, Italy is building a set of 79 hinged, inflatable steel gates to protect **Venice** from rising waters. Costing an expected US $4.5 billion, the project is due to be completed by 2012. Normally lying flat on the seafloor, the gates will be pulled up as needed in a 30-minute process that would block three inlets from the Adriatic and help keep the city safe from increasingly threatening surges triggered by Mediterranean storms. Minor floods are now so frequent in this gradually subsiding city that visitors and locals alike often find themselves facing streets ankle-deep with water.

Venice's gates would be employed for any surge expected to exceed 110cm (3.6ft), and they're reportedly capable of handling surges up to 200cm (6.6ft). In 1996 alone, over 100 floods topped the one-metre mark, and a severe 1966 flood reached 180cm (6ft). Alas, it wouldn't take much of an increase in global sea-level rise to push a repeat of that 1966 flood over the new barriers. Some researchers are already pondering the idea of a "Great Wall of Venice" that might eventually surround the city in order to preserve it.

Tornadoes: an overblown connection?

One of the staple ingredients of many books and articles about climate change is a photo of a menacing twister, usually with a caption that goes something like this: "Tornadoes and other forms of extreme weather are expected to increase in a warming climate." Actually, there is no sign that the most violent **tornadoes** are on the increase globally. The future could always surprise us, and an important scientific maxim – "the absence of evidence does not indicate evidence of absence" – may apply here. Still, there's no compelling reason why scientists would expect tornadoes to become much more frequent or intense in the wake of human-induced greenhouse warming.

The popular confusion appears to arise because tornadoes often get lumped under the heading of "extreme weather." It's true that extremes of temperature and precipitation are already on the increase and expected to increase further in a warming climate. But tornadoes are an entirely different animal. Twisters are spawned by **thunderstorms**, so they're dependent on the same warm, moist air masses that lead to thunder and lightning. But that's not enough in itself to produce a tornado. To get the most vicious type of twister – the kind that occur with regularity in only a few parts of the globe, most commonly in east India, Bangladesh, and the central and eastern United States – you need a peculiar type of thunderstorm known as a **supercell**. These long-lived storms thrive on a particular blend of conditions, including a layer of dry air about 2000–3000m high (6,600–9,800ft), as well as winds that increase and shift direction with height. Many supercells have an area of rotation, called a **mesocyclone**, from which tornadoes may emerge.

The rarity of all these conditions coming together in one place helps explain why so few parts of the world are prone to violent tornadoes (those with winds topping about 320kph or 200mph). Some have been reported in Europe, but generally you need a location where warm, dry air (from areas such as the US Desert Southwest or India's Thar Desert) can easily clash with warm, moist air (from such sources as the Gulf of Mexico or the Bay of Bengal). Since the geographic variables that lead to supercells won't be changing anytime soon, it's unlikely we'll see much change in the preferred stomping grounds of *violent* tornadoes, although it's conceivable that their US range could shift northwards and perhaps affect the southern tier of Canada more often.

Another possibility is that the period considered tornado season (generally early spring in the US South and late spring to summer in the Midwest) will shift a bit earlier in the calendar as the climate warms. This may already be happening: more of the largest US outbreaks have been occurring outside the boundaries of spring from the 1990s onwards (even accounting for the tornado "inflation" discussed below).

There's a bit more uncertainty with **non-mesocyclonic tornadoes** – the far weaker, far more common variety. These tornadoes don't require the

A tornado destroys a house in Mulvane, Kansas. Thankfully, there's little reason to believe that such twisters will increase as the climate warms
Eric Nguyen/Jim Reed Photography/Corbis

exotic blend of ingredients above: as long as the updraft in a thunderstorm is powerful enough, it may be able to produce a weak, short-lived tornado (though very few do). With the advent of video cameras and the growth of weather hobbyism, people have been reporting more and more tornadoes. The 1997 Hollywood blockbuster *Twister* appears to have played no small part in this boom, says Nikolai Dotzek of the German Aerospace Center (DLR). Combing through these reports, he and other scientists have confirmed that, in many places – including England, Ireland, South Africa and France – tornadoes are more frequent than we once thought.

Whether the global incidence of all types of tornadoes is truly on the increase is difficult to know, but experts believe that most or all of the upward trend in reported twisters is simply because more people are reporting them. Statistics from the US bear this out. The average yearly tornado count in the US ballooned from about 600 in the 1950s, just as the nation was implementing its watch and warning system, to 800 in the 1970s and around 1100 in the 1990s, when tornado videos became all the rage. However, US reports of violent tornadoes – the kind that were hard to miss even before storm chasing became common – haven't changed significantly in the entire century-long record, holding firm at around 10–20 per year. As for Europe, the continent as a whole, including the UK, reports around 160–180 tornadoes per year, according to a 2003 survey by Dotzek. He believes the number of reports could eventually top 300 as Europeans become more tornado-savvy. The 2007 IPCC assessment notes the jump in tornado reports worldwide, acknowledging the issues mentioned above, but it steers clear of any projections on how the localized conditions that spawn tornadoes might change.

Scientists are now using leading-edge computer models to get a better sense of possible global changes in severe thunderstorms, the type that produce gale-force winds and hailstones (and, sometimes, tornadoes). A US–Europe network of researchers led by Harold Brooks of the US National Oceanic and Atmospheric Administration is looking at how well global models reproduce the current patterns of severe local weather. These models are too coarse to pinpoint individual thunderstorms, so Brooks and colleagues are sifting through the global projections and hunting for the ingredients noted above – such as warm, moist air and **wind shear** (winds changing with height). Thus far, the models seem to be doing a good job of reproducing where severe weather happens now.

The next step is to extend the analysis to projections of the coming century's climate. Some initial work using a midrange emissions scenario hints at an increased US prevalence of unstable air but a reduction in wind shear. Normally this would lead to poorly organized thunderstorms that produce little more than heavy rain and lightning. However, Jeffrey Trapp and colleagues at Purdue University are finding that the instability may increase enough to compensate for weaker wind shear. As a result, they say, cities such as Atlanta and New York could see an increase in the number of days each year conducive for severe thunderstorms – perhaps a doubling in some locations – by the late 21st century. Of course, even when conditions are ripe, thunderstorms don't always form. Using a similar technique to gauge the risk for the central and eastern US, NASA's Tony DelGenio and colleagues also project a future with higher instability and weaker wind shear overall. However, their work shows the most favourable zones of instability and wind shear coinciding more often, leading to a jump in the potential frequency of the most intense thunderstorms. If further work supports these early results, then some of North America's most populous areas might have to add severe weather to their list of climate-change concerns.

Ecosystems & agriculture

The future of flora, fauna and farming

If all that global warming did was to make life a bit steamier for the people who consume the most fossil fuels, then there'd be a karmic neatness to it. Alas, climate change doesn't keep its multitude of effects so nicely focused. A warming planet is liable to produce a cascade of repercussions for millions of people who have never started up a car or taken a cross-country flight. Many animal and plant species will also pay a price for our prolific burning of fossil fuels.

Food, and the lack of it, could be where a changing climate exerts some of its most troublesome impacts for society. While the changes could affect ranching and grazing as well as arable farming, much of the research to date has focused on croplands. Because of longer dry spells, hotter temperatures, and more climatic uncertainty, the next century is likely to see major shifts in the crops sown and grown in various regions. Well-off countries of the North might break even or even benefit from the changes, if they can keep a close eye on the process and adapt their agriculture early and efficiently. Sadly, the same may not be true of the tropics, where most of the world's food crops are grown. The most problematic impacts on agriculture may wind up occurring in the poorest countries, those with the least flexibility and the most potential for catastrophic famine.

Humans use about a third of Earth's land surface for farming and other purposes. What global warming does to the other two-thirds of the land – the world's natural ecosystems – could be even more wrenching than the effects on managed lands. In his 2002 book *The Future of Life,* famed biologist E.O. Wilson warns that, if current trends continue, half of Earth's

"It will be difficult to anticipate future threats to biodiversity and ecosystem dynamics, even if we could know future climate change with perfect accuracy."

Jonathan Overpeck, University of Arizona

species could be gone by the year 2100. Pollution, development and other side-effects of civilization have already put uncounted species at risk. Now climate change threatens to make the situation far worse.

A landmark study of the extinction risks from climate change, led by Chris Thomas of the University of Leeds and published in *Nature* in 2004, looked at regions that together span a fifth of the planet's land surface. The report found that 15–37% of plant and animal species across these areas could face extinction by the year 2050, should a middle-of-the-road scenario for increased emissions come to pass. If emissions are on the high side, the range jumps to 21–52%.

The IPCC's 2007 report only reinforced this grim picture. It concluded that some 20–30% of species assessed to date are at increased risk of extinction should global temperatures rise more than 1.5–2.5°C (2.7–4.5°C) above recent values. "Current conservation practices are generally poorly prepared to adapt to this level of change", the IPCC warns. Throw in the risk of climate surprises, and there is every reason to be concerned about the potential impact of greenhouse gases on the life forms around us.

The canaries of climate change

Miners once brought canaries into their dank workplaces to test whether the air was safe to breathe; if the canary died, it was time to get out. Similarly, some creatures can't help but reveal the mark of a shifting climate through their very ability – or inability – to survive.

Amphibians and **reptiles** offer some of the most poignant examples of species at risk. As cold-blooded creatures, they must keep themselves within a fairly restricted range of air temperature, and they've got to crawl wherever they go. Those are tough constraints where the climate is warming and where highways and other human construction make it difficult to migrate. For many species of **turtle**, there's another complication: more females than males are hatched as the ambient temperature in their nesting area goes up. If a given species can't migrate quickly enough to avoid regions that are warming, then the gender imbalance could become a threat to survival. This only adds to the many other threats facing beach-nesting turtles, from commercial trade to pollution.

More than 30% of amphibians were found to be vulnerable, endangered or critically endangered in a 2004 survey. A 2007 study focusing on the protected rainforest of La Selva Biological Station in Costa Rica found that amphibian populations had dropped by roughly 75% since 1970 across the range of species present there, including many that had previously been considered robust. **Frogs** are among the most threatened species at many locations around the globe. A number of factors are probably involved in their decline, including the effects of pesticides and other chemicals on reproductive success. However, climate has been implicated in several cases of depletion or extinction. The atmosphere sometimes does this dirty work indirectly: climate shifts can help fungi to attack the amphibians more effectively. High atop the US Cascade mountains, intensified dry spells have been linked to mass die-offs of the **western toad**. The droughts make for shallower ponds, which allows dangerous ultraviolet light to penetrate more fully and weakens the pond-nurtured embryos of western toads. (Ozone depletion may also be allowing slightly more UV light in, although it doesn't appear to be the main culprit in the creatures' demise.)

A golden toad. The species was wiped out by a disease made more prevalent by climate change
Smith, Charles H/ US Fishery and Wildlife Service

Further south, twenty out of fifty species of frog in the **Monteverde Cloud Forest Preserve** in Costa Rica appear to have vanished completely after an unusually warm, dry spring. One special species found nowhere else, the exotically coloured **golden toad,** numbered at least 1500 in the La Niña spring of 1987. Within two years, it could be found no more. Resident scientist Alan Pounds suspected that climate was involved, but the link wasn't crystal-clear: the fungus believed to have killed the embryonic toads, **chytridiomycosis,** tends to prosper in cool, moist weather, yet the Monteverde forest has

"Disease is the bullet killing frogs, but climate change is pulling the trigger."

Alan Pounds, Monteverde Cloud Forest Preserve

The Apollo – one of the many butterfly species threatened by warming
Kenneth Lilly/DK Images

been warming overall. It took until 2005 before Pounds had solved the puzzle. As is the case in many parts of the world, the bulk of the warming at Monteverde was occurring at night, whereas days had actually cooled slightly (probably due to increased cloud cover). The reduced day-to-night range kept conditions closer to optimal for the villainous chytridiomycosis.

Butterflies are another excellent indicator of climate change on the march, partly because these widespread and aesthetically pleasing creatures have been carefully studied for centuries, especially in Britain. Both drought and flood can cause a butterfly population to crash, as happened to five of 21 species studied during a dry spell in the mid-1970s across California. Many other butterflies are threatened by slowly rising temperatures and other pressures that restrict their range. The **Apollo butterfly,** whose translucent wings are often seen across high-altitude areas in Europe, is no longer found at elevations below 838m (2749ft) in France's Jura mountains. Although they're skilled fliers, Apollo are finding themselves with less and less liveable terrain as they're forced to climb in elevation.

Even a creature as humble as the **dormouse** can tell us something about global warming. Made famous by Lewis Carroll's *Alice in Wonderland*, this tiny, nocturnal mammal now ranges only about half as far across Britain as it did in Carroll's time. Part of this is due to the fragmentation of the trees and hedgerows where it likes to live. However, mild winters interfere with its hibernation patterns, and warm summers cause additional stress. Were it not for so much civilization in the way, the dormouse might easily be able to migrate north to escape the warm-up.

Trouble also brews when interdependent species respond in differing ways to a warming atmosphere. Those who use temperature as their cue to breed or migrate may shift their timing as warming unfolds, while those whose behaviour is driven more by the waxing and waning of sunlight may not. At the Netherlands Institute for Ecology, Marcel Visser has been studying **disrupted synchrony** – the mismatched timing that can occur when creatures find that a food source has departed outside of its usual seasonal schedule. In studying a bird called the Dutch great tit, for example, Visser and colleagues found that the birds' egg-laying behaviour hasn't kept up with the earlier springtime emergence of caterpillars, its main food source. Similarly, in California, a Stanford University team found that the progressively earlier seasonal die-off of plantains on a nearby ridge left the bay checkerspot caterpillar without its favourite food. The butterfly eventually disappeared from the ridge.

Do examples like the ones above add up to a **global shift**? In an extensive survey published in *Nature* in 2003, Camille Parmesan of the University of Texas and Gary Yohe of Wesleyan University found that 279 out of 677 species examined in large-scale, long-term, multi-species research showed the telltale signs of human-induced climate change. On average, these species had moved roughly 6km (3.7 miles) northward, or about 6m (20ft) up in elevation, per decade. One of the most telling clues was the "sign switching" of eight butterfly species in North America and Europe. All of them moved northwards during the warm period of the early twentieth century, shifted back south during the mid-century cool-down (see graph, inside front cover), and then headed north again most recently. None of the butterfly species contradicted this pattern.

Yohe, an economist, remained somewhat sceptical at first. Even after his own paper with Parmesan was published, he refused to claim 95% certainty that human-induced climate change was affecting ecosystems on the large scale. But in 2006, after more research had accumulated, Yohe came around: "I now feel that we have crossed the 95% confidence threshold. We can conclude with very high confidence that we have detected the fingerprint of climate change."

The big squeeze

Simply keeping up with a changing climate will be a challenge for many animals and plants. It's hard to overstate how quickly the chemical makeup of the atmosphere is changing relative to Earth's history. The unprecedented increase in greenhouse gases translates into a warming projected for the next century that may seem gradual to us but will be warp-speed by geological standards (see The Long View, p.193).

In the Arctic, where the range of life forms that can survive the harsh environment is relatively narrow, it's easy to identify the animals most at risk, chief among them **polar bears** (see p.80). Elsewhere around the planet, ecologists are keep-

"We are going to get species going extinct. That's obvious."
Camille Parmesan, University of Texas

ing a close eye on **biodiversity hotspots**, where climate and geography team up to nurture a vast variety of species – far more than one might expect in such small areas. The 25 hotspots identified by researchers are scattered throughout the tropics and mid-latitudes, largely on coastlines and mountain ridges. The Monteverde cloud forest already mentioned is part of one such example. Most hotspots have large contrasts in temperature and moisture across small areas, which enables many different types of flora and fauna to find their niches. But the flip side of this concentrated contrast is that a seemingly small shift in climate could have major effects on plants and animals accustomed to a very specific environment. And because many hotspots are so localized – often surrounded by more uniform, human-dominated landscapes – plants and animals may not be able to easily shift their ranges outward.

Up, up and away?

For animals and plants in mountain homes that have grown too warm for comfort, there's nowhere to go but up – and that's not easy when you're already near the top. Some small mammals are at particular risk. For example, **pika** – the endearing, hamster-sized creatures familiar to hikers across the high country of the US West (and which may have inspired Pokémon) – have dwindled in number across both North America and Asia. With a metabolism that can't handle air much warmer than room temperature, pikas typically hide below rocks and ice to cool off, and they can't migrate easily. The US Geological Survey reports that American pikas have vanished from more than half of 25 traditional haunts in the US Great Basin since the early twentieth century. It's unclear whether heat has killed them directly or indirectly (by cutting down on their foraging hours), or whether other factors might be involved, such as a lack of insulating winter snow. In Australia, a comparable victim is the mountain pygmy possum, whose numbers have dropped as bush rats and feral cats move into its terrain.

When it comes to **plants**, the story includes both winners and losers. The losers are those attuned to the highest, most rarefied conditions: as temperatures continue to climb, many of these species seem fated to die off. In Europe, high-mountain territory covers just 3% of the landscape but accounts for some 20% of all native plants. However, as in the Arctic, a warming climate could produce a larger zone of plant-friendly territory overall. Treelines have risen some 150m (500ft) over the last century in Sweden's Scandes Mountains and up to 80m (260ft) in Russia's Ural Mountains. Swedish scientists were surprised to find large herbs at heights more than 100m (330ft) above their 1950s range, popping up in moraines only recently left behind by retreating glaciers. Some grasses may prove even more opportunistic, providing tough competition against other cold-adapted alpine plants. The potential "grassification" of high-altitude regions – hinted at by computer models, especially across the Alps – is a topic of keen research interest.

What about the rest of the tropics and mid-latitudes? These areas are often classified by geographers into **biomes**, regions that share similar traits of climate and landscape. Among the commonly used groupings for land areas are:

- **Forests** Covering roughly a third of the planet, these include the great boreal forests of Canada, Alaska and Russia; the evergreen and deciduous forests of mid-latitudes; and the tropical forests that support amazing biodiversity despite nutrient-poor soil.
- **Grasslands** The savannah that covers much of Africa and Australia and the temperate grasses that sprawl across the mid-sections of North America and Eurasia.

American pika
Joe McDonald/Corbis

Beyond anecdotal reports, there's a pressing need for a more thorough survey of flora and fauna in a changing high-altitude world. That type of census may soon emerge from GLORIA (Global Observation in Alpine Environments). Launched by the Austrian government with support from the European Union, this network of observing sites now includes scientists at more than 47 sites in Europe, the Americas, Asia, Australia and New Zealand.

As climate change pushes species upward, it may also open doors on plant and animal life of the past. Two tourists in 1991 discovered Ötzi, the Alps' now-famous 5300-year-old man, in a receding glacier. On a survey of Quelccaya in 2004, glaciologist Lonnie Thompson (see p.98) found a small, shrubby plant that he expected might have been around 5000 years old. Carbon dating soon proved it to be a moss that had been under ice for most of the last 50,000 years – and probably twice that long, making it a likely remnant of life before the last ice age.

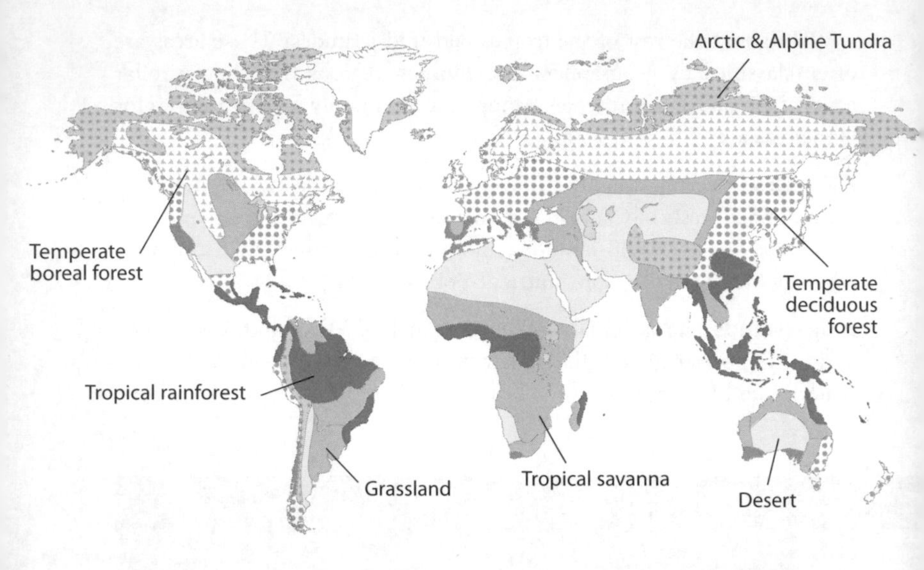

▶ **Deserts** About a fifth of Earth's land surface. Many of the largest deserts are clustered in the zones around 30°N and 30°S, where air tends to sink and dry out. Most of Antarctica is also considered desert.

▶ **Tundra** The cold lands of high altitudes and the Arctic that support only shallow-rooted, low-rising plants.

The edges of many of these biomes are shifting already. Some studies in Europe have shown a northward push of up to 120km (75 miles) in the north edge of certain biomes over the last century. The tropics themselves – as demarcated by high-level jet streams – widened by about 225km (140 miles) between 1979 and 2005, according to satellite data analysed at the Universities of Washington and Utah. The study suggests a polewards push of subtropical deserts. Likewise, one of the largest-scale shifts expected over the next hundred years is the expansion of **boreal forest** polewards into fast-thawing tundra and, in turn, the transformation of the southern edge of some boreal forests into deciduous forests or grasslands. But forests take time to establish themselves, and climate may outrun the process, leaving vast areas in transition until soils and plants can adapt to a new climate regime. To make matters worse, any climatic state we envision for the future is itself a moving target, since the greenhouse gases we're adding to the air now will be warming the climate for a century or more. Will new forests be able to take root in these ever-shifting conditions?

As for land-based creatures in the tropics and mid-latitudes, it might seem as if they can easily migrate poleward and upward as needed. But most of the planet's terrain outside hotspots and other protected areas is pockmarked with towns and cities and laced with highways, cultivated fields and other potential barriers to migration. Those creatures adapted to higher altitudes, like the pika (see p.152), may find themselves stranded on island-like patches of terrain, where climate will only constrict their range further over time.

If the exotic creatures of hotspots and the Arctic are the biggest losers in climate change, the biggest ecological winners may be the most humdrum. Animals that are common and robust could spread even further, such as the various species of deer now thriving across rural and semi-rural parts of North America and Eurasia.

A landscape of shifting climate may also offer a nearly ideal platform from which non-native plants, animals and insects can spread like wildfire. **Invasive species** are nothing new, of course, but in this era of mass global travel, it's easier than ever for unwanted life forms to tag along. Climate change could make particular landscapes more vulnerable to invaders than they might be otherwise. The most successful invasive species are often less picky than native species, able to tolerate a wide range of conditions, including heat and/or drought. One famous example is **tumbleweed**, or Russian thistle. Introduced to the US Great Plains in the 1870s, tumbleweed quickly spread across the US frontier, eventually becoming a symbol of its new home. Similarly, **rabbits** multiplied like, well, rabbits after 29 of them were brought to Australia in 1859 by rancher Thomas Austin. To say they spread is an understatement: according to the Australian government, rabbits remain the nation's most widespread and destructive pest. Today, the United States is dealing with a major invasive pest that's ravaged both the countryside and populated areas and even made its way to the Southern Hemisphere: the destructive and sometimes deadly **fire ant** (see p.157).

For details on the effects of climate on coral reefs and other marine creatures, see p.125.

Bugged by a changing climate

It may be **insects** – adaptable creatures *par excellence* – that take the best advantage of changing climate in the coming century. Many bugs respond strongly to small shifts in temperature, rainfall and humidity. The overall warming now under way may already be affecting insect distribution, and future warming should favour the spread of a number of species. In 2006, a spider-eating French wasp showed up in Sandy, Bedfordshire, the first of its species ever observed in Britain. On a more troublesome note, a 2002 review by Drew Harvell and colleagues at Cornell University found that **parasitic diseases** are likely to become more widespread and/or severe as the planet warms and moistens.

History gives us a sneak preview of what can happen. The so-called Medieval Warm Period (see p.220) – the several centuries of mild weather across much of the Northern Hemisphere between about 950 and 1300 AD – may have helped rats and fleas to thrive across Europe, paving the way for the **Black Death** that killed a third of Europeans in 1347–52.

Global warming is good news for malaria-carrying mosquitoes
US Department of Agriculture

Mosquitoes, and the diseases they bring, appear to be responding to the warm-up now under way. **Malaria** is moving into previously untouched parts of Africa's central highlands, possibly due to a scarcity of cool, mosquito-killing temperatures, although the cause is still being studied. Almost half the world's population is at risk of the disease, which now infects an estimated half a billion people each year – that's one in twelve of the world's population. Amazingly, this figure is four times higher than it was in 1990. Increased settlement in the highlands may be a factor in the upward curve, but it's likely that climate change is a key driver. And with as many as three million people dying each year from malaria, it's clear that any rise in infection rates carries a huge toll in terms of human suffering. Indeed, when World Health Organisation (WHO) scientists estimated that 160,000 people already die each year from the indirect effects of climate change, malaria was one of the main factors they pointed to. In places like Europe and the United States, where malaria was endemic for centuries, it's considered unlikely to make a comeback – even with warmer temperatures – as long as control measures and public-health systems remain strong. The 2007 IPCC report noted that short-term variations in rainfall are a huge factor in malarial risk across many regions,

March of the fire ants

It sounds like Hollywood horror at its most lurid: a voracious, supercharged pack of bright orange bugs chewing their way across America. But it's happening not on the silver screen but in real life, and a warming climate isn't helping matters. The red imported fire ant, which arrived in the United States from South America in the 1930s, has conquered large swathes of the South, from North Carolina to Texas. Since the 1960s the ants have expanded their territory each year by an area the size of New Hampshire, gradually pushing north and west towards their climatological limits.

The ants have many qualities going for them. Multiple queens can team up to help launch a colony (though workers eventually kill all but one of them). The ants reproduce like mad, often pushing out native ants completely and preying on bird hatchlings. They'll also attack mammals, and have even killed a few people, most of whom have been unlucky enough to fall on or near a nest due to an accident or injury. The ants can even survive floods: one of the surreal spectacles from Hurricane Katrina was the sight of fire ants floating atop the waters in dense clots, a technique they can sustain for up to two weeks.

Temperatures much below -12°C (10°F) spell doom for fire ants, but the gradual warming of winter nights could allow their northward US spread to continue this century. They're already expected to head west into California and up that state's coast. One projection shows that if carbon dioxide increases by 1% a year over the next century, the ant could increase its US range by over 20%, populating a belt from Oklahoma to Delaware. In the meantime, the ants have been discovered near Brisbane, Australia (in 2001) and in New Zealand (2004). Thus far, they seem to be thriving in their new homes, according to a 2005 study led by Robert Sutherst of Australia's Commonwealth Scientific and Research Organization. He says the fire ant "has the potential to impact widely on agriculture and the native fauna of Australia."

US Department of Agriculture

such as the drought-prone Sahel in Africa, making it difficult to assess the role of climate change with precision.

Dengue fever – caused by a group of four potentially fatal, mosquito-borne viruses – is also on the increase, with around two-fifths of the world's population now living in affected areas, according to the WHO. The 2007 IPCC report labels it the world's most important vector-borne viral disease. There are an estimated fifty million cases of Dengue each year, of which around 500,000 require hospitalization. A large proportion of these serious cases are accounted for by children. One study estimates that a 1°C (1.8°F) rise in average global temperature could raise the number of people at risk of dengue by up to 47%, but the complex factors that influence transmission make it hard to know if the disease would actually spread that quickly. Drought may be an important cofactor: it can increase the risk of dengue fever if it spurs people to hoard water in places where mosquitoes like to breed.

Unusual summertime warmth and drought helped another mosquito-borne illness make quick inroads into North America. The West Nile virus first showed up on the continent around New York City in 1999. Its mosquito host thrives in hot, dry weather, as protein builds up in stagnant pools that attract birds on which the mosquitoes feed. The virus spread regionally and then, during the record-breaking heat of 2002, it swept across much of the United States and into Canada. Up to 2005 it had killed more than one thousand North Americans.

Insects can also cause trouble for people indirectly, by attacking the food we eat, for instance. The Colorado beetle, a striped, centimetre-long US native, is a serious threat to potato crops in Europe. Each year a few of these beetles hitchhike into the UK, but to date the bug has failed to establish a foothold, thanks to vigilant inspections, public awareness efforts and eradication campaigns by the UK government. However, a study by Richard Baker of the UK Central Science Laboratory found that if Europe were to warm by an average of 2.3°C (4.1°F) by 2050, this pest could more than double its potential range in Great Britain, putting virtually all potato-farming areas at risk.

Shrinking forests

As we've seen, rainforests and their loss can have a huge impact on the global climate (see p.11), but how does global warming affect the rainforest itself? That's not so clear. It's precipitation more than temperature that drives the seasonal rhythms of the rainforest, so anything that affects when and how hard it rains can potentially threaten the forests' health. El Niño tends to produce drought across much of the Amazon, so a key question will be how El Niño evolves in the future – still an open question, although the recent trend has been in favour of more and/or stronger El Niños (see p.118). A sharp warming across the Atlantic has also been identified as a possible factor in the Amazon's severe drought of 2005. But the various global climate models considered in the 2007 IPCC report aren't yet in agreement on whether a sustained drying is on the cards for the Amazon. If climate change should push the Amazon towards more drought, the trend could be exacerbated by land clearing, which causes local-scale drying due to loss of the moisture-trapping forest canopy. The fear is that these and other positive feedbacks could lead to massive rainforest loss across the Amazon, although that scenario is far from a given.

There's less uncertainty about what climate is doing to forests further towards the poles. Mountain and boreal forests are one of the largest ecosystems already being strained by a rapidly changing climate. Permafrost melt (see p.83) is one big source of high-latitude forest stress, but there are other major factors. Across western North America, the intersection of pests and a warming atmosphere has led to the one of the world's largest **forest die-offs**. A series of major droughts and warm winters since the 1990s from Mexico to Alaska has transformed the landscape across huge swathes, destroying millions of hectares of forest through **beetle invasions** and **forest fires**. The extent of both is unprecedented in modern

Forest fires, like this one in Big Sur, California, are expected to become more frequent and severe as the planet warms
Frans Lanting/Corbis

records. In 2002 alone, British Columbia lost 100,000 square km (39,000 square miles) of lodgepole pines to fire and disease. That's more than enough forested land to cover Belgium and the Netherlands. Some 80% of British Columbia's pines will be gone by 2013, according to the Canadian government.

The main insect behind this devastation is the **bark beetle**, which comes in various species across North America. This hardy creature burrows into trees, lays its eggs, then girdles the tree from within, blocking the nutrient flows between roots and branches. Climate has given North America's bark beetles a carte blanche for explosive growth in recent decades. The spruce bark beetle, one of the most common varieties in colder forests, normally requires two years to reach maturity, but an extremely warm summer can allow a generation to mature in a single year.

The northward flow of maple syrup

The shift of a forest can be just as significant psychologically and economically as it is biologically. Take maples, for example. They're so closely identified with Canada that they're the centrepiece of the national flag, but for generations of Americans, it's New England that's synonymous with maple syrup. This piece of Americana appears to be in jeopardy, according to the US National Assessment of Climate Change. To produce the best syrup, maples need a series of freezing nights and milder days, together with a few prolonged cold snaps in late winter. But since the 1980s, winters across New England haven't lived up to their past performance. There have been fewer stretches of bitter cold, and nights are staying above freezing more often, even in midwinter.

Further north, maple production in Quebec was long limited by the lack of daytime thaws, as well as by the deep, sustained snow cover that kept maple harvesters from their trees. Now, improved technology allows the Quebecois to gather syrup more easily in deep snow, and the climate appears to be shifting in the Canadians' favour as well. Not only is maple production moving out of New England, but the long-term survival of maple trees themselves is in question across much of the US Northeast, according to the National Assessment. That would be a potential hit to the spectacular autumn foliage that attracts thousands of tourists each year.

Perhaps New England's best hope of keeping its stunning foliage, and its syrup, is the chance of a shift in North Atlantic currents (see p.119) that could chill the region – or at least keep it from warming as much as it otherwise might. But the chances of this happening are slim.

Conversely, it takes two bitterly cold winters in a row to keep the population in check. The combination of milder winters and warmer summers since the early 1990s has allowed the beetle to run rampant.

Meanwhile, **mountain pine beetles** – once limited to lower-elevation tree species such as lodgepole and ponderosa – have moved uphill as temperatures warm. They're now invading whitebark pine in Utah at elevations up to 3000m (10,000ft). "The outbreak has progressed at a truly astounding rate", says Jesse Logan of the US Forest Service. And in Canada, the beetles now threaten to invade **jack pine**, a critical move that could allow the insects to cross the continent and lead to massive infestations across the forests of eastern North America.

Drought weakens the ability of trees to fight off these aggressive bugs, because the dryness means a weaker sap flow that's easier for beetles to push through. Across parts of New Mexico, Arizona, Colorado and Utah, where a record-setting drought took hold in 2002, pine beetles and other drought-related stresses have killed up to 90% of the native **piñon** trees, turning the needles a sickly red before they fall off. A drought in the 1950s had already killed many century-old piñon, but the more recent drought produced "nearly complete tree mortality across many size and age classes", according to a 2005 study led by David Breshears of the University of Arizona. Even assuming the climate supports piñon regrowth, it will take decades to re-establish the landscape. Breshears sees the event as a sign that climate-driven landscape change could be far more rapid and widespread than even the experts expect.

Another destructive invader, the **spruce budworm**, has carved new territory for itself as the climate has warmed. It's now chewing up buds and needles on spruce trees as far north as Fairbanks, Alaska, where it first took hold in the early 1990s. Tree rings show no sign of any previous outbreaks in that area since the mid-1800s, and the Little Ice Age reigned for hundreds of years before that. In the decades to come, various budworm species – all of which benefit from warmer, drier summers – could ravage large stretches of boreal forest across Alaska and Canada.

Certainly, both insects and fire are part of the natural ecosystems of North America. Some species of trees actually rely upon fire to propagate. Jack pine seeds are tucked into tightly sealed resinous cones that only open when a fire comes along and melts the resin, which then allows the seeds to spread and sprout in the newly cleared landscape. The question is how climate change might amplify the natural process of forest fires, especially across the high latitudes of North America and Asia. For example, some studies project an increase of up to 100% by the end of this century in the amount of land consumed by forest fire each year across Canada. These fires are a major source of pollution, and they pack a double whammy on the carbon budget: the fires themselves add significant amounts of carbon dioxide to the air (not to mention carbon monoxide, methane and ozone-producing oxides of nitrogen), and the loss of trees means less CO_2 being soaked up from the atmosphere (at least until new growth occurs).

As for the beetles and budworms, most major infestations tend to run their course after a few years, followed by new epidemics a few decades later as the next crop of tree victims matures. But nobody knows how that natural ebb and flow will intersect with the climate-driven shifting of the insects' potential range across the US and Canadian West. And it's not out of the question that a decades-long **megadrought** could set in. It's happened before, at least across the US Southwest, and a 2007 study led by Richard Seager of the Lamont-Doherty Earth Observatory warns that by 2050 that region may be in the grip of a warming-driven drought that's more or less permanent. Los Angeles recently experienced its driest year in more than a century of record-keeping, with a mere 82mm (3.21") of rain from July 2006 through June 2007, setting the stage for catastrophic wildfires in and around LA that autumn.

All of these issues complicate long-running debates on how best to manage the forests of western North America. The overarching question is how to allow natural processes to play out while fending off the kind of large-scale, fire- or bug-produced devastation that could permanently alter the landscape.

Crops and climate: a growing concern

Global warming may not seem like a risk to farmers, who, you might think, can simply change what they cultivate to suit the climate. However, it's one thing to plan for a climate different from the one you and your ancestors have dealt with for centuries. It's another thing entirely when there is no single, fixed "new" climate to plan for, but a climate in constant flux.

That said, it's clear that the news is not all bad for agriculture in a warming world. One of the more prominent US organizations stressing the potential benefits of a world with more carbon dioxide is the Greening Earth Society. Since the 1980s, they've promoted the view that the extra CO_2 in the air will help crops and other vegetation to grow more vigorously. Moreover, as they point out, the tendency of global warming to be most pronounced in the winter and at night should only lengthen growing seasons on average across much of the world.

The fertilization effect of carbon dioxide has been underscored in many climate reviews, including the IPCC's, and is supported by research in the laboratory and in the field, albeit with a few caveats (see box overleaf). As for warming itself, scientists expect a hotter climate to enhance the overall potential for cereals and some other crops across large parts of northern Europe, Russia, China, Canada and other higher-latitude regions. This should lead to an overall expansion of the world's land areas favourable for agriculture. To take best advantage of these potential gains, however, farmers will need to be fast on their feet. When crops mature more quickly – as they tend to do in a warmer, CO_2-heavy atmosphere – they generally accumulate less biomass. Thus, unless a second crop is planted (or the farmer switches to different crops), the result can be a net drop in seasonal production. Not only does biomass tend to go down with the extra CO_2, but it appears that nutrients may become more scant as well. More than thirty studies to date show that CO_2-enhanced crops are significantly depleted in zinc, magnesium, or other micronutrients, perhaps because there aren't enough trace elements from the soil entering the plant to keep up with the photosynthesis boost from CO_2.

Among non-agricultural plants, woody vines appear to thrive on the extra CO_2. In part, it's because their clingy mode of growth allows the plant to devote more energy to photosynthesis as opposed to building structure to keep it standing. A 2006 study at Duke University found that one noxious vine – poison ivy, which sends over 350,000 Americans to doctors each year – grew at three times its usual pace with doubled CO_2. To make matters worse, the ivy's itch-producing chemical became even more toxic than before.

The most recent analyses have brought down the overall agricultural benefit we can expect from carbon dioxide increases, in part because the negative influence from **near-surface ozone** will be larger than scientists once thought. One of the open-air studies carried out as part of the Free Air CO_2 Enrichment program (see box overleaf) tested the effects of enhanced carbon dioxide and ozone on soybean fields. A doubling of atmospheric CO_2 boosted soy yields by about 15%, but that benefit went away when ozone levels were increased by as little as 20%. Another open-air study, this one in Wisconsin, obtained similar results for aspen, maple and birch trees. (To put a positive spin on the matter, some US studies have shown that reducing low-level ozone by less than half could preserve billions of dollars in crop value.)

Climate change itself will have an impact on crops, of course, some of it favourable. Modelling that fed into the 2007 IPCC assessment shows how **frost days** (the number of 24-hour periods when temperatures drop below freezing) might evolve over the next century. A study by Gerald Meehl and Claudia Tebaldi of the US National Center for Atmospheric Research found that the number of frost days should drop in many parts of the world, most dramatically across far western North America and northwest Europe. This is due to an expected shift in the jet stream that would pull more warm marine air onto these coastlines, helping to keep

Will greenhouse gases boost plant growth?

Scientists have long recognized one potentially positive side effect of fossil fuel use: the stimulation of forests, crops and other vegetation by the carbon dioxide we're adding to the atmosphere. Initial work in the 1980s suggested that a doubling of CO_2 could increase crop yields by as much as 35%. However, much of the initial work involved plants in open-top chambers, which allow fresh air in through the top but don't fully replicate how plants would grow outdoors. A more true-to-life model would be to add CO_2 to the air around vegetation in a natural setting and see what happens. This is the idea behind a series of projects called Free Air CO_2 Enrichment (FACE). More than a dozen large-scale FACE sites have been established in the US, Switzerland, New Zealand, Germany, Italy and Japan since the early 1990s. At each site, a circle of vent pipes injects CO_2 into the atmosphere surrounding a plot of land 8–30m (26–100ft) in diameter. Through computer-controlled settings, the CO_2 levels can be adjusted for winds and other factors.

In a 2005 review of more than a hundred FACE studies, Elizabeth Ainsworth and Stephen Long of the University of Illinois found that the crop benefits from enhanced CO_2 fell short of expectations. For example, in studies where CO_2 was increased over present-day levels by about 50%, the yields rose by about 7% for rice crops and 8% for wheat. This is only about half the effect one would have expected by extrapolating to the doubled-CO_2 levels in the earlier chamber studies, according to Ainsworth and Long. Other FACE studies indicate that the benefits may tail off for at least some crops as CO_2 levels continue to increase. Unlike crops, trees have fared better than predicted in the FACE studies. Ainsworth and Long found that the extra CO_2 caused the total amount of dry biomass produced above ground by trees to increase by an average of 28%.

Perhaps the biggest bonus from enhanced CO_2 could be an increase in the ability of some plants to deal with drought. Ainsworth and Long found that when conditions were dry, the boosted CO_2 levels improved yields by an average of 27%; during wet conditions, the extra CO_2 had no significant effect. One possible reason for this is that the extra CO_2 constricts a plant's pores, which also helps it retain moisture.

In order to take advantage of higher CO_2 levels, plants will need to draw on other nutrients to support their increased growth. That's possible but by no means certain. Nitrogen is a particular question mark. It's a critical part of industrialized more nights above freezing. Parts of the western United States could see ten or more fewer days with frost per year by the 2080s, if the model rings true. That, in turn, should lead to a somewhat longer growing season. In fact, some of the modelling for the 2007 IPCC report supports a boost of one to three days in growing-season length across most of North America and Eurasia, although it's the year-to-year timing of frost days – a hard-to-model quantity – that matters most.

Tim Meis of the University of Illinois setting CO_2 and ozone levels for a FACE study
Scott Bauer/US Department of Agriculture

agriculture: crops are fertilized with it so commonly, and so inefficiently, that most of the nitrogen bypasses the crop and enters the atmosphere, where it can travel thousands of kilometres before it's taken up by the soil. Nitrogen is also generated by microbes within the soil itself. It's unclear whether these two sources will provide enough nitrogen to allow natural vegetation to thrive on the extra carbon dioxide.

One of the largest and longest-running FACE studies suggests that, for grasslands at least, the extra nitrogen would be critical in order for plants to take advantage of any benefit from enhanced CO_2. Scientists from five US universities reported on the study in the journal *Nature* in 2006. On a Minnesota prairie, the team studied six plots that contained sixteen native or naturalized types of legumes and grasses. Along with boosting CO_2 levels, they added extra nitrogen to some of the plots. The plots without enriched nitrogen fell progressively further behind their nitrogen-enriched counterparts in making use of the extra CO_2.

Winners and losers in farming

Geography will play a huge role in how any agricultural benefits from climate change are partitioned. Currently, most of the world's food crops are grown in the tropics. In many of these areas, dry periods are expected to grow more intense and/or more long-lasting, interspersed with periods of intensified rainfall (a paradox explored in Floods & Droughts, p.58). These heightened contrasts are stressful for many crops. For example, the **groundnut** – a staple across western and southern India – produces far less fruit when temperatures consistently top 35°C (95°F) over a few days. Many other crops have similar temperature thresholds. **Wheat** that's exposed to 30°C (86°F) for more than eight hours produces less grain, and **rice pollen** become sterile after only an hour of 35°C heat.

One major study commissioned by the United Nations for 2002's World Summit on Sustainable Development compared the relative winners and losers in agriculture for a midrange scenario of global emissions increase by the 2080s. Among the findings:

▶ **Losses in the tropics** Between 42 and 73 countries, many of them in Africa and Asia, could experience declines of at least 5% in their potential to grow cereal crops. Between one and three billion people would be living in countries that could lose 10–20% of their cereal-crop potential.

▶ **Gains in the north** In contrast, most of the world's developed countries would experience an increase in cereal productivity of 3–10%.

▶ **Agricultural GDP** In terms of agricultural gross domestic product, the biggest winners are likely to be North America (a 3–13% increase) and the former Soviet Union (up to 23%). By contrast, Africa could lose 2–9% of its agricultural GDP.

While these projections look rosy for some areas, they don't necessarily factor in the potentially destructive role of extremes, such as the agricultural damage produced by Europe's 2003 heat wave. Years such as these are more likely to contain days with higher temperature and lower water availability, with negative impacts on a variety of crops. In its 2007 report, the IPCC stressed that recent work has "highlighted the possibility for negative surprises, in addition to the impacts of mean climate change alone". Some of the benefits projected for developed areas could be offset by thirsty soils and a resulting spike in irrigation demand, as seen in Australia's recent drought.

Wheat harvests in the US (such as Washington, shown here) may get a boost from CO_2, but yields are expected to drop in much of the poor world
US Department of Agriculture

There's still much work to be done in the realm of crop-climate modelling. Most studies of individual crops have been relatively small-scale, while analyses of future climate are at their strongest in depicting patterns across large sections of the globe. A natural intersection point is regional climate models, which zero in from global models to portray climate across a specific area in finer detail. For example, Andrew Challinor (Leeds University) and colleagues have developed a **General Large Area Model (GLAM)** for annual crops. GLAM is on a much larger scale than most crop models, but it's shown the ability to predict local variations (including the extremes that can harm many crops) while it's yoked to the UK Hadley Centre's regional climate model. According to recent work with GLAM, typical groundnut yields could fall by up to 70% by 2100 if agricultural practices aren't adapted to the new climate. Adaptation could reduce this damage by more than half, says Challinor.

Agricultural specialists are also working on hardier strains of key crops that can deal with potential climate shifts. In the Philippines, the International Rice Research Institute – famed for its work in bolstering yields in the 1960s Green Revolution – is working on drought- and heat-resistant strains of rice. Half the world's population eats rice as a staple, so its fate on a warming planet is critical to world foodstocks. A 2004 study from the IRRI showed that a 1°C (1.8°F) rise in average daily temperature can reduce some rice yields by 15%. The trick in formulating crops for a warming world will be to ensure that they stay as nutritious as their predecessors, given the effects of CO_2 on micronutrients noted above. Otherwise, the silent threat of "hidden hunger" – nutrient deficiencies that now jeopardize the health of billions – could only get worse.

PART 3

THE SCIENCE

How we know what we know about climate change

Keeping track

Taking the planet's temperature

It's customary to see polar bears, forest fires and hurricane wreckage when you turn on a TV report about climate change. What you're unlikely to see are the dutiful government meteorologist launching a weather balloon, the grandmother checking the afternoon's high temperature in her backyard, or a satellite quietly scanning Earth's horizon from space.

These are among the thousands of bit players – human and mechanical – who work behind the scenes to help monitor how much the planet is warming. There's no substitute for this backstage action. Day by day, month by month, year by year, the people and instruments who measure Earth's vital signs provide the bedrock for the tough decisions that individuals, states and companies face as a result of climate change.

The rise of climate science

The practice of assigning a number to the warmth or coolness of the air has royal roots. Galileo experimented with temperature measurements, but it was Ferdinand II – the Grand Duke of Tuscany – who invented the first sealed, liquid-in-glass thermometer in 1660. The Italian Renaissance also gave us the first barometers, for measuring air pressure. By the 1700s, weather observing was all the rage across the newly enlightened upper classes of Europe and the US colonies. We know it was a relatively mild 22.5°C (72.5°F) in Philadelphia at 1pm on the day that the Declaration of Independence was signed – July 4, 1776 – because of the meticulous records kept by US co-founder and president-to-be Thomas Jefferson.

Some of the first sites to begin measuring temperature more than three hundred years ago continue to host weather stations today. Most reporting sites of this vintage are located in Europe. A cluster of time-tested stations across central England provides an unbroken trace of monthly temperatures starting in 1659 – the longest such record in the world (see graph) – as well as daily temperatures from 1772. Like the remnants of a Greek temple or a Mayan ruin, these early readings are irreplaceable traces of a climate long gone. Such stations represent only a tiny piece of the globe, however, so on their own they don't tell us much about planet-scale changes. However, they do help scientists to calibrate other methods of looking at past climate, such as tree-ring analyses. They also shed light on the dank depths of the **Little Ice Age**, when volcanoes and a relatively weak Sun teamed up to chill the climate across much of the world, especially the Northern Hemisphere. A few regional networks were set up in Europe during those cold years, starting in the 1780s with the Elector of Mannheim, who provided thermometers and barometers to any interested volunteers.

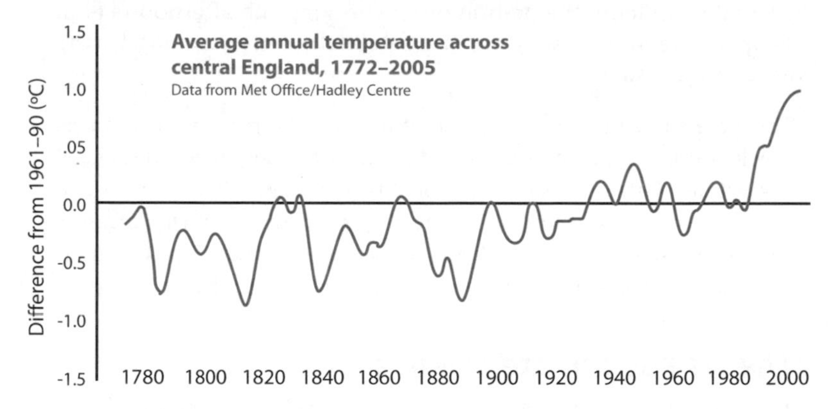

The more thorough assembling of temperature data began in earnest after the arrival of the telegraph in the mid-eighteenth century, which made possible the rapid-fire sharing of information. Overnight, it seemed, weather mapping changed from a historical exercise to a practical method of tracking the atmosphere from day to day. The embryonic art of weather forecasting, and the sheer novelty of weather maps themselves, helped feed the demand for reliable daily observations. **Weather services** were established in the United States and Britain by the 1870s and across much of the world by the start of the twentieth century. Beginning in the 1930s, the monitoring took on a new dimension, as countries began launching weather balloons that radioed back information on temperatures and winds far above ground level. Most of these data, however, weren't shared among nations until well after World War II.

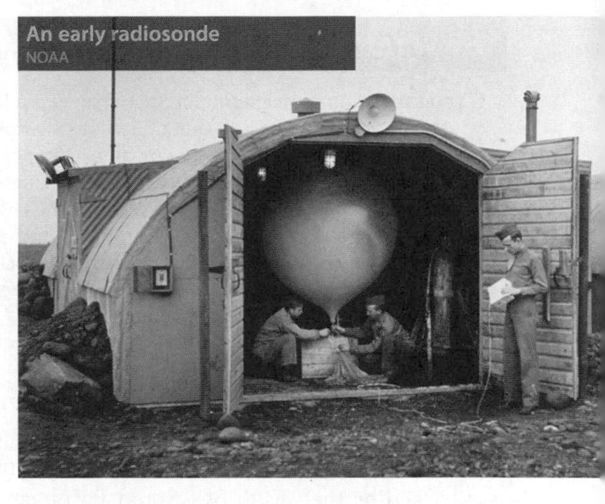
An early radiosonde
NOAA

One of the products of the postwar drive for global unity was the **World Meteorological Organization**, founded in 1951 to link national weather agencies and later folded into the United Nations infrastructure. In 1961, US president John F. Kennedy urged the UN General Assembly to consider "further cooperative efforts between all nations in weather prediction and weather control." The latter goal proved to be far-fetched, but the WMO did begin work on getting the world's weather services to agree on observing practices and data protocols. Before long, a global weather observing system was in place, a model of international collaboration and data exchange that stood firm against the tensions of the Cold War and continues to this day.

During this mid-century whirlwind of progress, the focus was on improving daily weather forecasts. What about the temperature of the planet as a whole? Amazingly, hardly anyone was tracking it. The pioneering climatologist **Wladimir Köppen** took an early stab at it in 1881. There was a brief flurry of activity in the 1930s when, in separate studies, the US Weather Bureau and British engineer Guy Stewart Callendar drew on the limited data available – mostly from North America and Europe – and found that Earth had warmed since the late 1800s. Little did they know that the warming itself happened to be focused in North America and Europe, a fact that emerged only decades later. "But for this accident, it is not likely that people would have paid attention to the idea of global warming for another generation", notes climate biographer Spencer Weart.

> **"Climate was described as 'average weather' and climatology was looked upon merely as the dry-as-dust book-keeping branch of meteorology."**
>
> **Hubert Lamb, 1959, referring to the interwar period**

Interest in global temperature had cooled off – as had the planet itself – by 1961, when Murray Mitchell

So what is Earth's temperature, anyway?

If you live in Rome, Melbourne or Shanghai, you might not realize how close your climatic experience is to that of Earth as a whole, at least in the average. Each of these cities has an annual mean air temperature within two degrees Celsius of 14.4°C (57.9°F), which is roughly the planet-wide average. A hundred years ago, the global average was closer to 13.6°C (56.5°F). The current number hides a substantial difference between north and south. It's apparently taking longer for greenhouse gases to warm the ocean-dominated Southern Hemisphere; the annual average temperature there is more than 1°C (1.8°F) cooler than in the Northern Hemisphere.

of the US Weather Bureau found a cooling trend that dated back to the 1940s. His was the most complete attempt up to that point at compiling a global temperature average. However, like his predecessors, Murray was tricked by the wealth of US and European weather stations. In the sparsely observed Southern Hemisphere, the mid-century cool-down was far weaker than it was in the north. It took until the late 1970s for temperatures to began inching back up and concern about global warming to resurface. In the early 1980s, several groups began to carefully monitor the annual ups and downs of global air temperature across the entire planet, eventually including the oceans. These data sets are now among the most important pieces of evidence brought into the courtroom of public and scientific opinion on climate change.

Measuring the global warm-up

Among those who find human-induced warming not much to worry about, there's a fairly wide spectrum of thought. Many now accept that Earth is warming up but argue that the rise isn't catastrophic and that any further rise should be well within our ability to adapt. At the other end, a few die-hards maintain that the planet really hasn't warmed up measurably in the last century – a position dismissed by virtually all climate scientists.

Still, the die-hards' position begs an interesting question: how do the experts come up with a single global surface temperature out of the thousands of weather stations around the world? It's a far more involved process than simply toting up the numbers and averaging them out. For starters, you need to include the best-quality observations available. Modern weather stations are expected to fulfil a set of criteria established by the WMO. For example, thermometers should be located between 1.25

and 2m (49–79") above ground level, because temperature can vary strongly with height. On a cold, calm morning, you might find frost on the ground while the air temperature at station height is 2°C (36°F) or even warmer. Also, an official thermometer should be housed in a proper instrument shelter, typically one painted white to reflect sunlight and with louvres or some other device that allows air to circulate (see photo). Anyone who's been in a car on a hot day knows how quickly a closed space can heat up relative to the outdoors.

Checking the temperature at a surface station
Bob Henson

Consistency is also critical. Ideally, a weather station should remain in the same spot for its entire lifetime. But as towns grow into cities, many of the oldest stations get moved. That means the station is sampling a different microclimate, which can inject a cool or warm bias into the long-term trend. One of the biggest agents of station change in the mid-twentieth century was the growth of air travel. Countless reporting stations were moved from downtown areas to outlying airports, where regular observations were needed for flight purposes and where trees and buildings wouldn't interfere with wind and rainfall measurement. Even when stations stay rooted, the landscape may change around them and alter their readings. Many urban locations are vulnerable to the heat island effect (see box overleaf). Large buildings help keep heat from radiating to space, especially on summer nights, and this can make temperatures in a downtown area as much as 6°C (10°F) warmer than those in the neighbouring countryside.

On top of all this, gaps can emerge in the records of even the best-sited, longest-term stations. Natural disasters, equipment or power failures, and wars can interrupt records for days, months or even years, as was the case at some locations during World War II. Problems may also emerge after

Bright lights, big cities, bogus data?

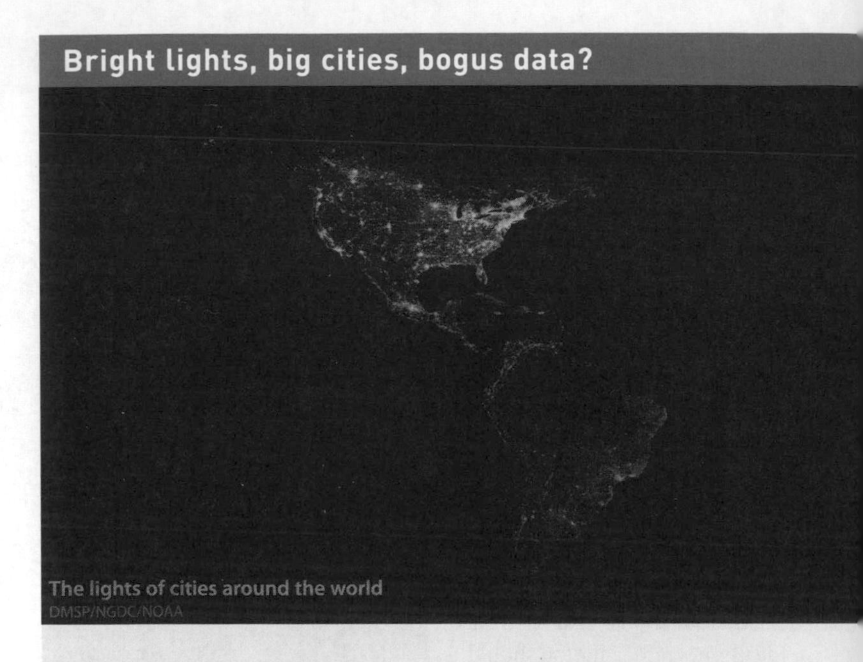

The lights of cities around the world
DMSP/NGDC/NOAA

One of the most calcified critiques of climate-change science is the accusation that the warming indicated by global observations is largely an artefact of urban growth. Nobody doubts the existence of the heat-island effect, by which the dense buildings and paved areas of cities absorb and trap heat in various ways. Indeed, urban heat islands are a form of climate change in themselves, and those who live in a growing urban area would probably experience a gradually warming climate even if we weren't pumping the atmosphere full of greenhouse gases.

However, mainstream climate scientists don't consider it kosher to fold the localized effects of heat islands into the planet-wide warming produced by fossil fuel emissions. Thus, the big three data compilers described in this chapter use a variety of methods to filter out urban effects from their data sets. The first challenge is working out which stations should count as urban, and which should count as rural.

James Hansen's group at NASA started out by using population data to identify urban stations, but they soon discovered this approach couldn't always distinguish whether a reporting station was sited in the heart of a big city or just outside town. In 2001 they applied a new strategy: identifying urban areas through the lights detected by US defence satellites on clear nights (especially during new moons). Hansen found that many rural US areas generated enough night-time light to put their free-of-heat-island status into question. Eventually, the group removed about 80% of their US reporting stations from the global average. That left roughly 200 bona fide rural sites – still enough to provide a reasonable picture of US climate, says Hansen.

In other parts of the world, where rural analogues for urban stations are scarce or where development covers entire regions, it's harder to weed out the city influence. But some recent studies indicate that urban heat islands may not be affecting the measured global average as much as scientists once thought. Thomas Peterson of the US National Climatic Data Center found that, from 1880 to 1998, the rural subset of the centre's 7500-station global record actually warmed a touch faster than the full record – hardly a sign of urban contamination. Peterson also took a fine-toothed comb to 289 US stations, using the satellite-based method, and found no statistically significant urban effects. He speculated that the siting of many US weather stations in open, park-like settings (including airport grounds) helps avoid the influence of core city areas, where it might well be warmer. "We need to update our understanding of urban heat islands", says Peterson. "This phenomenon is more complex than widely believed by those not immersed in the field."

Urban heat islands are strongest on calm nights, when there's no breeze to help disperse the city-trapped air. This inspired David Parker of the UK Hadley Centre to see whether calm nights might be warming up more than windy nights globally, which would presumably be a sign of heat-island bias in the climate record. As Parker reported in *Nature* in 2004, that doesn't seem to be the case. For a set of 264 stations across the globe during the period 1950–2000, he found identical warming trends for nights categorized as windy, calm or lightly breezy. Parker's study adds to the overall body of work showing that heat islands are a poor scapegoat for planet-wide warming. One other clue: the recent warming has manifested not only on land, but over the oceans, which aren't known for their urban sprawl.

the data are sent: electronic or software glitches may corrupt the numbers in ways that aren't immediately obvious.

In a perfect world, every temperature record would be free of the blemishes noted above. That's not the case, of course, but climate scientists have found clever ways to work with what they've got. By using the oldest and best data sets available, they can cull biased stations, adjust for incomplete records, weigh each station by the area it represents and produce a homogeneous picture of long-term temperature trends.

One major task is to find out how much the global reading for a given year has departed from the recent average (which is typically calculated from data covering a thirty-year period, long enough to transcend short-term dips or peaks). The three major climate centres that carry out yearly analyses of global temperature each have their own way of arriving at an average.

▶ **University of East Anglia** Phil Jones and colleagues, at the Climatic Research Unit of the University of East Anglia (UEA), pioneered a technique called the "climate anomaly method". As the name implies, this method starts with the anomaly observed at each station in a given year – how much warmer or cooler it is relative to a thirty-year base period, such as 1961–90. UEA does this calculation for about 4200 stations worldwide. It averages those station anomalies across grid boxes that span 5° each of latitude and longitude, then by hemisphere, and finally for the globe as a whole, with the UK's Hadley Centre providing the ocean analyses.

▶ **The US National Aeronautics and Space Administration** A group led by James Hansen of NASA devised the "reference station method". Instead of anomalies, it uses temperatures observed at about six thousand stations, then calculates averages across a much smaller number of grid boxes than UEA (about eighty in all). The longest-term record in each grid box is dubbed the reference station and considered the gold standard for that box. Records covering a shorter timespan are adjusted so that their trends are consistent with those from the reference stations.

▶ **The US National Oceanic and Atmospheric Administration** NOAA began using a newer technique, the "first difference method", in the late 1990s. It calculates the amount of temperature change from one year to the next at each station. If data are missing for some length of time at a certain site, the difference can be taken over a longer period without much effect on the final result. Among the three groups, NOAA

Amassing the figures

Phil Jones in his office, surrounded by climate data
Bob Henson

"It's the best thing we've ever done", says Phil Jones of the global temperature index he and colleagues created at the Climatic Research Unit of the University of East Anglia (UEA). The idea emerged during a pub lunch. It was 1979, and the group knew that others had tried to take Earth's temperature in a systematic way but hadn't achieved truly global coverage. With support from the UK and US governments, Jones and Tom Wigley (now at the US National Center for Atmospheric Research) led the charge to sift through stacks of data and correct as many sources of bias as possible.

In New York, James Hansen and colleagues at NASA's Goddard Institute for Space Studies were on the same wavelength. Hansen, a pioneer climate modeller, set out to maximize the value of climate data in poorly sampled parts of the globe. His group's technique allowed the most isolated stations to represent conditions within a radius of 1200km (750 miles), which helped expand the analysis across the station-scarce Southern Hemisphere and near the poles.

As the 1980s unfolded, the NASA and UEA groups took turns making headlines with a series of landmark papers. NASA led off in 1981 with the startling announcement that the world was once again warming – and, in fact, that it had been since the 1970s. UEA concurred in a 1982 paper, and in 1986 Jones and colleagues detailed their techniques in the most thorough global analysis to date. By the time Hansen testified before the US Congress in 1988 (see p.250), the US National Oceanic and Atmospheric Administration had begun its own program to monitor global temperature, led by Thomas Karl.

"There are many flaws in the input data for the temperature change analyses, especially in the early years, and the effect of these flaws cannot be fully removed", noted Hansen in 2005. "Despite these problems, however, the reality of global warming in the past century is no longer at issue."

calls on the largest number of stations: about 7200. In 2006 NOAA introduced a new method that fills in data-sparse areas, calling on longer-term patterns observed in nearby areas, and then statistically filters the results.

Each one of these techniques contains a lot of redundancy – on purpose. According to some experts, it takes no more than about a hundred strategically placed points around the world to create a decent first cut at global temperature. Everything beyond that refines and sculpts the estimate; important work, to be sure, but nothing that would change the result drastically.

Which of the three methods is best? That's an impossible question to answer, because there's no absolute record to compare the results against. Each centre's approach appears to have its strengths and weaknesses but, reassuringly, there doesn't seem to be all that much difference among the results. What look like important contrasts – for instance, NASA initially ranked 2005 as the warmest year on record, while UEA put it in second place – relate more to tiny differences in particular years, although some sceptics have taken the occasional reshufflings of the rankings as a sign of sloppy research and/or insignificant warming (see p.256). There's strong agreement on the bigger picture: the atmosphere at Earth's surface warmed nearly 0.8°C (1.44°F) from 1900 to 2005.

Improving the global thermometer

As critical as they are to climate-change science, traditional weather stations and balloon-borne radiosondes were never designed to measure the subtle trends of long-term climate change. Just as a hi-fi that's fine for heavy metal might not convey all the nuance of a symphony orchestra, a weather station might capture large day-to-day temperature changes well enough but contain tiny biases that become evident only after years have passed. To address the need for better long-term data, several UN agencies teamed with the International Council for Science in 1992 to launch the Global Climate Observing System. Through a broad web of activities, GCOS fosters the improvement of all types of data collection on climate change and its impacts, particularly on ecosystems and sea level.

With more and better data in the queue, the job of connecting these observational dots looms larger. In 2005 nearly sixty governments and the European Commission endorsed a ten-year plan to build a Global Earth Observation System of Systems (GEOSS). The name itself signals the multi-layered complexity of the task. GEOSS must find ways to incorporate upcoming satellite systems and new ground-based tools, while working to maintain the integrity of the observational network already in place. That network suffers from multiple ailments: gaps in coverage across space and over time, inadequate archiving, and a lack of certainty that data from valuable yet time-limited satellite missions will continue. The GEOSS ten-year plan calls for "targeted collective action" to address these and other observational concerns.

Estimates of full-globe temperature include **oceans**, which are a bit tricky to incorporate. Although few people live there, the oceans represent 70% of the planet's surface, so it's critical that surface temperature is measured over the sea as well as over land. There's no fleet of weather stations conveniently suspended above the sea, unfortunately, but there are many years of sea-surface data gathered by ships and, more recently, by satellites. Because the sea surface and the adjacent atmosphere are constantly mingling, their temperatures tend to stick close to each other. This means that, over periods of a few weeks or longer, a record of sea-surface temperature (SST) serves as a reasonable proxy for air temperature.

Until the mid-twentieth century, SSTs were mainly measured by dipping uninsulated canvas buckets into the ocean and hauling them back to deck (with the water often cooling slightly on the trip up). Around the time of World War II, it became more common for large ships to measure the temperature of water at the engine intake. This tended to produce readings averaging about 0.5°C (0.9°F) warmer than the bucket method. Some ships measured air temperature directly on deck, but these readings had their own biases: the average deck height increased through the last century, and daytime readings tend to run overly high due to the Sun heating up metal surfaces on the ship. It wasn't until the 1990s that routine measurements from ships and buoys were gathered promptly, processed for quality control and coupled with newer estimates derived from satellites, which can infer SSTs by sensing the upward radiation from the sea surface where it's not cloudy. Together, all these sources now help scientists to calculate temperatures for the entire globe, oceans included. As with the land readings, there are differences in how the three groups compile and process ocean data, but these don't appear to influence the long-term trends for global temperature too much.

Heat at a height

When people refer to global warming, they're usually talking about the air in which we live and breathe – in other words, the air near ground or sea level. But the atmosphere is a three-dimensional entity. Conditions at the surface are affected profoundly by what's happening throughout the **troposphere** – the lowest layer of the atmosphere, often called the "weather layer". The troposphere extends from the surface to about 8–16km (5–10 miles) above sea level. It's tallest during the warmer months and towards the tropics – wherever expanding air is pushing the top of the troposphere upward. In fact, global warming appears to be literally raising the troposphere's roof. In a 2003 study, Benjamin Santer and colleagues at the US Lawrence Livermore National Laboratory showed that the average depth of the troposphere had increased by about 200m (660ft) from 1979 to 1999.

There are two ways to measure temperatures high above Earth's surface. The first – as used by Santer's team – is to use **radiosondes**, the balloon-borne instrument packages launched from hundreds of points around the globe twice each day. Sending data via radio as they rise, the devices measure high-altitude winds, moisture and temperatures. As useful as they are, radiosondes have their limits for tracking climate change: they're launched mainly over continents, with large distances

How the greenhouse turns

The troposphere's name stems from the Greek *tropos*, meaning to turn – which is exactly what the troposphere does. Heated largely from below, as sunlight strikes Earth, this layer of air churns and bubbles like a boiling pot of water, with calm conditions the exception more than the rule. Usually the air is blowing faster horizontally than vertically, but no matter how you look at it, the troposphere is an ever-changing domain. All this motion – up, down and sideways – helps gives rise to the variety of weather we experience at ground level.

Because the troposphere is so well mixed, there should be a close connection between how much it warms near the surface and higher up. But there's not a one-to-one relationship, because greenhouse gases aren't spread through the troposphere as evenly as you might expect, and their effects vary with height.

▶ **Shorter-lived greenhouse gases**, such as low-level ozone, are chemically transformed or washed out by rain or snow within a few days. Thus, they tend to stay within the same general latitudes as their sources – often riding the jet stream from Russia to Canada or from eastern North America to Europe during their short time aloft.

▶ **Water vapour** is most prevalent at lower altitudes, especially above the oceans, but it's been increasing higher up, thanks in part to high-flying aircraft. (Planes also add fossil-fuel emissions to the thin, cold air.)

▶ **Longer-lived greenhouse gases**, such as carbon dioxide, are thoroughly mixed across the troposphere, both horizontally and vertically. That's why a single station high atop Hawaii's Mauna Loa (see p.30) can tell us how much CO_2 the global atmosphere holds.

In general, the higher in the troposphere a greenhouse gas is, the more powerful its effect on the troposphere as a whole. This is because the amount of energy a greenhouse molecule emits is directly related to its temperature. At high altitudes, where temperatures are low, the molecules emit less of the heat separating their paths; they're produced by more than one company, and so don't behave uniformly; and they're prone to small biases that can make a big difference when assessing long-term trends. The second technique is to use **satellites**. These provide a more comprehensive global picture, but they've also been at the centre of a recently resolved issue that was, for years, one of the most heated scientific debates in climate-change science.

Shortly after climate change burst into political and media prominence in 1988, Roy Spencer and John Christy of the University of Alabama at Huntsville began using data from a series of satellites to infer global temperature in three dimensions. Throughout the 1990s and into the new

energy they receive from Earth into space, resulting in more heat remaining in the troposphere.

This picture reverses in the stratosphere, where – oddly enough – human actions have led to *cooling* rather than warming. Carbon dioxide doesn't absorb much heat energy from Earth at these heights, but it continues to radiate heat to space, thus acting to cool the stratosphere. Meanwhile, the partial loss of sunlight-absorbing ozone over the last twenty or so years has exerted a cooling influence. As a result of these and other factors, temperatures in the lower stratosphere have plummeted to record-low levels.

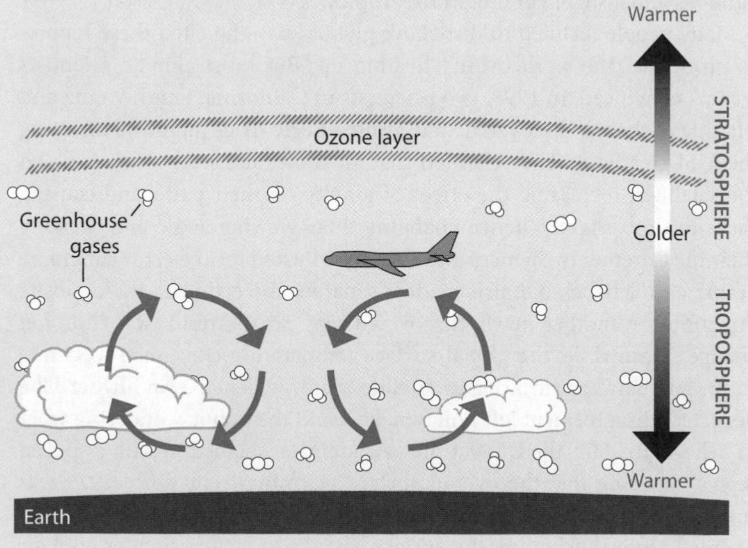

millennium, their findings roiled the world of climate research and lent ammunition to global-warming sceptics. Their data suggested that the troposphere wasn't warming much if at all, despite the observed warming at ground level. Unless the fundamental understanding of lower atmosphere was off-base, then one set of measurements had to be wrong.

Spencer and Christy's data are gathered primarily by instruments called **microwave sounding units** (MSUs), stationed aboard NOAA satellites. The MSUs measure the temperature of gases at various altitudes below them by measuring the energy that they emit. Unlike a surface station or weather balloon, the MSUs provide data that's inherently averaged over large areas. The formulas are complicated, but in general, each individual temperature measurement covers anywhere from about 8000 to 38,000 square km (3100–14,700 square miles) through a depth of a few kilometres and a time period of twelve hours. All told, the satellites cover more than 90% of the planet's surface.

The fireworks began with the first Spencer-Christy paper, published in the journal *Science* in 1990. It showed no discernable temperature trend since 1979 in the lowest few kilometres of the atmosphere. By comparison, surface temperatures had warmed roughly 0.2°C (0.36°F) during the 1980s. Updated through the 1990s, the Spencer-Christy data continued to show virtually no warming. What's more, radiosonde analyses tended to agree with the satellites, with relatively minor warming in northern midlatitudes and a slight cooling in the tropics.

Many people inclined to disbelieve global warming cited these reports as proof that the world wasn't heating up. But most climate scientists weren't convinced. In 1998, two scientists in California, Frank Wentz and Matthias Schabel, discovered that **atmospheric drag** might be affecting the MSU readings. They claimed that air molecules were slowing down the satellites, increasing the effect of gravity on their paths and causing them to drop slightly, hence changing their viewing angle and skewing their measurements. Spencer and Christy adjusted for the error caught by Wentz and Schabel, but also made a separate correction in the opposite direction for another newly discovered bias, so the result was little net change. Meanwhile, the global surface temperature continued to climb, so the gulf between surface and satellite trends widened even further. The persistent disagreement left policymakers and the public wondering what to believe. In 2000, the US National Academies weighed in with a special report, agreeing that the overall surface warming trend since 1979 was "undoubtedly real" but noting the possibility that high-altitude air *had* warmed more slowly than the surface, thanks to factors such as cool air filtering down from the record-cold stratosphere, sulphate pollution and volcanic emissions from 1991's eruption of Mount Pinatubo.

Since then, events have conspired to diminish the satellite data as an argument against global warming. For one thing, the years since 2000 have been notably warm compared to the 1980s and 1990s, both at the surface and aloft. Meanwhile, several other groups gained access to full MSU data, and their analyses (including new techniques for separating stratospheric cooling from tropospheric warming) tended to show even stronger satellite-based warming trends, closer to surface observations. Radiosonde trends have also crept upwards in some studies.

Overall, these and other corrections brought the surface and satellite estimates in much closer agreement with each other. In 2006, a report of the US Climate Change Science Program came to the same conclusion, as did the 2007 IPCC assessment. The IPCC noted that both surface and satellite data capture the year-to-year ups and downs associated with such events as El Niños and volcanoes. It's still unclear whether the troposphere above the tropics is heating or cooling, but for the globe as a whole, the IPCC concluded that the surface warming since 1979 (0.16–0.18°C or 0.29–0.32°F per decade) now concurs well with the MSU-derived tropospheric warming (0.12–0.19°C or 0.22–0.34°F per decade). With the dust settled, one thing is obvious: the world is warming up.

The effect of clouds on climate change is complex and not perfectly understood
Bob Henson

Global dimming

To get a clear idea of how our climate is changing, you've got to measure more than temperature. **Rain and snow** are an important part of the story, and they're uniquely challenging to assess (see Floods & Droughts, p.58). Other variables, such as **cloud cover** and **water vapour,** have their own complexities. Satellites can measure both, but only with strict limitations. For example, some can't sense lower clouds if high clouds are in the way. A set of globe-spanning satellites called COSMIC, launched in 2006 by a US-Taiwan partnership, is now providing fuller three-dimensional portraits of temperature and water vapour. Other new satellites are keying in on cloud cover. Even so, it's likely to be years before we can assess with confidence how these features are changing.

What's in the sky has everything to do with how much sunlight reaches Earth. Sunshine and its absence weren't part of the dialogue on global warming until very recently. A 2003 paper in *Science* gave the phenomenon of **global dimming** its first widespread exposure, and by 2005 it was the star of a BBC documentary. Global dimming is real: from the 1950s through the 1980s, the amount of visible light reaching Earth appears to have tailed off by a few percent. In terms of climate, that's quite significant, though it would be difficult for the average Jane or Joe to detect it next to the on-and-off sunlight cycles that occur naturally each day. It also wasn't enough to counteract the overall global warming produced by greenhouse gases. In fact, the timing of global dimming's rise to fame is actually a bit paradoxical, since the phenomenon appears to have already reversed. As noted in the 2007 IPCC report, "'Global dimming' is not global in extent and it has not continued after 1990." However, as we'll see, the recent brightening brings its own set of worries.

Global dimming offers a good example of how a set of little-noticed measurements, collected diligently over a long period, can yield surprising results and help answer seemingly unrelated questions. Back in 1957, as part of the worldwide International Geophysical Year, a number of weather stations across the globe installed **pyranometers**. These devices measure the amount of short-wave energy reaching them, including direct sunshine as well as sunlight reflected downwards from clouds.

Slowly but surely, the amount of energy reaching the pyranometers dropped off. Gerald Stanhill (who later coined the phrase "global dimming") discovered a drop of 22% across Israel, and Viivi Russak noted similar reductions in Estonia. Although they and several other researchers published their findings in the 1990s, there was hardly a rush to study the problem. With the spotlight on global warming by that point, scientists might have found it counterintuitive to investigate whether less sunlight was reaching the globe as a whole. Climate models, by and large, didn't specify such dramatic losses in radiation.

Yet the data cried out for attention, especially once they were brought together on a global scale. Using a set of long-term readings from the highest-quality instruments, Stanhill found that solar radiation at the surface had dropped 2.7% per decade worldwide from the late 1950s to early 1990s. Two other studies examining much larger data sets found a smaller global drop – around 1.3% per decade from 1961 to 1990. Even these studies failed to jolt the climate research community until 2003, when two Australians, Michael Roderick and Graham Farquhar, linked the dimming to a steady multi-decade reduction in the amount of water evaporating from standardized measuring pans around the world. Warmer temperatures ought to have evaporated more water, not less, they reasoned. However, they realized that a reduction in sunlight could more than offset this effect. It's also quite possible that global evaporation actually rose in spite of the pan-based data, as noted in the 2007 IPCC report. Pan-based sensors can only show how much water might evaporate from a completely wet surface, but a warmer, wetter and dimmer world could still send more total moisture into the air – through the leaves of plants, for example. The lifetime of moisture in the atmosphere, which could change with the climate, is another factor that makes the evaporation issue especially hard to unravel. Still, the notion of dimming through much of the latter twentieth century is now fairly well accepted.

Amazing but true: industrial pollution reflects sunlight away from Earth, offsetting some of the effects of global warming
Barry Lewis/Corbis

The plane truth about contrails

Can those sleek, wispy lines of cloud that trail behind jets really affect the future of our climate? Researchers have already found that contrails can produce surprisingly large effects on a regional basis. The most dramatic real-world example came after the terrorist attacks on New York and Washington on September 11, 2001. With all US commercial air traffic shut down for several days, the skies were virtually contrail-free for the first time in decades. The result was slightly warmer days and cooler nights, according to David Travis of the University of Wisconsin–Whitewater. He found that the average range in daily temperature across the US (the span between each day's highest and lowest reading) increased by more than 1°C (1.8°F) during the aviation-free days. In other words, the lack of contrails provided an open door for additional sunlight to reach the surface by day and for extra radiation to escape at night.

The 9/11 study provided a rare vantage point on an vexing issue. Contrails don't have a single, easy-to-summarize impact on climate, in contrast to the tone of some press and TV reports. "This is a very complicated problem, as there are multiple competing effects going on", Travis says.

Contrails are a special type of **cirrus**, which are high, thin clouds composed mainly of ice crystals and often resembling streaks or sheets. Contrails form as the water vapour and particles spewed out by an airplane's exhaust stream create water droplets that soon crystallize into ice. Pollutants in the exhaust plume serve as focal points for additional ice crystals within the contrail as well as in neighbouring clouds. (Some of these pollutants are greenhouse gases themselves, perhaps adding a few percent to overall global warming – an issue discussed separately on p.186.)

Most types of clouds cool Earth's surface as they screen out sunlight. However, cirrus clouds have an overall warming effect. That's because cirrus usually allow a good deal of sunlight through but effectively trap radiation flowing up from Earth. When a contrail first forms, it behaves more like a lower-level cloud, as the water droplets reflect ample amounts of sunlight and exert a cooling effect. Gradually, the contrail spreads, thins out and freezes, and its net effect soon shifts from cooling to warming. Many contrails evaporate within an hour or so, but some evolve into much larger cirrus clouds, adding to the warming influence.

The plot thickens depending on what time of day it is. At night, contrails can provide only a warming effect (since there's no sunshine to be reflected). Since night flights appear to be gaining in popularity, that could help tilt the scales towards a net warming effect. On the other hand, when sunlight hits a contrail at a low angle – in the early morning or late evening, or even at midday across high latitudes in winter – the contrail's reflectiveness appears to be boosted. On top of all this, contrails aren't globally uniform; they're focused where people happen to be flying. Right now contrails are most prevalent across North America, western Europe, and eastern Asia. Nobody knows whether increasing affluence will spawn more contrails above China and India, or how cut-rate airline traffic will evolve amid increasingly steep fuel costs.

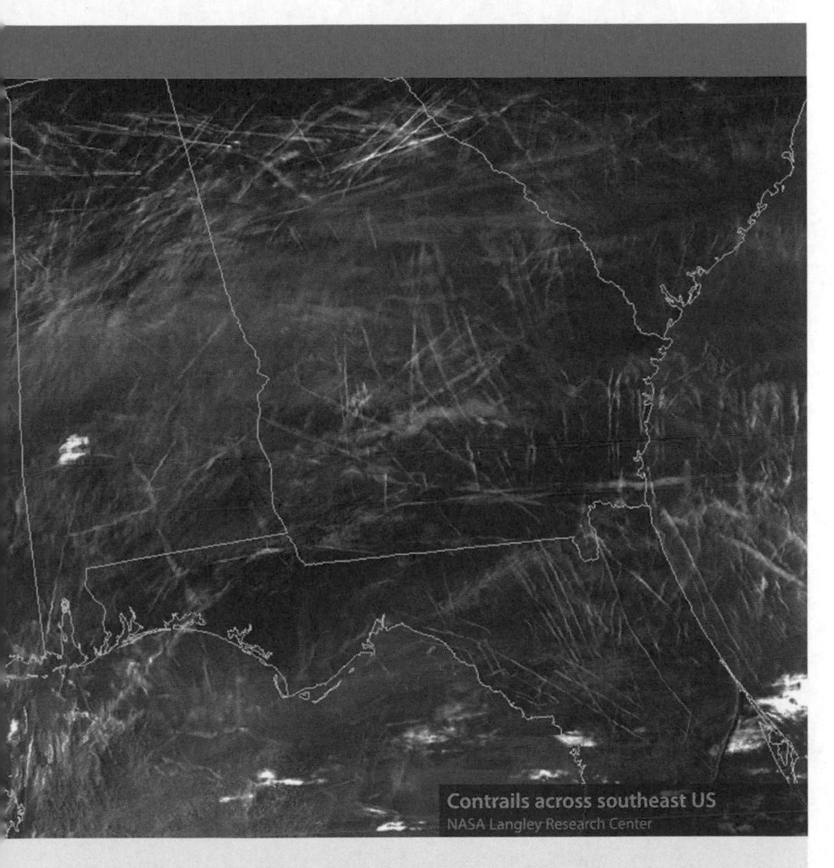

Contrails across southeast US
NASA Langley Research Center

Stir together the above ingredients, and you have a confusing mess indeed. Computer models tend to come down on the side of an overall warming effect in the regions where contrails are focused. Travis believes the models are generally on the right track, although he argues for caution (as does the 2007 IPCC report). The models may not be accounting for all of the above effects accurately, and the few observational studies thus far (including his own) don't show a significant temperature boost from contrails. "I do tend to believe that they ultimately have a warming effect. We just haven't yet found any empirical data to support this", says Travis.

Globally, the 2007 IPCC report estimates that contrails add 25–60% to the warming from aviation-related greenhouse gases. As the clouds spread out and persist, they may trap several times more additional heat, though the IPCC is less confident of this range. Whatever the total impact, it looks set to rise as air travel continues to grow, adding further to the impact of what is already a very climate-unfriendly form of transport (see p.37).

Some areas – especially urban, industrialized parts of the Northern Hemisphere – showed several times more dimming than other parts of the globe. This regional patchwork makes sense given what we know about the causes of dimming. The prime suspects are soot and other airborne particles, or **aerosols**, produced when fossil fuels are burned. For hundreds of years, people in London had first-hand experience with the sky-darkening effects of soot. Aerosol pollution does this type of darkening in several ways:

▶ **Attracting water** Many aerosols are hydrophilic, or water-attracting. When conditions are right for clouds to form, a highly polluted air mass can produce many times more cloud droplets. The moisture is spread more thinly, though, so each droplet is smaller on average. The upshot is that this larger array of droplet surfaces can reflect more sunlight upwards before it has a chance to reach the ground. From space, the clouds look brighter; from Earth, they're darker.

▶ **Extending cloud lifespans** In general, it takes longer to get rain or snow out of a cloud full of many small particles, as opposed to fewer, larger ones. For this reason, the addition of aerosols seems to lengthen the lifespan of clouds and thus increase their reflective power over time.

▶ **Reflecting light** Even on a cloudless day, aerosols themselves reflect sunlight to space. A major study of pollution across India and the adjacent ocean found that the subcontinent's thick veil of winter pollution can trim the amount of sunlight reaching the surface by as much as 10%.

Global warming itself may be responsible for some of the dimming – perhaps around a third – as it increases the amount of water vapour in the atmosphere (which makes more moisture available for cloud formation) and rearranges the distribution of clouds (lower and/or thicker clouds tend to reflect more than higher, thinner ones). As noted above, it's not entirely clear how cloud cover is evolving on a global scale, though increases of 1% or more per decade in water vapour have been recorded since the 1980s.

The big issue ahead may not be dimming but brightening. Aerosols are more visible and noxious than greenhouse gases, yet easier to control, so their reduction can be an easier political sell. Most of the world's highly industrialized nations began cleaning up their smokestacks and tailpipes by the 1970s, and the economic downturn of the 1990s across the former Eastern Bloc further reduced aerosol pollution. Although the breakneck

Good evening earthshine

Earthshine project, Big Bear Solar Observatory

Measuring how much sunlight hits Earth is one thing, but how much does Earth reflect back to space? Even a tiny change in that percentage – the albedo, or reflectivity – could have huge implications for climate. Clouds are the main reflectors of the planet: as a whole, they bounce about half of the sunlight they receive back to space. The oldest and most complete satellite observations of cloud cover are part of a program called the International Satellite Cloud Climatology Project. ISCCP can't track every cloud, but it does measure their collective impact on radiation and albedo. ISCCP data show a drop of 1–2% in Earth's albedo in the 1980s and 1990s, with a slight rise thereafter.

There's another way to measure how much light is bouncing off Earth: by measuring the amount that makes it to the Moon. This is earthshine, the faint light that illuminates the dark part of a less-than-full Moon. Leonardo da Vinci recognized the source of earthshine some five hundred years ago, and in the 1920s French astronomer André-Louis Danjon devised a clever way to measure it. By using a split-view telescope, Danjon dimmed the Sun-brightened part of the Moon (right) until its apparent brightness was the same as the part lit by earthshine (left). The stronger the earthshine, the less adjustment needed; thus, the adjustment itself served as a measure of earthshine strength.

Danjon and followers collected earthshine measurements in France through the 1950s. Since the mid-1990s, the Big Bear Solar Observatory has carried the torch, gathering data for global change studies as part of NASA's Project Earthshine. High in the mountains east of Los Angeles, Big Bear enjoys more than three hundred cloudless nights a year, and it experiences little of the atmospheric turbulence that can spoil sensitive measurements. Earthshine readings at Big Bear can rise or fall by a few percent over several hours as sunlight bounces off the sequence of oceans, continents, clear areas or cloudy patches presented by Earth's turning. Seasonal variations can run on the order of 10%. Over the long haul, the Big Bear earthshine data vary somewhat from satellite-derived readings, a topic now under debate in science journals.

You couldn't pick a better single spot from which to measure albedo than the Lagrange-1 point. That's the point in space about a million miles towards the Sun from Earth, where the two bodies' gravitation fields are balanced. A satellite stationed there could keep a continuous eye on the fully-lit side of Earth. Such was the idea behind Triana, one of Al Gore's most ambitious proposals as US vice president. Renamed DSCOVR after Gore left office, the project languished for years and was finally terminated in 2006 after NASA had spent more than $100 million building the satellite.

pace of development across China and India has kept aerosol emissions high in those areas, the global concentration of aerosols appears to have dropped by as much as 20% since 1990, according to a 2007 NASA report. At the same time, some areas have reported an increase in solar radiation of anywhere from about 1% to 4% per decade since the early 1990s. Beate Liepert (Lamont-Doherty Earth Observatory) suspects this may reflect a recovery from the Sun-shielding effects of Mount Pinatubo's 1991 eruption and from the extensive cloudiness generated by the frequent El Niño events of the early and mid-1990s. All this, in addition to greenhouse gases, may help explain why global temperatures rose as sharply as they did in the last decade of the twentieth century.

Our clean-up of aerosols comes at an ironic price. Few would complain about a world that's brighter and less fouled by aerosol pollution. Yet the overall cooling impact of aerosols will diminish as we reduce their global prevalence. Thus, in gaining a cleaner atmosphere, the world stands to lose one of its stronger buffers against greenhouse-gas warming. As the planet warms and brightens, it remains to be seen whether policymakers will be alert to this risk.

It's possible, though, that policymakers won't have much say in the matter. Commentators such as James Lovelock have raised the concern that a severe global economic downturn (caused by climate change or anything else) could lead to massively reduced aerosol emissions over a relatively short period, causing the planet to warm faster than ever just when we're least able to put money into doing anything about it.

The long view

A walk through climate history

During the four and a half billion years of our solar system, Earth has played host to an astounding array of life forms. Much of the credit goes to our world's climate, which is uniquely equable compared to conditions on the other major planets. Even if they had water and other necessities, Venus would be too hot to sustain life as we know it, and Mars and the planets beyond it too cold. But in fact, there's evidence that both planets once harboured oceans and may have been mild enough to support life before their climates turned more extreme. Likewise, during its lively history, Earth's climate has lurched from regimes warm enough to support palm-like trees near the poles to cold spells that encased the present-day locations of Chicago and Berlin under ice sheets more than half a mile thick.

The changes unfolding in our present-day climate make it all the more important to understand climates of the past. Thankfully, over the last fifty years, technology and ingenuity in this field of study have brought dramatic leaps forward in our knowledge. The clues drawn from trees, pollen, ice, silt, rocks and other sources leave many questions still to be answered. Taken as a whole, though, they paint a remarkably coherent picture. We now have a firm grasp of the changes that unfolded over the past three million years – a period ruled mainly by ice ages, with the last 10,000 years being one of the few exceptions. We also know a surprising amount about what transpired even earlier, including many intervals when Earth was a substantially warmer place than it is now.

Could those ancient warm spells serve as a sneak preview of what we might expect in a greenhouse-driven future? We know that there's already more carbon dioxide and methane in our atmosphere than at any time in at least 800,000 years, and perhaps much longer than that. Carbon dioxide has increased by more than 30% in only a century, an astonishing rate by the standards of previous eras. At the rate we're pumping greenhouse gases into the atmosphere, they could be more prevalent by the year 2050 than they've been in at least ten million years. That's one reason why **paleoclimatologists** (those who study past climates) are taking a close look at how climate and greenhouse gases interacted during previous warm intervals. The analogies are admittedly less than perfect. As we'll see, Earth's continents were positioned quite differently 200 million years ago than they are now (even ten million years ago, India was still drifting across the Indian Ocean and North and South America had yet to join up). As a result, the ocean circulations were very different as well. Also, most of the species that were then swimming, crawling or flying across our planet are now extinct. Many other aspects of the environment at that time are hard to discern today.

In spite of our less-than-ideal hindsight, we've gained enormous ground in our understanding of how climate works by piecing together the past. And new advances continue to arrive at a rapid clip in the still-young field of paleoclimatology. Until recently, for example, the oldest ice-core records of past atmospheres extended back about 440,000 years. In 2004, a new ice core from Antarctica provided a record that's almost twice as long (see box on p.215). There's hope that a future core could yield data going back nearly 1.5 million years, including the point when ice ages switched their on-and-off tempo. If there's one message that's become increasingly clear from the ice cores, it's that natural forces have produced intense and rapid climate change in the past. That's something worth thinking about as we contemplate our future.

What controls our climate?

The great detective story that is paleoclimatology hinges on a few key plot devices. These are the forces that shape climate across intervals as brief as a few thousand years and as long as millions of years. Greenhouse gases are a critical part of the drama, of course, but other processes and events can lead to huge swings in climate. These vary greatly in terms of time-span. An unusually large volcano may cool the planet sharply, but most of the impacts are gone after a couple of years. In contrast, climate change due to slight shifts in the Sun's orbit takes tens or hundreds of thousands of years. That's still relatively fast in geological terms. The slow drift of continents, another major player in climate variation, can operate over hundreds of millions of years.

The overlap among these various processes creates a patchwork of warmings and coolings at different time intervals, some lengthy and some relatively brief. It's a bit like the mix of factors that shape sunlight in your neighbourhood. On top of the dramatic and regular shifts in sunlight you experience due to the 24-hour day and the transition from summer to winter, there are irregular shifts produced as cloudiness comes and goes.

Below is a quick summary of the biggest climate shapers. We'll come back to each of these later in the chapter.

▶ **The Sun** is literally the star of the story. At its outset, the Sun shone at only about three-quarters of its current strength; it took several billions of years for its output to approach current levels. One fascinating question (discussed later) is how Earth could be warm enough to sustain its earliest life forms while the Sun was still relatively weak. In another five or six billion years the Sun – classified as an "ordinary middle-aged star" – will reach the end of its life. At that point it should warm and expand dramatically before cooling to white-dwarf status.

▶ **Variations in Earth's orbit** around the Sun are hugely important. These small deviations are responsible for some of the most persistent and far-reaching cycles in climate history. Each of these orbital variations affects climate by changing when and how much sunlight reaches Earth. As shown in the graphics overleaf, there are three main ways in which Earth's orbital parameters vary. Several visionaries in the mid-1800s, including James Croll – a self-taught scientist in Scotland – suspected that these orbital variations might control the comings and goings of ice ages. In the early 1900s, Serbian mathematician Milutin Milankovitch quantified the idea, producing the first numerical estimates of the impact of orbital variations on climate. While his exact calculations have since been refined, and there's still vigorous debate about what other factors feed into the timing of climate cycles, the basic tenets of Milankovitch's theory have held firm.

James Croll
NOAA Paleoclimatology Program/Courtesy John Imbrie

Round and round we go (and so does our climate)

Everyone who lives outside the tropics knows the power of the four seasons. But a surprising number of people – including most of the Harvard graduates interviewed at random in a classic documentary – can't explain why the seasons occur. Many assume it's because Earth is closer to the Sun in summer and more distant in winter. But that doesn't wash: it's true that Earth's orbit is slightly asymmetric, but we're closest to the Sun on 4 January, when it's summer in Sydney but winter in London. The real reason for the seasons is that Earth is tilted about 23.4° relative to its orbit around the Sun (see graphic). That tilt aims each hemisphere towards the Sun in an alternating pattern that produces summers, winters, springs and autumns.

What does this have to do with climate change? Each of the factors just alluded to – Earth's tilt, the eccentricity of its orbit, and the timing of its closest approach to the Sun – change slightly over time in a cyclical manner. This is due mainly to the intersecting gravitational effects of the Sun, the Moon and the other planets in our solar system. Together, the orbital cycles produce dramatic swings in climate over tens of thousands, or even hundreds of thousands, of years. By studying how these natural climate shifts come and go, scientists also find clues about how Earth might respond to change that's human-induced.

▶ **Earth's tilt** goes up and down, ranging from about 21.8° to 24.4° and back over about 41,000 years (the range widens or narrows a bit with each cycle). The tilt is now around 23.4° and on the decrease. When the tilt is most pronounced, it allows for stronger summer Sun and weaker winter Sun, especially at high latitudes. Ice ages often set in as the tilt decreases, because the progressively cooler summers can't melt the past winter's snow. At the other extreme, increasing tilt produces warmer summers that can help end an ice age.

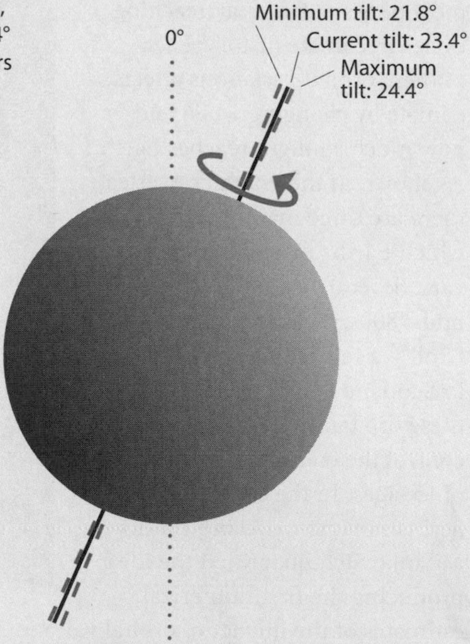

▶ **Earth's orbit around the Sun** is not precisely circular, but elliptical, with the Sun positioned slightly to one side of the centre point. Currently this brings Earth about 3% closer to the Sun in early January (perihelion) than in early July (aphelion), with about 7% more solar energy reaching Earth at perihelion. The **eccentricity** or "off-centredness"

of the orbit varies over time in a complicated way. The net result is two main cycles, one that averages about 100,000 years long and another that runs about 400,000 years. When the eccentricity is low, there's little change through the year in the Earth-Sun distance. When eccentricity is high, the sunlight reaching Earth can be more than 20% stronger at perihelion than at aphelion. This either intensifies or counteracts seasonality, depending on the hemisphere and the time of year that perihelion and aphelion occur. The last eight ice ages have come and gone roughly in sync with the 100,000-year eccentricity cycle, but it's not yet clear how strongly the two are related (see p.224).

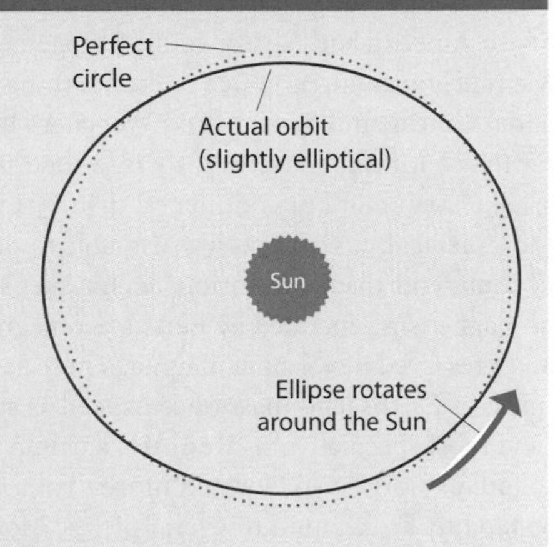

▶ **Earth's axis** rotates slowly around an imaginary centre line, like a wobble in a spinning top. The main cycle of this process – known as **precession** – takes about 26,000 years, and it shifts the dates of perihelion and aphelion forward by about one day every 70 years. Thus, in about 13,000 years, Earth will be closest to the Sun in July instead of January. This will intensify the seasonal changes in solar energy across the Northern Hemisphere and weaken them in the south. The stronger summer Sun is likely to bolster the African and Asian monsoons, as was the case around 10,000 years ago (see box, p.60).

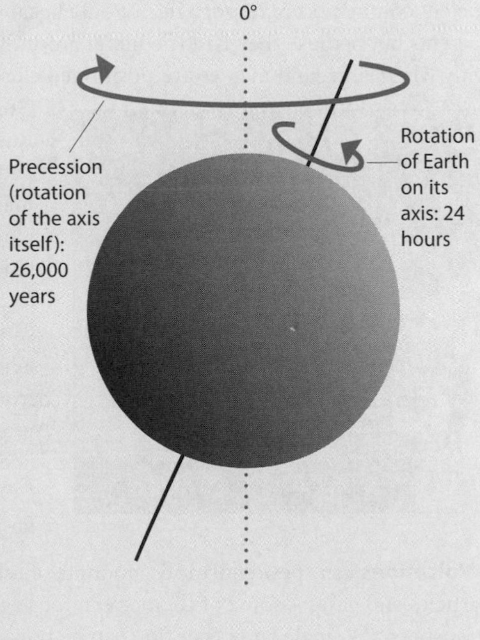

▶ **The changing locations of Earth's continents** also play a huge role in determining climate. Francis Bacon observed in 1620 how easily South America and Africa would fit together if they were puzzle pieces. Many schoolchildren notice the same thing long before they're taught about **continental drift**, Alfred Wegener's bold 1915 hypothesis that Earth's continents moved slowly into their present positions. Today, satellites can monitor continental drift by tracking minuscule land movements. But scientists are also able to deduce the long-ago locations of continents thanks to various techniques – including analyzing traces of magnetism generated by Earth's geomagnetic field over the millennia and preserved in volcanic magma. These and other clues tell us that much of Earth's land mass once existed as **supercontinents**. The first (and most speculative) is **Rodinia**, a clump that bears little resemblance to today's world map. Formed more than a billion years ago, Rodinia apparently broke up into several pieces. Most of these were reassembled as **Gondwana**, which drifted across the South Pole between about 400 and 300 million years ago. Gondwana became the southern part of **Pangaea**, an equator-straddling giant. In turn, Pangaea eventually fragmented into the continents we know now. The locations of these ancient continents are important not only because they shaped ocean currents but because they laid the literal groundwork for ice ages. Only when large land masses are positioned close to the poles (such as Gondwana in the past, or Russia, Canada, Greenland and Antarctica today) can major ice sheets develop. However, high-latitude land isn't sufficient in itself to produce an ice age. The southern poles harboured land for most of the stretch from 250 to 50 million years ago, yet no glaciation occurred, perhaps because CO_2 concentrations were far higher than today's.

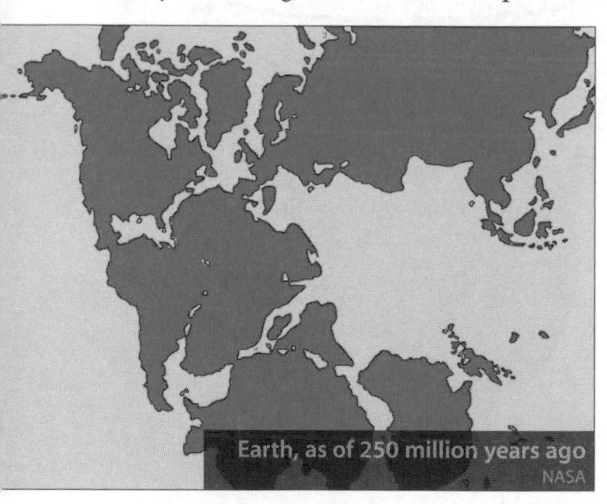
Earth, as of 250 million years ago
NASA

▶ **Volcanoes** can spew enormous amounts of ash, soot and other particles and gases – some of them greenhouse gases – into the atmosphere. A single large eruption can be powerful enough to loft

The eruption of Mount Pinatubo on June 12, 1991
USGS/Cascades Volcano Observatory

Sun-shading particles into the stratosphere, as was the case when Mount Pinatubo blew its top in 1991. When this happens, the high-altitude debris can cool the planet by more than 1°C (1.8°F) for a year or more, until gravity pulls the particles back to Earth. Tropical volcanoes are especially effective at this cooling, since their stratospheric debris can easily spread across both northern and southern hemispheres. In the much longer term, volcanoes act as a warming influence, since they add perhaps 0.1–0.3 gigatonnes of carbon to the atmosphere per year on average. That's a substantial amount, though it's less than 10% of what humans now add through fossil fuel burning. At present it appears that the volcanic addition is balanced by other natural forces and feedbacks that remove carbon at roughly the same pace. However, it's possible that a frenzy of volcanism early in Earth's history pumped the atmosphere's greenhouse-gas levels far above anything that's occurred since. Later on, undersea eruptions associated with the spreading of continental plates may have been the main source of the carbon dioxide that led to Earth's remarkable warmth from about 250 to 50 million years ago (see p.211).

▶ **Celestial bodies** such as **asteroids** make cameo appearances in Earth's climate saga. Most are minor. Only the very rarest of giant asteroids, which typically arrive millions of years apart, can produce climatic effects comparable to those from a major volcanic eruption. As with volcanoes, a huge asteroid might cool climate for a year or two by blasting dust and soot into the stratosphere. However, a big enough asteroid could also produce an explosive wave of heat as it enters the atmosphere, enough to set much of Earth's vegetation ablaze. Such fires would send enough carbon dioxide into the atmosphere to cause a substantial global warming that could last centuries, far longer than the brief post-collision cooling. This lurid scenario apparently played out around 65 million years ago, when an enormous asteroid struck Earth. The most likely point of impact is an area on Mexico's Yucatán Peninsula now underlaid by a giant crater of pulverized rock. Scientists combing the globe have found widespread deposits of iridium and other substances that are far more prevalent in asteroids than on Earth, all apparently deposited at the same time. Also dating from the same period are clumps of charcoal-like carbon, a hint that widespread fires occurred. More than half of all species are thought to have perished in this cataclysm and its aftermath, including most dinosaur species. (A few that survived may be the predecessors of modern-day birds and crocodiles.)

How scientists read the clues

The technology used to analyse materials and glean climate clues from them has advanced at a phenomenal rate in recent decades. The traces of past climates can show up in animal, vegetable and mineral forms, providing a wealth of evidence for paleoclimatic detective work. Below are the four main types of paleoclimate **proxy data** – records that can be used to infer atmospheric properties such as temperature or precipitation.

▶ **Biological**: tree rings, pollen, corals and fossils of plants, animals, insects and microscopic creatures such as diatoms. Their locations, concentrations and conditions point towards the atmospheric and oceanic conditions under which various life forms prospered. Many biological materials can be dated using radioactive decay rates or the presence of stable isotopes (more on these later). The chemical makeup of these samples can also yield important information.

▶ **Cryological**: cylindrical samples called ice cores, collected by drilling through dense, deep ice sheets or glaciers where air bubbles and dust are trapped within yearly layers of snowfall. The bubbles preserve the greenhouse gases and other atmospheric constituents present at a given time, and the ice cores can be analysed through radiometric dating (see p.202). The dust layers and the character of the ice itself also reveal climate processes at work.

What trees tell us about climate

It was an astronomer – who happened to be looking down instead of up – who launched the science of **dendrochronology**. In the 1920s, Andrew Ellicott Douglass pondered the rings in tree stumps he encountered while hiking through the forests of northern Arizona. Douglass discovered that the width of the tree rings in this arid area was correlated with the precipitation amounts that occurred during the trees' lifetimes. He went on to found the Laboratory of Tree-Ring Research at the University of Arizona in Tucson in 1937. Among his noteworthy accomplishments was developing a method using tree rings to determine the age of wood beams in prehistoric structures across the US Southwest.

From its roots in Arizona and elsewhere, dendrochronology has branched out across the world, with past climates now profiled at thousands of sites. To create a typical data set, the ring patterns obtained from several trees in the same area are matched, cross-dated and analysed to produce an annual, localized set of tree-ring growth indices known as a site chronology. For some types of living trees, such as bristlecone pines, these chronologies can go back thousands of years. (Tree lovers take note: the slender cores can be extracted from live trees without harming them.) If a series of progressively older dead trees can be found whose lifespans overlap with living trees, the sequence can be extended back even further. In Germany, the University of Hohenheim has created a 12,480-year record using oak and pine trees. How much further back can dendrochronologists go? There is at least one place in the world where trees from the depths of the last ice age are preserved in relatively pristine shape. That's on the North Island of New Zealand, where bogs may hold Kauri pine dating back as far as 50,000 years.

It takes far more than counting in order to use tree rings as guides to past climates. Each site has a unique blend of characteristics – temperature, precipitation, soil conditions and the like – that affect how trees grow. For example, trees at arid forest borders are limited by moisture availability, so the variations in their rings will reflect ups and downs in precipitation. It's also possible to deduce catastrophic events, such as when a tree is shunted by an advancing glacier or partially knocked over by a storm. One clue is the asymmetric ring pattern produced as the tree attempts to grow upright after the injury.

Some of the most dramatic findings of dendrochronology relate to the region where the discipline was born. Connie Woodhouse of the US National Oceanic and Atmospheric Administration went to the Arizona lab in the early 1990s, and later made headlines when she and colleague Jonathan Overpeck compiled evidence, based on tree rings and other proxy records, of what Overpeck dubbed "megadroughts". These persistent dry spells, lasting several decades each, occurred across large portions of the central and western US prior to 1600 AD. Megadroughts can affect the course of entire civilizations (see p.220), even though there may be occasional years of near-normal rain amid the dry years.

An ancient bristlecone pine
NOAA/Jonathan Pilcher

- **Geological**: rock, sand dunes, ocean sediments, glacial debris, stalagmites and other materials sampled from sea and lake beds and dry land. Again, radioactive decay within these materials helps establish their age. Volcanic rock can also be dated by the traces of Earth's geomagnetic field. The location and condition of telltale geological features helps establish when sea levels rose and fell or when glaciers scoured the land surface. Many ocean sediments are laden with the remnants of shells built by sand-sized creatures, including bottom-dwelling **foraminifera** as well as **plankton** that prefer near-surface waters. The prevalence and type of these shells can reveal the amount of carbon and other elements that entered the sea at various points.
- **Historical**: written records of atmospheric conditions and biological phenomena tied to climate. The latter might include the timing of spring blooms for a given type of tree or the poleward extent of a particular species of insect. In some areas, including England and Japan, such records go back hundreds of years with almost daily precision.

Establishing dates for climate proxies can be a challenge. Trees are among the easiest: each ring indicates a year's worth of growth, and its width hints at the state of that year's climate (see p.201). For trees that are still alive, or ice cores drawn from the surface downward, it's a simple matter to count backwards from the present and establish dates in the past, using multiple samples to ensure accuracy. For many other proxies, such as fossilized trees, you also need a benchmark, something you can use to establish the time period in which the proxy was created. A typical starting point is **radiometric dating**, which revolutionized the paleoclimatic dating game in the 1950s.

Radiometric techniques rely on the fact that elements such as **carbon** and **uranium** exist in several different forms, or **isotopes**, each with a different number of neutrons. Some of these isotopes are unstable, and they decay over time at measurable and predictable rates. For instance, neutrons hitting the upper atmosphere are continually converting nitrogen into an unstable carbon isotope known as ^{14}C. This sifts down into Earth's atmosphere and gets absorbed by living things, decaying over time as new ^{14}C isotopes are created higher up. Given a certain amount of ^{14}C, half of it will have decayed back into nitrogen after about 5700 years (its **half-life**). Knowing this, scientists can measure how much ^{14}C has decayed in a given substance and use that information to determine how old the substance is. Although these unstable isotopes are rare compared to stable carbon atoms, they're well distributed throughout Earth and its ecosystems. This allows geochemists to date a wide variety of substances, from carbon dioxide in an ice-core air bubble to the shells of marine creatures. One limit to using ^{14}C is that it has a relatively short half-life of around 5700 years. Since there's very little carbon left to measure after a few half-life cycles, most carbon dating only goes back about 50,000 years. For older substances, **mass spectrometry** dating is used. Uranium is the material of choice for this, because it decays into lead at a very slow rate. Uranium's half-life is about 700 million years for the isotope ^{235}U and an astounding 4.5 billion years – the age of Earth itself – for ^{238}U.

As with any measurement technique, errors can creep into radiometric dating. This happens mainly when a substance is contaminated by other materials of a different age through chemical reactions, erosion or some other process. Careful cross-calibration with similar materials from other locations can help alleviate this problem.

Isotopes don't have to be unstable to be useful in dating. Stable isotopes provide another window on climate's past. The most extensively studied is ^{18}O, a form of oxygen. Because ^{18}O is heavier than standard oxygen, it condenses more readily. The difference between the two rates depends on temperature, ocean salinity, and other factors. So, for example, coral reefs hold differing amounts of ^{18}O based on the sea-surface temperature present when they formed. In ice cores (and other places where ancient water is trapped) the relative amount of ^{18}O can help reveal the temperature at which the moisture condensed. This can help delineate annual layers in the ice as well as reflecting longer-term temperature trends.

Computer models are another important part of the paleoclimate toolbox. In theory, any model that simulates future climate ought to be able to reproduce the past as well, given enough information to start with. Indeed, many of the most sophisticated global models designed to project future climate are tested by seeing how well they reproduce the last century. Another useful test is to see how well the models handle modern-day seasonal transitions: in mid-latitudes, at least, going from summer to winter can serve as a microcosm of a long-term overall climate shift. For prehistoric times, there's far less data to check a model against, but even in these cases simple models can be helpful in assessing what climates might be plausible. There's often a back-and-forth between paleoclimate modelling and data: observations might point to a particular climate scenario that can be tested through models, and the model results might confirm or deny a particular interpretation of the data, sending scientists back to the drawing board (or the field).

The really big picture: from Earth's origins to the expansion of life

It takes quite a mental stretch to fully comprehend the total lifespan of Earth's atmosphere – unless, that is, you're a biblical literalist (see box opposite). To carve up this vast expanse of time into manageable portions, paleoclimatologists use the same naming system as do geologists. The box on p.206 shows the major geological eras, the intervals they denote, and the analogous lengths of time they'd occupy if Earth's 4.5 billion-year lifespan were compressed to 24 hours. In this imaginary day, dinosaurs wouldn't come on the scene until after 10pm, and the period since the last ice age ended – the time when most human civilizations emerged – would occupy only the last fifth of a second before midnight.

Earth itself offers few clues to help us figure out what climate was up to during the planet's first two billion years or so. In fact, a good deal of our knowledge about Earth's earliest climate comes from space. By studying the behaviour of stars similar to the Sun, astronomers have deduced that the solar furnace (a continuous nuclear reaction that forms helium out of hydrogen) took a long time to reach its modern-day strength. It appears the Sun's luminosity started out only about 70–75% as strong as it is now, gradually increasing ever since. With the Sun relatively weak for so long, you might envision early Earth as a frozen, lifeless wasteland. Amazingly, this wasn't the case. Fossils of tiny bacteria show that life existed as far back as 3.5 billion years. Rocks from around that time show the telltale signs of water-driven erosion, another sign that Earth wasn't encased in ice. How did Earth stay warm enough for life during those early years? That's the crux of what paleoclimatologists call the **faint young Sun paradox**.

Greenhouse gases are perhaps the most obvious solution to the paradox. Put enough carbon dioxide in the air, and Earth can stay warm enough to sustain life even with a weaker Sun. The amount of carbon now stored in fossil fuels and in the oceans is more than enough to have provided such a greenhouse boost if it were in the air during the weak-Sun period. Volcanoes can inject huge amounts of greenhouse gas into the atmosphere in a short amount of time, so it's believed that the intense volcanism of Earth's earliest days likely added enough carbon dioxide to keep the planet from freezing. However, that begs another question: how did the carbon dioxide levels gradually decrease over billions of years at just the right tempo to keep Earth warm, but not too warm? It's possible, but unlikely, that volcanic activity diminished at exactly the pace needed.

Earth's age: 4.5 billion years or 6000 years?

Virtually all physical scientists accept the basic chronology of Earth's history summarized in this chapter. However, millions of Americans beg to differ. These Christian fundamentalists generally take the Bible at its word: that Earth was created in six 24-hour days about 6000 years ago. A cottage industry of experts, including some PhD scientists, promulgates this view through such enterprises as the Institute for Creation Research. They often employ the Bible's story of a giant flood to explain the formation of the Grand Canyon and other geological features. (The official gift shop at the canyon's national park sells a book along these lines called *Grand Canyon: A Different View*.) In recent years many US school boards have pondered the teaching of intelligent design (ID), a somewhat watered-down version of creation science that seeks evidence for the existence of a creator without specifying whom that might be. Most of the ID movement's energy is focused on biological rather than climate science, although its proponents seem to be philosophically inclined towards scepticism about human-induced global warming. (That's not the case for all US evangelicals – see p.268.)

Public opinion polls suggest a cognitive split among many Americans that allows them to accept Earth science that contradicts the Bible even as they draw the line at the more discomforting idea of human evolution. A 2006 report by the US National Science Foundation shows that about 75% of Americans and 85% of Europeans surveyed in 2004–05 agree with the statement, "The continents on which we live have been moving their location for millions of years and will continue to move in the future." However, US surveys have shown consistently that nearly half of Americans believe God created humans in the last few thousand years.

Another possibility is that some other climate factor kicked into gear to keep greenhouse gases within a range tolerable for life on Earth.

One candidate is the steady expansion and elevation of continents during much of Earth's history. **Land masses** reduce the amount of carbon dioxide in the air through a multimillion-year process known as chemical weathering, which occurs when rain or snow fall on rocks that contain silicate minerals. The moisture and silicates react with carbon dioxide, pulling the CO_2 out of the air. Carbon and water then flow seaward in various forms, and most of the carbon ends up stored in ocean sediments that gradually harden into rocks. Over billions of years, this process has

Geological era and origin of the name	**Periods included** (oldest to youngest)	**Dates** (millions of years ago)
Hadean after Hades, Greek god of the underworld	n/a	4500–3800
Archaean from the Greek *arkhaios*, ancient	n/a	3800–2500
Proterozoic from the Greek *proteros*, earlier, and *zoe*, life	Paleoproterozoic Mesoproterozoic Neoproterozoic	2500–543
Paleozoic from the Greek *palaios*, ancient	Cambrian Ordovician Silurian Devonian Carboniferous Permian	543–248
Mesozoic from the Greek *meso*, middle	Triassic Jurassic Cretaceous	248–65
Cenozoic from the Greek *kainos*, new	Tertiary Quaternary	65–today

been responsible for some 80% of the carbon now stored underground. (Rain and snow also draw CO_2 out of the air when they leach limestone, but in that case the carbon heads seaward in a form used by marine creatures to build shells, a process that returns the carbon to the atmosphere.) Chemical weathering is such a powerful process that the growth of the Himalayas over the last 55 million years might have helped to pitch Earth into its recent series of ice ages.

If Earth's lifespan were a day (midnight to midnight), then this era begins at:	What's going on?
Midnight	Earth cools after its cataclysmic birth; surface remains liquid. Atmosphere likely a stew of toxins.
3:44am	Continents began to form; bacterial life develops. Atmosphere still hostile to present-day life forms.
10:40am	Continents solidify. Earth cools markedly 750–580 million years ago. Afterwards, the climate warms and life begins to diversify.
9:06pm	Plant and animal life flourishes. Supercontinent Gondwana forms and transitions to Pangaea. This period is mostly warmer than present; brief glaciation around 430 million years ago; longer cold interval with frequent glaciations from around 325 million years ago up to a mass extinction at the end of the period.
10:41pm	New plant and animal forms emerge, including dinosaurs. Pangaea separates into modern-day Africa, Eurasia and the Americas. The entire period is warmer than present, with no major glaciations. A meteorite triggers another mass extinction at the end of period.
11:39pm	Mammals flourish and humans evolve. The Indian subcontinent joins Asia; Australia breaks off from Antarctica. Climate remains warm until dramatic cooling begins around 50 million years ago. Antarctic ice begins to form at least 35 million years ago. The Northern Hemisphere glaciation starts around 2.7 million years ago and continues to present day, interrupted by relatively brief interglacial periods such as the one we're now in.

Apart from these vast changes, chemical weathering can also influence the atmosphere through a **negative feedback** process, one that tends to counterbalance climate change in either direction. In general, the warmer it gets, the more rapid chemical weathering becomes and the more carbon dioxide is pulled from the air. In part this is because the chemical reactions behind weathering operate more rapidly when it's warmer. Another reason is that warmer oceans send more water vapour into the air, and more rain and snow falls onto continents to stimulate weathering. Thus, the warmer the air gets, the more carbon dioxide is drawn out of the atmosphere through weathering, and that should help produce a cooling trend. Conversely, if ice starts to overspread the land, weathering should decrease and carbon dioxide levels ought to increase, thus working against the glaciation.

Earth's profusion of **plant life**, which began around 400 million years ago, serves as another brake on greenhouse gases. Plants and trees take in carbon dioxide through photosynthesis and release it when they die. Generally, plant growth increases on a warmer, wetter Earth – so the hotter it gets, the more plants there are to help cool things down. Indeed, the concept of Gaia positions life itself as the main negative or balancing feedback on the entire Earth system (see box opposite).

Of course, there are also many **positive feedbacks** between CO_2 and climate, which tend to amplify rather than dampen a change. For instance, as carbon dioxide increases and warms the planet, more water vapour – itself a greenhouse gas – should evaporate into the atmosphere and further increase the warmth. Similarly, oceans absorb less carbon dioxide the warmer they get, which leaves more CO_2 in the air to stimulate further warming. It's the balance between negative and positive feedbacks that generally determines which way climate will turn at any given point. A big concern about the 21st century is that positive feedbacks, such as those now taking place in the Arctic, may overwhelm negative feedbacks, some of which (such as chemical weathering) take much longer to play out.

Whether they're looking a thousand or a billion years into the past, paleoclimatologists take it for granted that greenhouse gas concentrations are intimately linked with the rise and fall of global temperature. Ice cores and other records bear out this tight linkage over the last million years, and there's no reason to think it should be absent earlier on. There's a subtlety in this relationship, though – one often pointed out by global-warming sceptics. Because the CO_2 changes are so closely coupled in time to the growth and decay of ice sheets, it's hard to tell exactly which came first and when.

For example, there appear to be periods in which undersea volcanism produced vast amounts of CO_2, after which temperatures rose dramatically, and other periods where orbital cycles triggered ice-sheet growth and colder temperatures, with CO_2 amounts dropping as a side effect. Paleoclimatologists are still sorting out these chains of events. It appears that orbital changes play a major role in kicking off and ending ice ages, but during the growth and decay stages, there's a web of positive feedbacks

Gaia and global warming

Though often associated with New Age thinking, the concept of Gaia isn't necessarily warm and fuzzy. The name itself, a reference to the Greek goddess of Earth, was suggested in 1969 by William Golding, author of the dystopic novel *Lord of the Flies*. Today, the father of Gaia theory, scientist **James Lovelock**, is deeply disturbed about the direction in which global warming – or as he calls it, "global heating" – is pushing the Earth system. His 2006 book *The Revenge of Gaia* warns that "we live on a live planet that can respond to the changes we make, either by cancelling the changes or by cancelling us."

Lovelock was inspired in the 1960s by the contrast between the lack of evidence for life on Mars and the fact that life evolved and prospered on Earth despite the slow ramp-up of solar energy (see p.224). "Together, these thoughts led me to the hypothesis that living organisms regulate the climate and the chemistry of the atmosphere in their own interest", writes Lovelock. He sees Gaia as a single, self-regulating entity that encompasses Earth's living things and the chemical and physical backdrop that sustains them. Though Lovelock doesn't view Gaia as an evolving organism in classical Darwinian terms, he does see the web of interacting feedbacks among greenhouse gases, vegetation and creatures on land and in the oceans as a system that preserves and perpetuates the conditions that foster life.

James Lovelock
Bruno Comby/www.ecolo.org

It's been a long slog for Lovelock to convince disciplinary scientists of the merit in his all-embracing theory. Partial vindication came in 2001 when a group of leading Earth scientists issued the Amsterdam Declaration on Global Change. It asserts that "the Earth system behaves as a single, self-regulating system comprised of physical, chemical, biological and human components". Lovelock goes further by insisting this system has a goal: "to keep the Earth habitable for whatever are its inhabitants".

Like any organism, Gaia can be pushed to its limits, says Lovelock, and warming the atmosphere is asking for such trouble. If Gaia could express a preference, Lovelock believes that "it would be for the cold of an ice age, not for today's comparative warmth." One example: both drought and flood are a greater risk in warm climates than in cool ones, because water tends to evaporate more quickly the hotter it gets. Another example: oceans warmer than 10°C (50°F) become stratified such that nutrients are quickly depleted in the warm surface layer. Lovelock fears that greenhouse gases could trigger an irreversible cascade of feedbacks that profoundly disrupt the self-regulating processes of Gaia. As for what to do about this, Lovelock thinks the standard renewable-energy goals of more wind farms and solar panels aren't nearly enough. He calls for an immediate ramp-up of nuclear power, as a form of energy triage, while we develop and refine alternative sources of power over the long haul. "It is much too late for sustainable development", he says; "what we need is a sustainable retreat."

at work among CO_2, ice and other climate elements. For instance, cooler oceans absorb more carbon dioxide, which acts to support further cooling. In any case, the fact that CO_2 can sometimes lag behind a major climate change doesn't mean that it can't help trigger one, which is the situation we're finding ourselves in now.

Close calls for life on Earth

US poet Robert Frost famously observed in 1920 that some people expect the world to go out in fire, while others think the end will be an icy one. Our planet edged towards those extremes at several different points in its history. Two examples brought vastly different consequences for ecosystems.

One of the coldest periods in early Earth history ran from about 750 to 580 million years ago. There's evidence that glaciers scoured most of Earth's large land masses, including some within 10° of the Equator. The implication is that much or all of the planet was covered with ice for millions of years – a scenario dubbed Snowball Earth in 1992 by geologist Joseph Kirschvink of the California Institute of Technology. Some researchers believe Earth was tilted more than 50° at the time; this would have given the Equator less intense sunlight than the poles. If large land masses were located near the Equator, the extra tilt would have supported the creation of low-latitude glaciers. Then, if the planet actually did freeze over entirely, the powerful positive feedback of a "white Earth" could have reflected most of the incoming sunlight and helped preserve the ice. How such a state would have ended isn't known, but a slow accumulation of carbon dioxide from volcanoes – on top of a cold-induced slow-down in chemical weathering (see p.205) – seems the most likely possibility. Whether or not ice actually covered the entire Earth, this period of widespread glaciation was followed by a sudden profusion of multi-cellular organisms. Some theorists speculate that a Snowball Earth and a subsequent warm-up could have prodded the rapid evolution of primitive life into more complex forms.

At the other end of the spectrum, an especially intense warming around 250 million years ago, after a long interval of glaciation, spelled doom for most of the life forms present at the time. The Permian/Triassic extinction drew the curtain on more than 90% of marine species and more than two-thirds of land-based creatures. The die-off unfolded quickly in geological terms, over less than a million years. Greenhouse gases likely soared to many times their present-day amounts, and high-latitude oceans warmed to as much as 8°C (14.4°F) above present-day readings. The warmer oceans likely enhanced the separation of surface waters from cooler deep oceans, reducing the usual mixing between the layers that distributes oxygen and nourishes many marine organisms. Researchers haven't settled on a single cause for the extinction. Among the possibilities: enhanced volcanic activity, a massive meteorite (although there's no sign of an impact crater dating from that era), an outpouring of the greenhouse gas methane in the form of methane hydrates released from the ocean floor into the atmosphere (see p.212), or some blend of these and/or other factors.

Hot in here: Earth's warm periods

What will the world look like if the more dire projections of climate change by the year 2100 come to pass? Computer models and basic physical theory tell us a lot about what to expect, but we can also gaze back into climate history for clues.

A good place to start is the **Mesozoic era**. It began after the world's greatest extinction episode, which occurred about 250 million years ago (see box, p.206), and ended with another vast extinction 65 million years ago. In between are three periods well known to dinosaur lovers: the Triassic, Jurassic and Cretaceous. All three intervals were notably warm, with no signs of any major glaciations or even much ice at all. In part this is because Earth's land masses were located away from the North Pole, mostly clustered in the warm supercontinent of Pangaea. However, parts of Pangaea extended to the South Pole, and the warmth continued even as Pangaea broke up and the Northern Hemisphere continents shifted closer to their present-day locations.

Two of the warmest intervals were the mid- to late Cretaceous and the early Eocene (part of the Tertiary epoch), which fall on either side of the asteroid-driven extinction that occurred 65 million years ago. In both periods, average global readings soared to 5°C (9°F) or more above present values. That's near or beyond the IPCC's high-end projections of global temperature for the year 2100. Proxy data for these past periods indicate that carbon dioxide was several times more prevalent in the atmosphere than it is today. This enhances scientists' confidence that the extra CO_2 stoked the ancient warmth and that the greenhouse gases we're now adding could do something similar in our future.

Where did the extra CO_2 come from? The most likely candidate is an active period of **undersea volcanism** (see graphic overleaf). Earth's tectonic plates clash most visibly along the world's volcanic coastlines, such as the Pacific's "ring of fire." However, there's additional volcanic activity far beneath the oceans. Most major basins have a mid-ocean ridge, where tectonic plates separate and vast amounts of magma and carbon dioxide surge upwards to fill the void. The magma eventually solidifies to form fresh crust, slowly elevating the centre of the ridge as the older material spreads outward. It appears that, on average, the plates were separating up to 50% more quickly about 100 million years ago than they are now. This would have allowed more CO_2 to bubble upwards and pushed the plates towards continents more vigorously, stimulating more activity (and more CO_2 emissions) at coastal volcanoes.

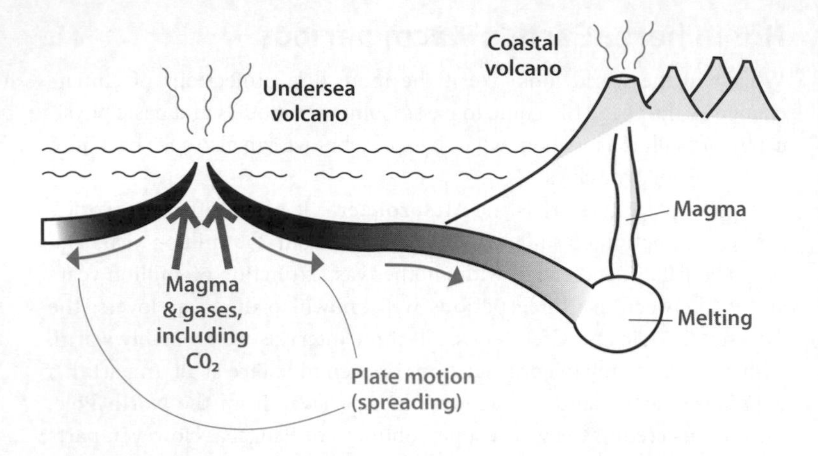

Another source of greenhouse gas might have provided even more of an influx at key times. Hundreds of billions of tonnes of carbon are believed to be locked in cold ocean seabeds in the form of **methane hydrates**, ice crystals that hold methane in their molecular structure (also known as methane clathrates). Methane hydrates remain stable if conditions are cold enough and/or if enough pressure is exerted on them (as is the case at the bottom of the sea). But a change in ocean temperature or circulation during a warm era might be enough to destabilize the methane and send it pouring into the atmosphere. This is a long-term concern for our future climate (see p.86), and it may have played a role in at least two massive warmings during already warm times. One was the temperature spike and mass extinction that kicked off the Mesozoic era 248 million years ago. The other is a distinct blip early in the already toasty Eocene epoch called the Paleocene-Eocene Thermal Maximum (PETM), which occurred about 55 million years ago. During the PETM, high-latitude air and ocean temperatures soared by as much as 7°C (12.6°F) over a few thousand years and remained high for almost 100,000 years. Isotope studies by NASA indicate that a vast amount of greenhouse gas was pumped into the atmosphere during the PETM, though over a longer period and perhaps at a slower rate than our current emissions.

One thing that's fairly clear about past warm periods is that mid- and high-latitude areas were downright balmy as compared to today. Coral reefs grew as far polewards as 40°, close to the latitudes of present-day Philadelphia and Melbourne. Crocodiles, dinosaurs and leafy trees flourished in polar regions, while other areas featured vast deserts covered with sand dunes. A 2006 study of sediments beneath the Arctic Ocean found that sea-surface temperatures during the PETM soared to as high as 23°C (73°F), which is far beyond earlier estimates. As for the tropics, it's generally believed that they weren't much if any warmer than they are now, although some startling evidence to the contrary has recently surfaced. Karen Bice of Woods Hole Oceanographic Institution reported in 2006 on results obtained from sediments and fossilized shells collected from the tropical North Atlantic off Suriname. These indicate water temperatures for the late Triassic period (100 to 84 million years ago) that ranged between 33°C and 42°C (91–107°F), exceeding even hot-tub levels. "These temperatures are off the charts from what we've seen before", said Bice.

The oceans were at a high-water mark during these warm times. For millions of years, **sea level** was more than 200m (660ft) higher than it is now. This put vast stretches of the present-day continents under water, including the central third of North America and all of eastern Europe from Poland southward. The oceans were so high and vast for several reasons. One is that there wasn't any water locked up in long-term ice sheets. Another is that water, like any other fluid, expands as it warms up, just as the oceans are doing now (albeit on a much more modest scale). A third reason is the shape of the oceans themselves. The faster separation of tectonic plates noted above would have produced larger mid-ocean ridges as new crustal material surged upward. This, in turn, would have raised the height of the seafloor and the resulting sea level by surprisingly large amounts – perhaps enough to account for most of the difference between the Triassic/Eocene sea levels and today's.

From greenhouse to icehouse: the planet cools down

The most recent 1% of our atmosphere's history is mainly a story of cooling. Starting in the mid-Eocene, around 50–55 million years ago, sea levels began to drop, continents shifted ever closer to their current locations, and global temperatures began to fall. The cooling appears to have temporarily reversed between about 15 and 25 million years ago, then resumed with gusto. This long process wasn't linear – there were sharp warmings and coolings in between, each lasting many thousands of years – but overall, Earth's temperature took a tumble of perhaps 4°C (7.2°F) or more. This eventually pushed the planet into the glacial/interglacial sequence that began less than three million years ago and continues today.

What caused the cool-down? Once again, it appears that carbon dioxide was the most direct culprit. CO_2 in the atmosphere went from several times its modern-day amounts in the mid-Eocene to barely half of its present-day concentration during the last few ice ages. The question of what drove down CO_2 levels is far more challenging. Mid-ocean ridges were separating more slowly when the cool-down began, a sign that volcanic activity was pumping less CO_2 into the atmosphere. The biggest decreases in volcanism apparently occurred prior to forty million years ago. That would correspond to the first pulse of global cooling, but the spreading rate had increased again by the time the second pulse began, leaving that one unexplained.

In their search for a solution to this riddle, one group of US scientists cast their eyes towards the **Tibetan plateau**. This gigantic feature began to take shape around 55 million years ago as the Indian subcontinent joined Asia, pushing the Himalayas and the plateau upwards. The pace quickened about thirty million years ago, and today most of the Alaska-sized plateau sits at elevations greater than 3700m (12,000ft). It's bordered by mountains that soar more than twice as high.

In the late 1980s, Maureen Raymo (of Woods Hole), Flip Froelich (now at Florida State University) and William Ruddiman (of the University of Virginia) proposed that the intense chemical weathering produced by the Tibetan plateau could have pulled enough carbon dioxide from the air to trigger the global temperature drop that led to the most recent ice ages. Since weathering is a negative (dampening) feedback, one might argue that there should be less weathering rather than more as the ice ages came on. But Raymo and colleagues claim that the Tibetan plateau is a special case, so vast and tall (perhaps the most massive above-ground feature in Earth's history) that its emergence overwhelmed the usual negative feedbacks associated with chemical weathering. In the absence of enough data to confirm or disprove the theory, the debate goes on.

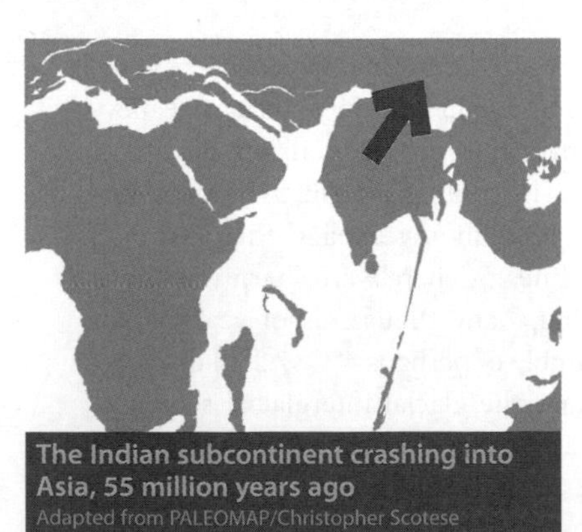

The Indian subcontinent crashing into Asia, 55 million years ago
Adapted from PALEOMAP/Christopher Scotese

There were other tectonic developments of interest during the long cool-down. **Antarctica** separated from South America between 20 and 25 million years ago. With air and water now able to circle around Antarctica unimpeded, the resulting tight circulation might have fostered the growth of the first major Antarctic ice sheet about thirteen

Core value: getting information out of ice

Our understanding of the last few ice ages has been boosted immeasurably by the rich vein of data obtained through ice cores from Greenland, Antarctica and several other glacier-studded areas. The task is laborious: each ice sheet is more than 3km (1.9 miles) thick, but each cylindrical core is only about 10cm (4") in diameter, and each segment of the core must be drilled and removed a few metres at a time. Weather only permits a few short months of work each year. Even in midsummer, daytime temperatures can be colder than -20°C (-4°F) atop the Greenland ice sheet; in central Antarctica that's a mild summer day.

The first major ice cores were retrieved from both Antarctica and Greenland as part of the International Geophysical Year (IGY) of 1957–58. In Greenland, the first cores were taken at camps far from the thickest part of the ice sheet. As interest in climate change intensified and researchers clamoured for more of the ice, two parallel projects set out to gather definitive cores from the top of the sheet. By the early 1990s, these US and European teams – separated by only about 32km (20 miles) – had drilled from the summit of Greenland's ice sheet to bedrock, a depth of just over 3km (9900ft). The two cores they obtained each spanned a little more than 100,000 years of climate.

Meanwhile, Russian scientists at Antarctica's Vostok Station, which was founded during the IGY, were engaged in their own decades-long drilling effort. By 1995, they'd reached a depth of around 3600m (11,800 feet) before they halted to avoid contaminating a vast lake beneath the ice. The Vostok core is only about 20% longer than those from Greenland. However, since the highest reaches of the Antarctic ice sheet get so much less snowfall per year than Greenland's summit, each metre of the Vostok core details a much longer stretch of climate. The total Vostok record covers about 440,000 years.

One of the most striking findings from both poles is that carbon dioxide and methane, two key greenhouse gases, rose and fell in near-lockstep with the comings and goings of ice ages. Typically, the concentrations dropped by 20–50% during a glaciation, then quickly recovered afterwards.

The new champion of ice archives is the core completed in 2004 by the EU-sponsored European Project for Ice Coring in Antarctica (EPICA). Retrieved at a site atop Dome C, about 500km (300 miles) from Vostok, the EPICA core has already yielded climate analyses extending back 800,000 years. Researchers with EPICA hope to use the core to look back nearly a million years – around the time when ice ages mysteriously transitioned to much longer intervals than before.

A "drill dome" in Greenland
NOAA/Mark Twickler

million years ago (though modelling studies have been inconclusive on this point).

Meanwhile, North and South America slowly approached each other, finally joining at the Isthmus of Panama nearly four million years ago. This closed the longstanding connection through which warm, salty water had flowed beneath trade winds from the tropical Atlantic into the Pacific. In a major readjustment, the Atlantic shifted into a mode much like the one we see today, with the tropical waters flowing northwards along an intensified Gulf Stream and on into the far North Atlantic. This influx of warm, saline water into the Arctic would have cut down on sea-ice formation. Ironically, it could have also furnished the moisture needed to build ice sheets over land and thus hastened glaciation.

Thanks to some combination of the factors above – plus, perhaps, others yet to be discovered – Earth inexorably slid into what some experts call a "deteriorating" climate. Antarctic ice grew in a start-and-stop fashion. Glaciers began to appear after about seven million years ago in the Andes, Alaska and Greenland. Finally, **ice sheets** began spreading across North America and Eurasia about 2.7 million years ago. Since then, more than two dozen glaciations have ground their way across much of present-day Canada, northern Europe and northwest Russia. In between each glaciation was a shorter, milder period, with ice extent and temperatures closer to those at present and even warmer in some cases.

At first, these **interglacial** periods occurred about every 41,000 years. This corresponds remarkably well to the 41,000-year cycle in the tilt of Earth's axis (see box, p.196). When Earth's tilt is maximized, the extra solar energy received at high latitudes in summer melts ice sheets more easily (even though winters are getting less Sun than usual). When the tilt is lower, summers are cooler, and ice sheets get a running start with unmelted snow from the previous year.

About a million years ago, however, the 41,000-year rhythm of ice ages ended. The ice ages got stronger and began to last much longer – about 100,000 years each, with interglacial periods on the order of 10,000 to 15,000 years long. What was behind this shift isn't yet clear, although it could be that a gradual thickening of the ice sheets made them more resistant to melting on the warm-summer side of the 41,000-year cycle. Earth's orbit also has a 100,000-year cycle in its off-centredness (see p.196). Most paleoclimatologists don't believe that this relatively weak cycle was enough in itself to wrench a change in the ice-age frequency. However, it may have interacted with some other cyclic process to provoke the timing. One of the more exotic hypotheses for the current timing of ice ages is that Earth's orbit may be passing through a disk of interplanetary dust about every 100,000 years.

The period just before the last ice age began, about 130,000–115,000 years ago, provides a sobering illustration of what climate change could accomplish over the next few centuries. That interglacial kicked off with a sharp warming most likely driven by orbital cycles. Coral-reef data indicate that sea-level was up to 6m (20ft) higher than present-day levels, even though greenhouse gases were less abundant than today. A study in 2006 by Jonathan Overpeck (of the University of Arizona) and Bette Otto-Bliesner (of the US National Center for Atmospheric Research) shows that major ice-sheet melting at both poles may have fed the sea-level rise. The same computer modelling shows that Arctic summer temperatures by 2100 will be close to those achieved in the last interglacial warming. Although it could take decades or even centuries for that warming to translate into a 6m sea level rise – which would swamp major coastal areas around the world – Overpeck worries that the process of ice-sheet melt could prove irreversible sooner than we think.

Emerging from the ice

By the time of its peak about 20,000 years ago, the last ice age had carved out a very different-looking world than the one we know now. A major effort in the 1970s to reconstruct this period's topography showed a North America and Europe half encased in ice, with global temperatures about 4°C (7.2°F) colder than today's average. South of these ice sheets, it was largely dry and windy, especially over continental areas. Tundra covered most of southern Europe, and spruce trees characteristic of high latitudes extended into the mid-Atlantic area of the present-day US. One place where rainfall exceeded today's norm is the US Great Basin: Nevada and Utah were freckled with dozens of large lakes that now exist only as salt flats. Sea levels were so low – more than 100m (330ft) below today's – that Australia extended northwards into New Guinea and Japan connected Korea and China.

The ice age may not have looked exactly like this through its entire 100,000-year lifespan. Until ice cores were plucked from Greenland and Antarctic, many scientists had pictured ice-age climates as being relatively static. Instead, the cores revealed that Earth swung through a number of distinct warmings and coolings during the start of the last ice age, about 115,000 years ago, up to and beyond its conclusion around 15,000 years ago. As the clarity and completeness of ice-core data improved in the

1990s, it was joined by analyses of ocean sediments from several dispersed basins and stalagmites from caves in China. The new evidence made it clear that warm and cold swings during the ice age were far more widespread, and could unfold far more rapidly, than scientists had previously thought (see box opposite).

There's little debate about what brought the last ice age to an end. Two of the orbital **cycles** discussed on p.196, involving Earth's tilt and its precession, synchronized to produce a strong peak in the summertime input of solar energy across the Northern Hemisphere. Starting about 15,000 years ago, ice sheets began to retreat dramatically. As the melting intensified, water occasionally poured out of glacial lakes in spectacular bursts, some of which were large enough to influence ocean circulation and trigger rapid climate shifts.

Perhaps the most dramatic was the **Younger Dryas** period, a frigid encore that brought a return to cold nearly as extreme as the ice age itself. The period is named after a tenacious alpine flower – *Dryas octopetala* – that thrives in conditions too austere for most other plant life. The flower's pollen, preserved in Scandinavian bogs, provides evidence of extremely cold periods. The most recent ("youngest") such period began nearly 13,000 years ago and persisted for some 1300 years. Ice cores from Greenland indicate that the Younger Dryas took as little as a decade both to set in and to release its grip. During the interim, average temperatures across the British Isles sank to as low as –5°C (23°F), and snowfall increased across many mountainous parts of the world. Drought and cold swept across the Fertile Crescent of the Middle East. This may even be what prodded the ancient Natufian culture to pull back from wide-ranging hunting and gathering and focus on what ranks as the world's earliest documented agriculture.

As for the cause of the Younger Dryas, most paleoclimatologists point to the massive draining of **Lake Agassiz**, a vast glacial remnant that covered parts of present-day Manitoba, Ontario, Minnesota and North Dakota. Such an influx of fresh water could disrupt the thermohaline circulation that sends warm waters far into the North Atlantic (see p.120). However, there's spirited debate about whether the Agassiz melt flowed in the right direction to affect the North Atlantic. An even more perplexing piece of the puzzle comes from the Southern Hemisphere, where records from Antarctica and other areas show that the cool-down there began a thousand years earlier – closer to 14,000 years ago.

The Younger Dryas was followed by a much shorter and less intense cooling about 8200 years ago. Once those were out of the way, the next

How fast can climate flip?

Evidence as far back as the 1970s hinted that climate during glacial periods might not have been as static as many believed. The pieces came together in the early 1990s, when scientists confirmed a sequence of major warmings and coolings that unfolded against the cold backdrop of the last ice age. Instead of furnishing a constant chill, it seems the 100,000-year ice age unfolded in a much more irregular fashion. Gerald Bond and colleagues at the Lamont-Doherty Earth Observatory drew on data from sediments in the North Atlantic to trace the evolution of the ice age in eye-opening detail. The kind of upwards and downwards spikes they found don't seem to occur in warm regimes like our present-day, post-glacial climate. But they serve as a reminder that climate can switch from one mode to another in the geologic equivalent of a heartbeat. The great ice sheets that coated much of Europe and North America never disappeared, but conditions across many ocean and continental areas varied sharply with these warmings and coolings, each named for paleoclimatic pioneers.

▶ **Going up** On 23 different occasions during the last ice age, air temperatures quickly climbed about half of the way back to their interglacial levels, then sank back to more typical ice-age readings. The warm-ups pushed average temperatures in Greenland up by as much as 16°C (29°F) in as little as forty years, while the much slower return to glacial cold took about a thousand years. These warmings – which appear to have been concentrated in the Northern Hemisphere – are called Dansgaard-Oeschger (D-O) events, after Danish geophysicist Willi Dansgaard and Swiss geochemist Hans Oeschger.

▶ **Going down** Less frequently – at six points during the ice age – climate lurched in the other direction. Vast fields of icebergs poured from North America into the North Atlantic, disrupting the ocean circulation and cooling climate. These so-called Heinrich events are named after German scientist Hartmut Heinrich, who discovered particles of Canadian soil scraped off by ice sheets and deposited in the North Atlantic by rafts of icebergs.

Together, the Heinrich and D-O events explain much of the variability in global climate that shows up in sediments and ice cores from the last 100,000 years. Scientists are still hunting for what might lie behind the timing and occurrence of both types of climate swings. For instance, each Heinrich event occurs after a series of three to five D-O events, implying that the extended warm spells might have progressively destabilized the growing ice sheets and led to a Heinrich-style iceberg armada. Another interesting facet is that many of the D-O events are separated by around 1500 years, with a few spaced at about 3000 and 4500 years. "This suggests that the events are triggered by an underlying cycle … but that sometimes a beat or two is skipped", notes Stefan Rahmstorf of the Potsdam Institute for Climate Impact Research. Other paleoclimatologists believe these sub-beats may be little more than random variations.

few thousand years were, by and large, a nourishing time for civilizations. Scientists who map out the climate for this period can draw on human history as well as other biological, physical and chemical clues. (There's still a bit of a bias towards Europe, the Middle East and North America, due in part to the concentrations of researchers and documented civilizations on those continents. And, as with any climate era, the proxies aren't perfect.)

Across northern Africa and south Asia, the peak in summertime sunshine stimulated by orbital cycles during the postglacial era produced monsoons far more intense and widespread than today's. From about 14,000 to 6000 years ago, much of the Sahara, including the drought-prone Sahel, was covered in grasses and studded with lakes that played host to crocodile and hippopotami. Traces of ancient streams have been detected by satellite beneath the sands of Sudan, while the Nile flowed at perhaps three times its present volume. Meanwhile, from about 9000 to 5000 years ago, most of North America and Europe were ice-free and summer temperatures there likely reached values similar to today, perhaps a touch warmer in some areas and at some times.

The good times couldn't last forever. Starting about 5000 years ago, global temperatures began a gradual cool-down, interspersed with warmings lasting a few hundred years each. This pattern is consistent with ice-core records from towards the end of previous interglacial periods, although it's far from clear when the next ice age will actually arrive (see box opposite). One of the earliest and sharpest signs of this global cooling is the prolonged drought that crippled Near East civilizations about 4200 years ago. Orbital cycles had already weakened the summer sunlight that powered bountiful monsoons across Africa and Asia. At some point, the gradual drying killed off the grasses that covered much of the Sahara, which quickly pushed the region from savannah towards desert.

From 1000 AD to today (and beyond)

Because of its context-setting importance for what lies ahead, the last thousand years of climate have drawn intense scrutiny. The last millennium began near the peak of a 300-year stretch of widespread warmth called the **Medieval Warm Period** (or the Medieval Climatic Optimum), when temperatures in some parts of the world were close to modern-day levels. It may indeed have been an optimal climate for Europeans, or at least a liveable one, but people in parts of the Americas suffered from intense drought that ravaged entire cultures.

Sowing a savoury climate

Whether or not we're due for another ice age relatively soon is a matter of intense debate. The last three interglacials each lasted about 15,000 years, which would imply that it's nearly time for a new glacial era to begin. Some paleoclimatologists believe that orbital cycles won't trigger the next era for as long as 50,000 years. However, if in fact the next ice age is due, then human activity could be postponing its arrival. Veteran paleoclimatologist William Ruddiman argues that the development of agriculture boosted greenhouse gases and delayed a glaciation that should have already begun. Ruddiman introduced his theory with a provocative paper in 2003 entitled "The Anthropogenic Greenhouse Era Began Thousands of Years Ago". He's since elaborated it for a broader audience in the book *Plows, Plagues and Petroleum*.

Ruddiman's theory is based on ice-core records that show how greenhouse gases responded during previous interglacial periods. He noticed that in most interglacials, greenhouse gas levels tend to spike quickly and then gradually drop, setting the stage for the next icing. However, the most recent interglacial hasn't obeyed the pattern. After an initial spike and the beginnings of a drop, carbon dioxide amounts rose again by about 8% about 8000 years ago, then levelled off until the Industrial Revolution. Methane concentrations also increased starting about 5000 years ago.

Ruddiman attributes these rises to the rapid and widespread growth of agriculture during these periods. The clearing of forests would have increased CO_2 levels, he says, and rice paddies would have emitted large quantities of methane. To bolster his case, Ruddiman points to the drop in agricultural activity and subsequent reforestation after plagues swept Europe in the 1300s and the Americas in the 1500s and 1600s. The timing, he points out, corresponds well with the reduction in both global temperature and carbon dioxide levels that occurred during the Little Ice Age.

William Ruddiman
Courtesy William Ruddiman

Ruddiman's theory has met both enthusiastic agreement and stout resistance. One twist came with the new EPICA ice core from Antarctica, which shows that methane levels rose in the midst of the lengthy interglacial about 400,000 years ago. Critics say this proves that human influence need not be invoked to explain the methane rise in the most recent interglacial. But EPICA doesn't rule out human influence, says Ruddiman, who believes the timing of the interglacials from the EPICA record has been misinterpreted and that an alternative timing would support his claims. Through all the critiques, Ruddiman's belief in his theory hasn't faltered: "I really have not had a serious day of doubt", he says.

One example is the **Mayans**, whose empire extended from present-day Mexico across much of Central America. After nearly 2000 years of development, many Mayan cities were abandoned between about 750 and 950 AD. Sediments in the nearby Caribbean reveal several strong multi-year droughts during this period. The Mayan collapse in eastern Mexico apparently coincided with that region's most intense and prolonged drought in over a millennium.

To the north, **Anasazi** settlements near the US Four Corners region experienced periods of erratic rain and snow during the Medieval Warm Period. Tree-ring studies indicate an especially intense drought near

Painting the Little Ice Age

Some of northern Europe's greatest artists used oil and brush to set the mood that many associate with the Little Ice Age: cloudy, snowy and dank. Pieter Bruegel the Elder may have used the frigid winter of 1565 as source material for the dull, greenish sky of *Hunters in the Snow*, part of his series of seasonal depictions. This was one of the first portrayals of a snowy landscape in European art, noted William Burroughs in the British journal *Weather*. Bruegel extended the wintry theme to other topics, including *The Adoration of the Magi in the Snow*. Many Dutch artists, notably Hendrick Avercamp, took to cold-weather depictions in the mid-1600s, another period of brutal chill across the region.

The northern Renaissance also spawned a new realism in sky portraiture. Back in the early 1400s, Flemish painter Jan van Eyck was one of the first to depict cloud types that a meteorologist today might recognize and label. Hans Neuberger quantified the treatment of clouds by US and European painters in an unusual 1970 study that appeared in *Weather*. Sampling 41 museums in nine countries, Neuberger examined more than 12,000 paintings produced between 1400 and 1967. He found that blue skies, which predominated up to 1550, gave way to low clouds in more than half of the post-1550 paintings. Neuberger didn't attempt to analyse how much of the trend was related to the Little Ice Age weather and how much to artistic fashion.

English landscape painters of the Little Ice Age held true to their island's cloudy climate. Every English sky examined by Neuberger had at least some cloudiness, and the sky was typically a pale blue at best. The English Romantic artist J.M.W. Turner specialized in foggy, misty tableaux as well as striking sunsets; the latter may have reflected the volcanic dust that added vivid hues to many sunsets in the early 1800s. Later in the century, the gigantic Krakatoa eruption of 1883 led to sunsets so striking they were noted in press reports in New York and London. According to astronomer Donald Olson of Texas State University, Krakatoa may also have inspired Edvard Munch's iconic masterpiece, *The Scream*. In describing what triggered the painting, Munch wrote of experiencing a "blood-red" sunset in present-day Oslo that resembled "a great unending scream piercing through nature" – though Munch didn't give a date for this experience. Although a full the end of the 1200s, when the culture's emblematic cliff dwellings were abandoned. Scholars in recent years have warned against pinning the Mayan and Anasazi declines solely on drought, pointing to signs of other factors such as migration, a high demand on trees and other resources, and power struggles within each culture and with neighbouring peoples. However, climatic stresses can't have helped matters.

After the Medieval Warm Period faded, the long-term cooling resumed, kicking in around 1300 and continuing into the mid-1800s. This was the Little Ice Age (see box), the coldest interval globally in thousands of years. Periodic plagues and famines ravaged Europe, and glaciers descended

Francis G. Mayer/Corbis

Dutch painter Hendrick Avercamp (1585–1634) was a specialist in winter scenes, especially skaters on frozen rivers, canals and flood waters

decade separates the eruption from *The Scream*, Olson believes that Munch may have encountered a Krakatoa sunset and waited years to depict it.

The legendary frost fairs held on the River Thames in London during occasional freeze-ups were captured in a number of paintings, including *A Frost Fair on the Thames at Temple Stairs* (1684) by Dutch painter Abraham Hondius. However, these festivals weren't as frequent as one might assume. Outside of the especially frigid mid-1600s, the Thames froze at London only about once every twenty or thirty years from the 1400s until 1814, when the last freeze-up was recorded. Moreover, it wasn't the end of the Little Ice Age that ended the frost fairs. When London Bridge was replaced in the 1830s, it allowed the tide to sweep further inland. This made it virtually impossible for the Thames to freeze at London, and it hasn't happened since.

from the Alps to engulf a number of villages. The chill wasn't constant: as historian Brian Fagan notes, the Little Ice Age was marked by "an irregular seesaw of rapid climatic shifts, driven by complex and still little understood interactions between the atmosphere and the ocean."

Another influence may have been a drop in **solar energy**. Isotopes of carbon in tree rings and beryllium in ice cores show a drop-off in solar

Four ways to think of past and future climate

The answer to whether Earth is warming or cooling depends in large part on what time frame you're considering. Each graph below includes a very rough schematic of past (and plausible future) temperatures.

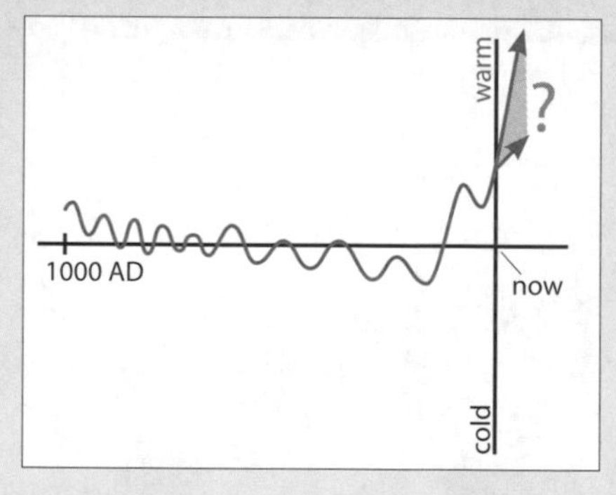

▶ **Over hundreds of years** Around 1000 AD, Earth was nearly as warm as it is now. After that, temperatures over much of the planet cooled by around 0.2–0.5°C (0.4–0.8°F), mostly during the period known as the Little Ice Age, from the 1300s to around 1850. The global average has since rebounded by nearly 1°C (1.8°F) as human-produced greenhouse gases have increased. This century is projected by the IPCC to warm by 1.4–5.8°C (2.5–10.5°F). Further warming after 2100 could be substantial if greenhouse emissions continue to increase through this century, and the delayed response of ice sheets and deep oceans to warming could produce major sea-level rise over several centuries.

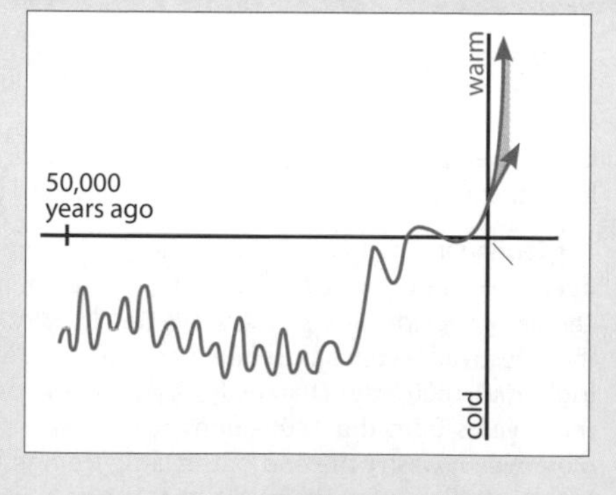

▶ **Over tens of thousands of years** The most recent ice age began about 115,000 years ago and ended about 11,500 years ago. Then came a dramatic warm-up, which lasted until about 3000 BC. Since then, Earth's temperature has changed relatively little, with a very slight cooling interrupted by warmer periods and punctuated by the last century's sharp temperature rise. More than a thousand years from now, after humans have exhausted fossil fuels and the resulting greenhouse gases have left the atmosphere naturally (mostly through slow absorption by the ocean), we may return to cooler times. If the length of the radiation during much of the Little Ice Age. Moreover, sunspot observations that began around 1610 show a near-absence of reported sunspots between 1645 and 1715 (the so-called Maunder Minimum). However, by 2006 the estimates of this solar effect had been considerably toned down.

Also in the mix are volcanoes, which seem to have erupted more frequently after 1500 than during the Medieval Warm Period. The 1815

last interglacial is any guide, we're due for another ice age in the next few thousand years, although some scientists believe it's more likely to take 20,000 years or more. It's even possible that a mix of polar melting, land-use changes and slowly waning greenhouse-gas concentrations could postpone the next ice age for an undetermined period.

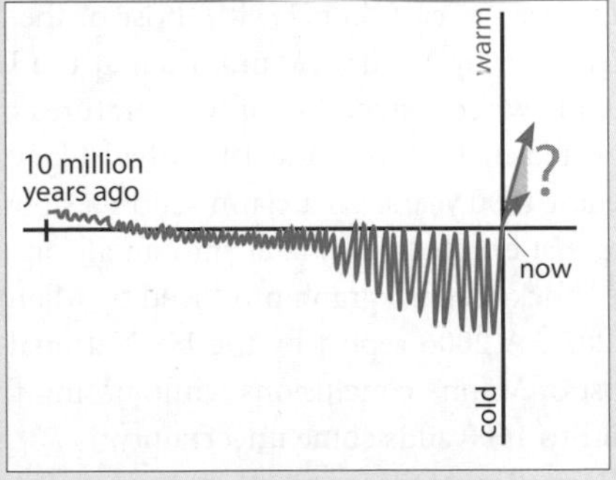

▶ **Over tens of millions of years** A gradual, though sometimes erratic, cooling trend has been under way for at least 55 million years, perhaps due in part to carbon dioxide removed by weathering atop the growing Himalayas. Northern Hemisphere glaciations began about 2.7 million years ago. Warm and cool periods come and go atop this overall cooling. Simple extrapolation would keep Earth on this extremely slow cooling trend, although there's no evidence to tell us how long that might last.

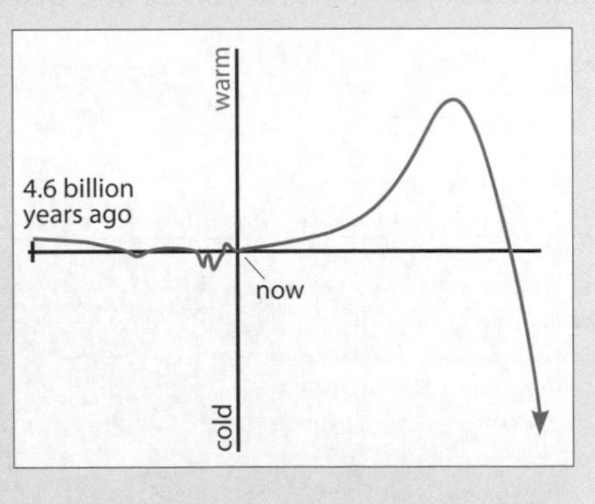

▶ **Over billions of years** On this scale, the details are blurry but the overall picture is clear. Changes in greenhouse-gas levels have so far kept our climate relatively stable (not too hot or cold for life) even though the Sun's output has risen by more than a third since the solar system was formed around 4.6 billion years ago. The Sun will continue to heat up and eventually undergo changes to its size and structure, producing a climate on Earth hot enough to evaporate the oceans and make life impossible. Eventually, the sun will shrink and start to fade, with the solar system reaching its cold "final configuration" in around twelve billion years.

eruption of Indonesia's **Tambora** – the most violent ever recorded on Earth – led to a disastrously cold summer across much of the globe in 1816. That "year without a summer" brought crop failures to northern Europe as well as snows in Vermont as late as early June. (It also spawned a monster: while spending part of that dreary summer at Lake Geneva, a teenaged Mary Shelley started work on the novel that became *Frankenstein*.) Both the Medieval Warm Period and the Little Ice Age appear to have been strongest over the Northern Hemisphere continents, although it's hard to completely eliminate geographic bias from the records. Some researchers argue that both phenomena were primarily regional events, as opposed to the global-scale warming under way now.

After the mid-1800s, Earth's climate took a decided turn for the warmer and by the end of the twentieth century it was clear that global temperatures were at least on par with those of the Medieval Warm Period. After a team led by Michael Mann, then at the University of Virginia, carried out a new reconstruction of temperatures over the last millennium, they reported in 1999 that the 1990s had likely been the warmest decade of the last 1000 years. That claim set off a slow burn among climate sceptics, one that erupted years later into an all-out firefight about the validity of the "hockey stick" graph produced by Mann and colleagues for the IPCC (p.287). A 2006 report by the US National Research Council supported most of Mann's conclusions while noting that the sketchiness of proxies prior to 1600 adds some uncertainty.

Regardless of these and other dust-ups, it's clear that the greenhouse-gas content of our atmosphere has entered territory never before explored in human history. Only time will tell how quickly the climate might follow.

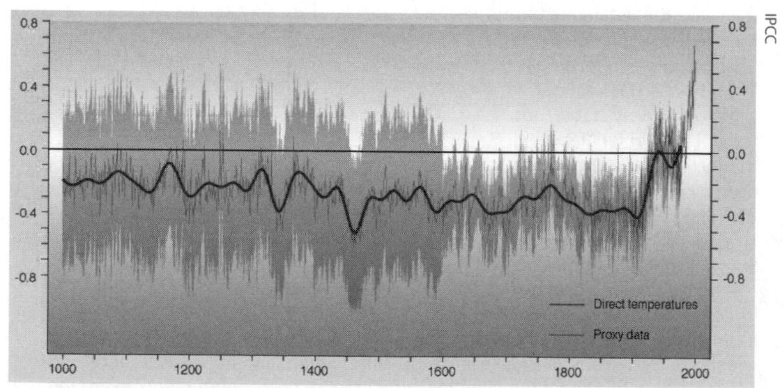

The IPCC's "hockey stick" graph, showing temperature (departures from the 1961–90 average) in the Northern Hemisphere during the past 1000 years

Circuits of change

Modelling the future climate

In 1957, Roger Revelle and Hans Seuss of the Scripps Institution of Oceanography described the intersection of fossil fuels and global climate change in words that proved far more resonant than the authors might have imagined at the time. Their landmark study, published in the journal *Tellus*, included this oft-quoted passage:

> …human beings are now carrying out a large scale geophysical experiment of a kind that could not have happened in the past nor be reproduced in the future. Within a few centuries we are returning to the atmosphere and oceans the concentrated organic carbon stored in sedimentary rocks over hundreds of millions of years.

Like any good scientists, they were curious about the outcome: "This experiment, if adequately documented, may yield a far-reaching insight into the processes determining weather and climate."

It may seem callous to describe our current situation – glaciers melting, cities baking, hurricanes churning, species disappearing – as an interesting experiment. It's also quite accurate. Each year our emissions push the chemistry of the atmosphere a little further into territory never before explored in human history. If there were a million Earths in our solar system, each with a different concentration of greenhouse gases, we'd have a better idea of just what to expect in the future. Unfortunately, as Revelle and Seuss point out, we can't replicate the experiment that we're now conducting on planet Earth. The next best thing is to simulate it, and the only place that's feasible is inside a computer.

Reproducing the atmosphere with **computer models** has been a key activity in computer science ever since the field was born, just after World

Climate, weather and chaos theory

One of the most common gripes about climate-change projections is the idea that a computer model can tell us anything about climate in the distant future. It's often phrased this way: "If they can't get the forecast right for next week, how can they predict the climate a hundred years from now?" Even Michael Crichton has used this tactic (see p.260).

But a weather forecast and a climate projection are two different beasts. Weather models can't see clearly beyond a few days because of inherent limits to the predictability of small-scale weather (ie what will happen in your neighbourhood, as opposed to the globe as a whole). Weather observing stations are situated a few kilometres or miles apart at best, and in much of the world you can go 100km (60 miles) between stations, not to mention the largely unsampled oceans. Small disturbances missed in the data can influence larger weather events over time – the "butterfly effect," as discovered in the 1960s by Edward Lorenz, the father of chaos theory. We'll never have enough weather stations to catch every one of these tiny weather makers. That's why forecast quality – while steadily improving – will always be problematic for days more than about two weeks in the future.

Climate models aren't interested in the vagaries of individual weather events so much as the influence of long-term climate shapers such as greenhouse gases, solar variations and polluting aerosols. Chaotic variations in weather normally average out over years and decades, so they don't corrupt a climate projection the way they might a weather forecast. The downside is that global models can't provide the precision in time and space that many policy makers and the public want to see, although progress is being made through techniques such as "downscaling" from global output to regional depictions.

War II. At first the emphasis was on weather prediction. By the 1960s, forecasters were using 3D models of atmospheric flow, or **general circulation models**, to gaze into the future of weather patterns several days out. Since then, breakthroughs in weather and computer science, together with Moore's Law (the maxim that computer processor speed doubles about every eighteen months), have allowed for steady improvements in weather forecasting models and thus in forecast skill. Revelle and Seuss might have been happy to have a solid three-day forecast; now, ten-day outlooks are commonplace.

Models focusing on climate long lagged behind their counterparts in the weather field, and with good reason. The most obvious weather changes emerge from the interplay among a fairly limited set of ingredients in the atmosphere: pressure, temperature, moisture, wind and clouds. Weather is also shaped by other factors in the surrounding environment, such as ocean temperature, vegetation and sea ice. But these latter factors don't change much over the course of a few days, so a weather model can focus on tracking the atmosphere while keeping the rest of the environment constant.

Climate models don't have that luxury. As the weeks roll into months, vegetation thrives and decays. Sea ice comes and goes. Greenhouse gases accumulate. The ocean gradually absorbs heat from our warming atmosphere, sometimes releasing it in giant pulses during El Niño events. In the very long term, even the topography of Earth changes. All of these variations feed into the atmosphere and influence weather – often so subtly that the effect isn't obvious until it's been playing out for years. For all of these reasons, it's extremely difficult to go beyond a weather model's focus on the atmosphere and to depict Earth's whole environment accurately in a **global climate model**.

Difficult, but not impossible. Climate modelling has undergone a rapid transformation in the last twenty years. As recently as the 1980s, the most sophisticated climate models portrayed the atmosphere and the land surface – and that was about it. Gradually, scientists have forged models that depict other parts of the Earth system. Many of these are incorporated in ever-growing global climate models. The best of the bunch now include land, ocean, sea ice, vegetation and the atmosphere, all interacting in a fairly realistic way. Global models are far from ideal; as sceptics often point out, there are still major uncertainties about climate processes that no model can resolve. However, the skill of recent global models in replicating twentieth-century climate, as we'll see below, is one sign that the models are doing many things right.

How computer modelling got started

The best analogy to a climate model may be a state-of-the-art computer game. Instead of cars tearing down a racecourse, or warlords fighting over territory, this game follows the heating, cooling, moistening and drying of the atmosphere. It's not quite as sexy but far more important to the future of the planet.

The action unfolds in a virtual space that's typically divided into a three-dimensional mesh. Imagine the steel latticework in a building under construction, but extended across the entire surface of the planet (see graphic on p.231). At the centre of each rectangle formed by the intersection of the imaginary beams is a **grid point**, where atmospheric conditions are tracked as the model's climate unfolds. In a typical global

From great ideas to global views

The computer modellers now helping the world decide how to confront climate change owe a big debt to L.F. Richardson, a far-seeing British scientist of the 1920s. The story begins in World War I, when a team of meteorologists in Bergen, Norway, came up with the first three-dimensional theory of how weather worked. They named the boundaries that separated cold and warm air "fronts", after the battle fronts then raging in Europe. Meanwhile, Richardson, then an ambulance driver for the French army, was hatching a scheme to calculate the future of the atmosphere.

After the war, Richardson returned to England and set to work. Building on ideas that emerged from the Bergen group, he came up with seven equations based on physical principles from Isaac Newton, Robert Boyle and Jacques Charles. If these equations could be solved, Richardson believed, one could not only describe the current weather but extend it into the future. He envisioned a "forecast factory", where hundreds of clerks would carry out the adding, subtracting, multiplying and dividing needed to create a forecast by numbers. Richardson and his wife spent six weeks doing their own number-crunching in order to test his ideas on a single day's weather. The resulting outlook was abysmal, but Richardson's test demonstrated to the world that one could – in principle, at least – carry out calculations and forecast the weather using more than intuition and rules of thumb that were standard in the 1920s.

Richardson's equations materialized again almost thirty years later in the world's first general-purpose electronic computer. The ENIAC (Electronic Numerical Integrator And Computer) was created in the US at the close of World War II. In order to show the machine's prowess, its developers hunted for a science problem that could benefit from raw computing power. Weather prediction filled the bill. Richardson's equations were updated and translated into machine language, and in Princeton, New Jersey on March 5, 1950, the ENIAC began cranking out the first computerized weather forecast. The computer took almost a week to complete its first 24-hour outlook, but ENIAC did a better-than-expected job of portraying large-scale weather a day in advance.

As weather modelling became established in the 1950s and 1960s, a small group of US climate experts began to adapt some of the weather models for climate and to create new ones from scratch. One centre of action was the Geophysical Fluid Dynamics Laboratory, based in Princeton. Another was the Courant Institute in New York and a third was the National Center for Atmospheric Research, newly established in Boulder, Colorado. In the mid-1960s, NCAR's Warren Washington and Akira Kasahara built one of the first general circulation models that spanned the globe. (Many models at the time were two-dimensional or covered only the Northern Hemisphere). Generations later, the seeds of that early model remain in NCAR's Community Climate System Model, used by university researchers around the world.

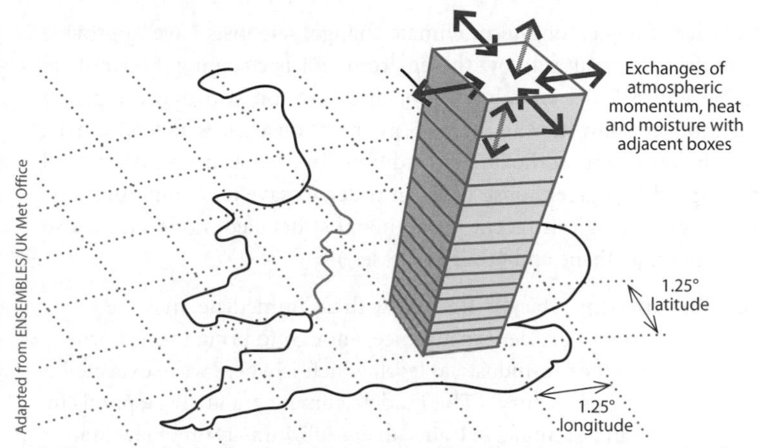

A climate model divides the atmosphere into 3D chunks and measures the change in each one as weather unfolds over days, years and decades of model time

model, each grid point accounts for a horizontal area of roughly 1.25° longitude by 1.25° latitude. At mid-latitudes (40°N), that represents an area of around 140km by 100km (87 x 62 miles) – about the size of Connecticut or roughly half the size of Sicily. Thus, each grid point represents the atmosphere over a fairly large area – much more so than for a weather model. The typical global model also includes perhaps 25 or so vertical layers. Each layer spans anywhere from 100 to 500 metres (330–1650ft) close to Earth's surface, with the vertical distance between grid points increasing as you go up.

A climate model operates in a parallel universe of sorts, one in which time runs far more quickly than in real life. Each model advances in **time steps** that move the simulated climate forward by anywhere from five minutes to a half hour, depending on the model's complexity. At each time step, the model calls on formulas derived from the laws of physics to compute how all the atmospheric and environmental variables have changed since the last time step – the amount of sunlight hitting each spot on Earth, the convergence of winds, the formation and dissipation of clouds, and so forth. Each step only takes a few seconds to process, but it can still take weeks of dedicated time on a supercomputer to depict a century of climate in detail. The output from all this number-crunching is saved in various formats along the way. After the processing is done, scientists can go back, sift through the results, calculate yearly averages and other statistics and produce graphics.

In order to depict long-term climate change, scientists have to provide the model with input on how the environment is changing. One critical variable is fossil-fuel emissions. Modellers rely on two major types of simulations to show how an increase in greenhouse gases affects climate. (In both cases, carbon dioxide is traditionally used as a stand-in for all human-produced greenhouse gases in order to save on computing time and expense, though in recent years the most detailed models have also incorporated methane and other key gases.)

▶ **Equilibrium runs** typically start off with an immediate, massive injection of carbon dioxide (for instance, enough to bring the airborne amount to twice the pre-industrial level, which is where we're expected to be by later in this century). The model's climate gradually responds to this extra CO_2 and eventually attains an equilibrium at some new (and presumably warmer) state. The idea isn't to see how long the climate takes to respond – since we'd never add that much CO_2 instantaneously in the real world – but where the climate ends up. To save on time and expense, these runs typically use simplified oceans.

▶ **Transient runs** more closely resemble reality. In these simulations, CO_2 is added in smaller increments that resemble the actual amounts added by human activity – typically 1% per year, compounded year over year. In this case, scientists are interested as much in the process as the outcome. For instance, does the model climate warm smoothly, or does the temperature go up in fits and starts? The various global models have tended to show closer agreement for transient runs than for equilibrium runs – an encouraging sign, since the transient scenario is closer to how greenhouse gases increase in real life.

Of course, you can't simply add CO_2 to a global model and leave everything else alone. As discussed in Ecosystems & Agriculture (p.147), a large part of that CO_2 gets absorbed by vegetation and the oceans. There are also other processes that kick in as carbon dioxide warms Earth. Consider the sea ice that melts as the atmosphere warms up. As we saw in The Big Melt (p.75), the loss of that ice means less reflection of sunlight and an amplifying, positive-feedback loop that leads to even more warming. The positive feedback from water vapour also has to be portrayed accurately: the warming produced by the extra CO_2 helps evaporate more moisture from the sea, and that water vapour raises global temperature further.

These environmental processes are quite different from the physics that drives day-to-day weather, and they can't be added to a climate model in an easy or simple way. The original global climate models of the 1960s and 1970s couldn't hope to account for these processes in a realistic fashion. But as computing power has improved, the models have grown steadily in sophistication and detail. They've incorporated a growing number of sub-models, dealing with each of the major climate-influencing elements of what scientists call the **Earth system**. When an atmospheric model is yoked to one or more of these sub-models, the new hybrid is called a **coupled model** (even when it has three or more components).

In a simple coupling, a sub-model runs side by side with a global model and passes it information – a one-way dialogue, as it were. For instance, a global model might be fed information on airborne nitrogen levels from a chemistry model, but the resulting changes in climate wouldn't feed back into nitrogen production, as they would in real life. More realistic are **integrated** couplings, which allow two-way exchanges of information between model and sub-model. However, these can take years to develop.

Models also grow in the level of detail they provide. Model **resolution** – the fineness of the spacing between grid points – has gradually increased over the years as computing power has grown. Today, some cutting-edge

Hot spots for climate modelling

It takes vast amounts of computing power to run a comprehensive global model at a reasonable pace. Also needed are dozens of physical and computer scientists to ensure that a model's software is accurate and efficient. As such, the world's major climate models are run at a fairly small number of institutions – most of them run by national governments. They include Australia's **Bureau of Meteorology Research Center** and its **Commonwealth Scientific and Industrial Research Organization**, both in the Melbourne area; the **Canadian Centre for Climate Modelling and Analysis**, based at the University of Victoria; Japan's **Frontier Research Center for Global Change** in Yokohama; the **Hadley Centre**, part of the UK Met Office, based in Exeter; Germany's **High Performance Computing Centre for Climate- and Earth-System Research** (Deutsches Klimarechenzentrum) and the **Max Planck Institute for Meteorology**, both in Hamburg; and the **Laboratory of Dynamic Meteorology** in Paris, part of France's Centre National de la Recherche Scientifique. In the US, there's NOAA's **Geophysical Fluid Dynamics Laboratory** in Princeton, New Jersey, NASA's **Goddard Institute for Space Studies** in New York City, several facilities of the **Department of Energy**, and the **National Center for Atmospheric Research** (NCAR) in Boulder, Colorado.

global models are beginning to resemble weather models. For instance, a team from Japan's Meteorological Research Institute recently simulated the frequency of tropical cyclones in a doubled-CO_2 world using a model with a resolution on the order of 20km (12 miles), with 60 vertical layers. To accommodate the sharp detail of these models, their time frame has to be restricted. Typically the models start out with a "future" atmosphere (such as one with doubled carbon dioxide) and simulate only a decade or so rather than a century or more.

Climate modellers have typically avoided incorporating current weather conditions when projecting the climate decades into the future, since there's so much day-to-day variability in weather. However, bringing current ocean conditions into the model could help nail down how global temperature is likely to unfold over the next decade. The UK Met Office's Hadley Centre used this strategy to issue a novel kind of forecast unveiled in a 2007 paper in *Science*. By running a global model in two ways – either including or omitting the patchwork of present-day ocean temperatures – the group showed that incorporating current ocean conditions, and the "memory" they bring to the atmosphere, can add measurable benefit to a climate outlook. The model captures realistic-looking peaks and valleys of global warm and cool spells caused by natural variability, but with an underlying rise that projects at least half of the period from 2009 to 2014 to be warmer than the record-setting year of 1998.

Especially when the focus is on a weather or climate feature with restricted geographic extent, such as hurricanes, it makes sense to limit the model's extra-sharp precision (and the related computing burden) to the areas where it's most needed. This can be done using a **nested grid**, whereby the globe as a whole is simulated at a lower resolution and the areas of interest are covered at higher resolutions.

Similarly, when studying how climate might evolve in a fixed area, such as parts of a continent, global models are often coupled to **regional climate models** that have extra detail on topography. Bringing aerosols into a model allows rainfall and other important features to be depicted with more precision. The output from nested global models and regional models is especially useful in creating the climate assessments that are sorely needed by policymakers. However, bumping up a model's resolution isn't a simple matter: it involves a careful look at the physics of the model to ensure that processes which don't matter so much in a low-resolution simulation, such as showers and thunderstorms, are realistically depicted at the new, sharper resolution.

Whatever its design, every global model has to be calibrated and tested against some known climate regime in order to tell if it's functioning properly. The easiest way to do this is through a **control run** with greenhouse gases increasing at their twentieth-century pace. By simulating the last century of climate, scientists can see if a model accurately depicts the average global temperature as well as regional and seasonal patterns, ocean circulations and a host of other important climate features. Once it's passed this test, a model can go on to project the climate of the future – or the distant past, the goal of paleoclimate modelling.

Another safety check is that the world's leading models are regularly checked against each other through such activities as the Program for Climate Model Diagnosis and Intercomparison, conducted at the US Lawrence Livermore National Laboratory. When the models agree, that bolsters confidence in the results; if one model deviates markedly from the others, it's a flag that more work is needed.

The limitations

In both weather and climate models, it's not easy to calculate every iota of atmospheric change. **Clouds**, for example, are still a tremendous challenge. If they change in a systematic way, clouds can have major implications for how much radiation gets reflected and absorbed from both the Sun and Earth. But clouds are far more varied and changeable in real life than any model can accurately represent, so modellers have to simplify them in their simulations. There's currently no gold standard for how to do this, and new techniques are being explored all the time. Typically, each grid point in a model will be assigned an amount or percentage of cloud cover based on how much water vapour is present and whether the air is rising or sinking (rising air is conducive to cloud formation). The type of cloud and its reflective value are estimated from the height and temperature of the grid point.

The most sophisticated models also include **aerosols** – pollutants, dust and other particles that can shield sunlight and serve as nuclei for cloud droplets and ice crystals. This allows scientists to see how changes in pollution might affect cloud formation and temperature, as in the case of the global dimming observed in the late twentieth century (see p.186).

In order to test whether their cloud schemes are working well, scientists run their models for present-day climate and then compare the overall distributions of cloud cover to the real-world situation observed by satellite. (These data sets have their own limitations, though, since most satellites can't yet delineate multiple cloud layers very well). The results are also

Modelling for the masses: ClimatePrediction.net

Until very recently, climate modelling was restricted to the supercomputers of a few universities and research labs. That all changed with ClimatePrediction.net, an online project that has set global models churning in more than 100,000 homes and offices worldwide. It's the brainchild of Myles Allen, a physicist and climate analyst at Oxford University. Allen was inspired in the late 1990s by the success of the SETI@home project, which has involved more than a million people around the world dedicating standby time on their PCs to assist in the search for extraterrestrial life.

The amount of computing time needed to analyse radio-telescope data for SETI is vast, but it can be parcelled out in small doses. Each volunteer agrees to download a program that runs on her or his PC at times when the computer would otherwise be resting (the software never interferes with other tasks). These "screensaver scientists" get the satisfaction of contributing to a grand scientific endeavour with a minimal commitment of time, as well as the fun of joining a like-minded online community. At least three couples have met and married through SETI@home.

Right away, Allen saw the potential value to climate science of this distributed approach to computing. On an up-to-date home computer, a fifty-year simulation of global climate could be run in six months, he figured. If thousands of volunteers were involved, the virtual horsepower would be far more than any research centre could afford on its own. Moreover, this setup would be an ideal way to run ensembles – sets of closely related simulations, each with slightly different starting conditions (see p.242). To drum up interest, Allen pitched the idea in a commentary for *Nature* in 1999. "Anyone respectable enough to sit on a peer review committee", he wrote, "will probably find the idea of getting schoolchildren to run full-scale climate models on their parents' PCs completely daft. But there will be others for whom the idea is as natural as Amazon.com."

Allen's idea, which was originally dubbed CASINO21 (an allusion to Agenda 21, the UN's action plan for sustainable development), didn't take off instantly. "It was very difficult", recalls David Stainforth. A physicist and colleague of Allen's at Oxford, he became the project's first recruit and is now its chief scientist. "Most people at the time thought it was stupid and implausible. I didn't." Gradually, the pieces fell into place. Allen and Stainforth obtained financial backing from the UK's National Environmental Research Council (NERC). They settled upon the name ClimatePrediction.net and launched the experiment in 2003.

At first there were technical problems that alienated many volunteers. In response, Allen and Stainforth went to the University of California, Berkeley, and arranged to use the same computing infrastructure that allowed SETI@home to work with its vast cadre of volunteers. The new version, launched in 2004, proved more durable, and since then, the project has gone from strength to strength.

The greatest participation in CP.net is in North America and Europe: over half of the 100,000-plus computers involved as of 2006 were in the US, UK and Germany.

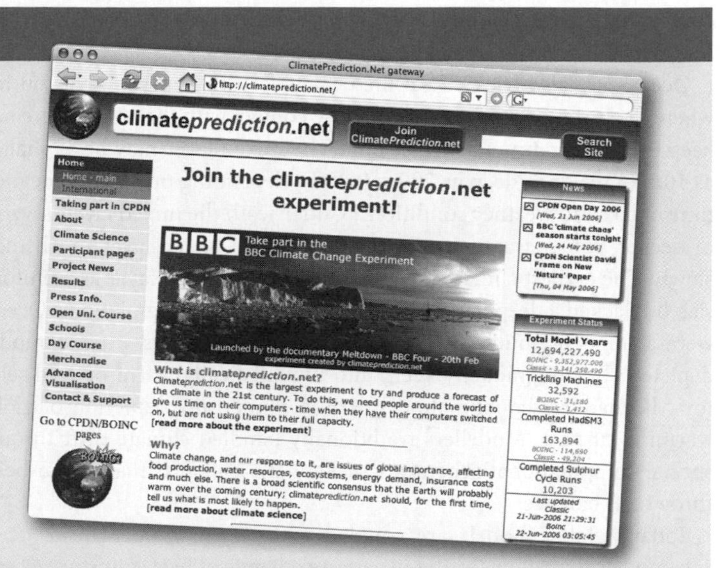

But there's a healthy sprinkling of users all across the globe, with participants in 139 nations. Among countries with at least a million residents, the Czech Republic has the highest per-capita participation rate, with one computer involved in the project for every 330 Czechs. The CP.net team works to engage volunteers with its extensive website, which includes background on climate science, techniques for comparing one volunteer's results against others, and instant-messaging tools for the volunteer community.

The first science goal for ClimatePrediction.net was to nail down sensitivity, the amount of global warming that a doubling of carbon dioxide would produce (see p.239). Volunteers were given a version of the UK Met Office's Unified Model, with an atmospheric component from the UK Hadley Centre linked to a model of the uppermost ocean layer (the so-called mixed layer). Each simulation ran for fifteen model years for calibration, fifteen years at pre-industrial levels of carbon dioxide, and fifteen years at doubled CO_2, enough model time for the climate to reach a new, warmer equilibrium temperature. The results grabbed headlines. Out of 2500 simulations, most were in the ballpark of previous studies (see p.239), but a few showed a climate sensitivity as high as 11.5°C (20.7°F). Allen believes the possibility of such high values was downplayed in previous research because traditional schemes for measuring sensitivity in any process (climatic or otherwise) don't portray the high part of the range well – just as a slow-functioning camera takes fuzzy pictures of a fast-moving scene. Or, as Allen says, "It's like trying to measure the speed of a car with a dodgy stopwatch".

From its initial work, CP.net has moved on to investigate such hot topics as the likelihood of a shutdown of the Atlantic thermohaline circulation (see p.56) and the impact of sulphate aerosols on climate.

compared to special cloud-resolving models that are too expensive to run routinely but provide a valuable cross-check.

Oceans are also tricky. Real-life oceans feature a **mixed layer** on top, where the turbulence of waves and close contact with the atmosphere help keep the temperature fairly constant. The mixed layer may be as shallow as 10m (33ft) or as deep as 200m (660ft), depending on season, location, time of day and weather conditions. Underneath the mixed layer is a zone of deeper, colder water that interacts with the mixed layer much more slowly. The earliest climate models depicted the mixed layer as a uniform slab but lacked a deep layer, a problem for long-term simulations. Oceans were a frequent source of **climate drift**, which occurs when a model-depicted climate gradually veers into unrealistically warm or cold territory. If unaddressed, climate drift could render any depiction of global warming suspect. Modellers traditionally handled climate drift through special computational schemes that kept the global climate from being thrown off balance.

Today's global models are coupled with much more realistic ocean sub-models that allow for the two-way exchange of heat, moisture, chemistry and momentum among the surface, the mixed layer and deeper layers. Still, the complicated circulations of the real ocean remain hard to depict. For instance, the North Atlantic's flow is often characterized as a simple loop, with warm water flowing north off the US East Coast towards Europe, cold water streaming back south towards Africa, and warmer waters flowing west with the trade winds towards the Caribbean. But in reality, the northward flow branches into several components as it approaches Greenland and the British Isles, and the water descends far undersea into a cold return flow that moves southward in several distinct layers. Where the contrasts in temperature and salinity are strong, small but powerful eddies can spin off. All this complexity is hard enough to measure, much less portray in a model.

The tropical Pacific remains one of the most difficult areas for climate models to depict in detail, not only because of ocean circulation but because of the gigantic areas of rainfall spawned by the warm waters, especially in and near Indonesia. This warm pool periodically expands eastward or contracts westward during El Niño and La Niña (see p.118). Most global models can now depict El Niño and La Niña, but the timing and geographic extent of these events are still hard to pin down. Even in neutral conditions, it's a challenge for models to realistically capture the amount of heat and moisture pumped into the upper atmosphere by tropical showers and thunderstorms.

A sensitive topic: predicting the warm-up

The question of what will happen in a greenhouse-warmed world is so essential to climate modelling that it's spawned an index that's widely used to assess model performance. This is the **sensitivity** of a global model, most often defined as the amount of temperature change produced in the model when carbon dioxide is doubled. The CO_2 might be introduced all at once, or it might be added gradually. Either way, the climate eventually settles down at a new global temperature, and the difference between that temperature and the old one is the sensitivity. It provides a convenient way to compare models and how readily they respond to greenhouse gas.

This definition isn't cast in concrete: one might consider a tripling or quadrupling of CO_2 instead, and various starting or ending points can be used, such as the peak of the last ice age. However, the doubling of CO_2 since pre-industrial times (generally taken as the mid-1800s) serves as a useful and time-tested convention familiar to all climate modellers.

At the current rate of increase, greenhouse gases will have roughly doubled in the atmosphere by the mid- to late 21st century (though we might be able to postpone the doubling, or perhaps even forestall it entirely, through a concerted effort to reduce greenhouse emissions). The doubling of CO_2 used in model sensitivity gives us a good idea of how much warming to expect once the atmosphere and oceans have adjusted to the extra greenhouse gas, which could take decades longer than the doubling itself.

Even though sensitivity is expressed as a single number, it's actually the outgrowth of all the processes that unfold in a model. That's why equally sophisticated climate models can have different sensitivities. That said, what's striking about sensitivity is how little the average value has changed as models have improved. Some of the earliest global climate models showed a sensitivity of around **2.5°C (4.5°F)**. As more models of varying complexity came on line, the range of possibilities was estimated to be **1.5–4.5°C (2.7–8.1°F)**. This range – something of a best guess at the time – was first cited in a landmark 1979 report on climate change by the US National Academy of Sciences. It was held constant by the IPCC in its first three assessments and was narrowed only slightly in the 2007 report (to 2.0–4.5°C or 3.6–8.1°F). Most of the major coupled models now show sensitivities clustered within a degree of **3.0°C (5.4°C)**, which is the best estimate pegged in the latest IPCC report. The fact that this value hasn't changed much after nearly thirty years of modelling is, arguably, testimony to how well modellers have captured the most important processes at work.

A sceptic might grouse that the models have simply been **tuned** – their physics adjusted – to produce the result that people expect. Indeed, modellers have to guard against making so many adjustments to a model that it gets a "right" answer for the wrong reasons. But there's also a physical rationale behind the 3.0°C number. It's fairly easy to show that, if the atmosphere had no clouds or water vapour, the warming from doubled CO_2 would be somewhere around 1°C to 1.2°C (1.8–2.2°F). It's believed that water vapour roughly doubles the impact from the extra CO_2. Together with smaller effects from other greenhouse gases, this brings the total change into the 3.0°C ballpark.

Be that as it may, we can't be assured that larger (or smaller) changes are completely out of the question. The small but real risk of a dangerously high sensitivity has been borne out by a number of studies based on observations, and more recently by the massive experiment in distributed climate modelling known as ClimatePrediction.net (see p.236). In its first study, the project ran more than 2000 variations of a sophisticated climate model from the UK Met Office. The results, published in 2005, showed that the sensitivity to doubled CO_2 could be higher than **8°C (14.4°F)**. If such a warming came to pass, the consequences could be catastrophic.

Most of the ClimatePrediction.net variations showed much less sensitivity, with a consensus value of **3.4°C (6.1°F)**. That's well within the range of previous work (and expected by the nature of the model used). It was the upper-end estimates that raised enough of a red flag to trigger follow-up research by modellers elsewhere in 2005 and 2006. Among the findings:

▶ **James Annan and Julia Hargreaves** of Japan's Frontier Research Center for Global Change used past climate data and statistical techniques to estimate a 95% chance that the true climate sensitivity lies below 4.5°C (8.1°F), the upper bound first cited in 1979.

▶ **Gabriele Hegerl** of Duke University examined more than 1000 simulations of the last millennium of climate using an energy balance model. Hegerl and colleagues reported in *Nature* that, with 95% certainty, the climate sensitivity was less than 6.2°C (11.1°F).

▶ **A team led by Reto Knutti** of the US National Center for Atmospheric Research used seasonal change as an analogue for climate change. They looked at how more than 2500 simulations from ClimatePrediction.net depicted the annual warm-up and cool-down caused by the yearly strengthening and weakening of sunlight. Relating that response to the model's response to increased carbon dioxide, they found a 95% likelihood that sensitivity was below 6.5°C (11.7°F), with the cut-off perhaps as low as 5°C (9°F) depending on model assumptions.

Clearly, there's still disagreement about just how sensitive climate might be. But all of the recent studies indicate that the most likely outcome of a doubling of carbon dioxide is a global temperature increase close to the one scientists have been projecting for decades: somewhere not too far from **3°C (5.4°F)**. None of this rules out a much more sensitive climate. It simply adds to the weight of evidence for an amount of warming that's

What exactly does probability mean?

There's an important difference between subjective and objective probabilities. In weather forecasting, a computer model might calculate a 60% chance of rain. That's an **objective** probability – it depends only on the physics or statistics in the model and the data it's working from. However, if a human forecaster raised the value to 80% based on information that the model didn't have, such as a glance out of the window that revealed gathering storm clouds, the new figure would be a **subjective** probability – that is, produced by a person, using her or his best judgment.

Many objective probabilities can be drawn from past experience. For instance, if decades of climate in a given city show that it rains on average about one out of every four days in June, then a summer-solstice festival has a 25% chance of getting dampened in any given year. (Those odds may change as the event nears, much like the example above.)

Scientists can't use purely objective probability when it comes to climate change, because by definition the climate is entering new territory – there's no set of past human-induced warmings on which probabilities could be based. One way around this is for humans to apply expert judgment, as the forecaster above did, and assign probabilities based on the evidence at hand. For example, in 2001 the IPCC stated that doubling carbon dioxide in the atmosphere would make a warming of 1.5–4.5°C "likely." Elsewhere in that report, the IPCC defines "likely" as a probability range of 66% to 90%.

While these ranges can be useful to policymakers, they aren't based on hard-and-fast scientific experiments. Now that climate simulations can be carried out en masse, it's possible to derive probabilities as if those simulations were a set of climates similar to ours. That's how various groups are coming up with percentages and ranges like the ones discussed on p.278. These are more objective than before, but they're still based on models rather than real-world climate, since we can never have an archive of thousands of global warmings to compare this one to.

certainly nothing to be cavalier about. A 3°C rise, after all, would bring Earth to its warmest point in millions of years. Plus, there's no reason that greenhouse gases couldn't increase well beyond the benchmark of doubled CO_2.

Listening to the models

It's up to policymakers and the public to decide how society ought to respond to the onslaught of data that climate models give us. Probabilities like the ones above can help immensely. It may seem confusing to think of a temperature range and a probability at the same time, but that's the direction modellers are heading, and it's the type of output that's most useful in making decisions. Once upon a time, weather forecasts would call for "possible showers" or "a chance of snow" without ever specifying a number. That made it hard to tell exactly how likely it was that your picnic would get rained out. It wasn't until computing become widespread in the 1960s that a new set of statistical weather models enabled forecasters to look at a wide range of outcomes and assign likelihoods. By the 1970s, "probabilistic" forecasts were the norm and people had grown accustomed to phrasings such as "a 30% chance of rain."

In much the same way, climate modelling is now affordable enough that some high-end models can be run a number of times. Each run might use a different rate of CO_2 increase, for example, to show the many ways the climate might unfold depending on how serious we get about reducing emissions. Or the model might simulate the same CO_2 increase a number of times, but with starting conditions that vary slightly, in order to see how natural variations in climate affect the results. These large sets of simulations are called **ensembles**, and they've become increasingly important in climate modelling. As in the examples cited above, some ensembles include more than one thousand simulations, which was unheard of as recently as the late 1990s. The sheer size of these ensembles allows scientists to calculate statistics and probabilities that a single model run can't provide.

The European Union is exploring how this technique might help guide policy in a huge multi-year project called ENSEMBLES. Through the end of the decade, scientists at 66 institutions in Europe, Australia and the US will be conducting hundreds of simulations of European climate, looking at changes that manifest globally, regionally and locally over seasons, years and decades. The results will feed directly into assessments of the impact of climate change on various economic sectors, from agriculture to insurance. Apart from creating ensembles, scientists will also be hard at work making other improvements to their models. Among the current areas of keen interest are weaving in more interactive chemistry, adding more detailed land cover and ecosystems, and capturing the various effects of aerosols more precisely.

One might ask why we ought to believe anything that global climate models tell us. Some contrarians pose this question often, and it's healthy for everyone to recognize what climate models can and can't tell us. For instance, models will always struggle with uncertainty about the starting-point state of the atmosphere, which affects shorter-term weather and climate outlooks the most. At the same time, our lack of foresight about greenhouse emissions becomes enormous when looking a century out. In between, though, is a "sweet spot" identified by Peter Cox and David Stephenson (University of Exeter) in a 2007 *Science* paper. They urge their fellow scientists to focus their research energies on the time frame from about 2040 to 2060, a period of keen interest for long-term business and government planning and one where the total climate uncertainty may actually be less than at the longer or shorter end.

One good reason to be confident in the general quality of today's climate models is their ability to simulate the past – especially the twentieth century, for which we have by far the most complete climate record. These three graphs,

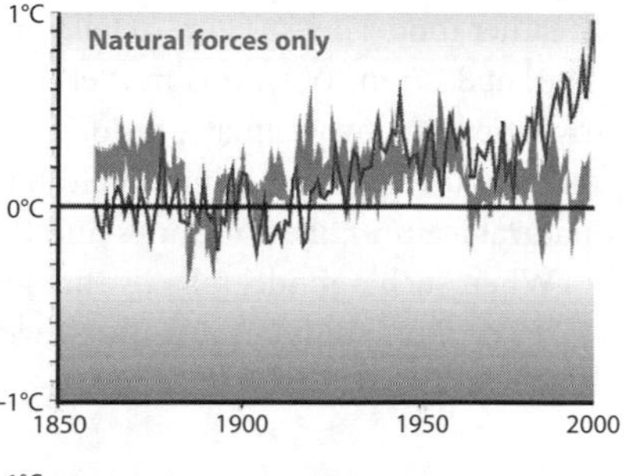

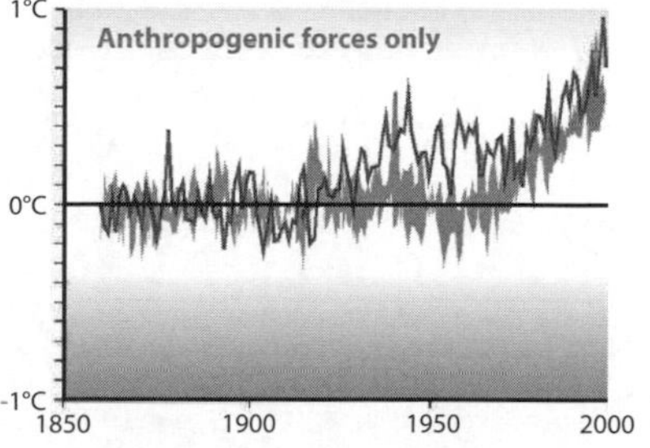

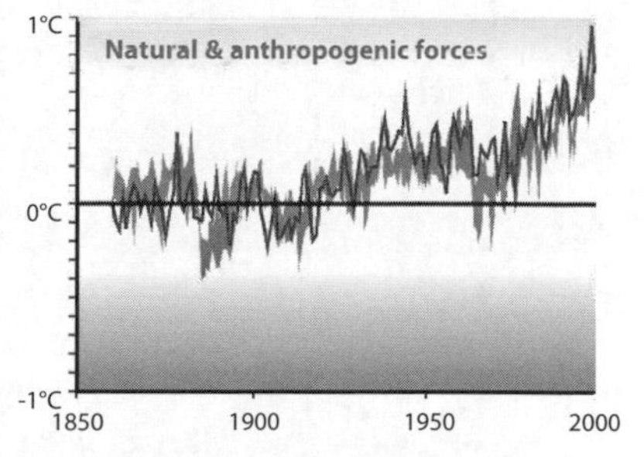

When models look only at natural climate forces (top) or human-produced ones, such as greenhouse gases and sulphate pollution (middle), they fail to accurately reproduce the last century's climate (the dark line in each graph). A model that incorporates both kind of forces (bottom) does a much better job. The grey shading shows the range of each model ensemble.

Graphs: IPCC

prepared for the 2001 IPCC report, show how one climate model captures the ups and downs in global temperature over the last century. It doesn't reproduce every year-to-year wiggle, but that's not the idea, just as a weather model doesn't aim to tell you the exact temperature in your backyard at 8:27pm. What this model does show accurately is the character of twentieth-century climate. It does so by including the gradual increase in greenhouse gases and Sun-shielding pollutants, as well as the effects of natural features like volcanoes and the Sun.

When such a model tells us that global temperatures could rise by 3°C (5°F) or more in the next century, we'd be foolhardy to ignore it.

PART 4

DEBATES & SOLUTIONS

From spats and spin to saving the planet

A heated debate

How activists, sceptics and industry have battled for column inches and the public mind

Every good scientist is part sceptic, ready to let go of a hypothesis if new evidence points another way. It's interesting, then, that in the world of climate change science the sceptic label has been focused on a tiny subset of contrarians – those who buck the mainstream and downplay the risk of global warming. Often quite vocal, this small group of scientists became familiar to anyone reading or hearing about climate change in the 1990s. That's when long-simmering questions about the future of our climate came to a boil.

It wasn't only sceptical scientists stirring the pot. Industry groups and their spokespeople had a huge say in the negotiations that led to the Kyoto Protocol in 1997, and they've continued to wield influence, although many corporations have since gone considerably greener. At the other end of the spectrum, environmental activists have argued for the defence of the planet and its ecosystems, often with colourful protests tied to key diplomatic moments. The ensuing debates have raged for well over a decade, with the public trying to grapple with an issue unlike anything that's come beforehand.

The early days

As we've seen, the greenhouse effect was discovered right back in the nineteenth century, but serious scientific and media debate about climate change didn't take off until much later. One of the first news stories came in

Is the World Getting Warmer?

By ALBERT ABARBANEL and THORP McCLUSKY

Was this past mild winter just part of a natural cycle? Will the old-fashioned blizzard someday disappear? Here are the startling facts revealed by world-wide weather surveys now going on.

This *Saturday Evening Post* article from 1950 was one of the very first stories on climate change in the popular press

1950, after global temperatures rose from around 1900 to the 1940s. *The Saturday Evening Post*, then one of America's biggest magazines, asked a question one might hear today: "Was this past mild winter just part of a natural cycle?" Their article "Is the World Getting Warmer?" rounded up a variety of anecdotal evidence, including "tropical flying fish" sighted off the New Jersey coast. As possible causes for the warm-up, the article cites solar variation and other natural factors. Greenhouse gases aren't even mentioned.

The scientific debate picked up in the 1960s. But at that time, there was plenty else to worry about – nuclear annihilation, for instance – so few people outside of scientific circles heard much about the risk of climate chaos. Things began to change in the 1970s, when the embryonic environmental movement called out air pollution as an example of humans' soiling of the planet. With early photos from outer space now highlighting Earth's stark aloneness, it was suddenly easier to believe that humans could affect the atmosphere on a global scale.

But what grabbed most of the press in the 1970s wasn't a global warming but a cool-down. Earth's temperature had been gradually slipping for some three decades, and a few maverick scientists speculated that dust and Sun-blocking sulphate particles could be responsible for the cooling. A British documentary in 1974 called *The Weather Machine* warned that a single brutal winter could be enough to plaster northern latitudes with a "snow blitz" that the next summer couldn't entirely erase, thus leading to continent-encrusting ice sheets within decades. If nothing else, climate had started to seem more fluid, more unstable, than most people had ever thought possible.

> **"Climatological Cassandras are becoming increasingly apprehensive, for the weather aberrations they are studying may be the harbinger of another ice age."** *Time* magazine, 1974

Even as reporters chattered about cold, many scientists were concerned about the long-term outlook for warmth. In a 1972 *Nature* paper entitled "Man-made carbon dioxide and the 'greenhouse effect'", J.S. Sawyer predicted a temperature rise of 0.6°C (1.0°F) for the rest of the twentieth century – a figure that was only slightly off the mark. A key 1975 paper in *Science* by Wallace Broecker (Lamont-Doherty Earth Observatory) asked if we were "on the brink of a pronounced global warming." Two studies late in the 1970s by the US National Academy of Sciences confirmed that the ever-increasing levels of CO_2 in the air should lead to significant warming. Computer models were improving quickly, and they continued to indicate that warming was on the way. Finally, the atmosphere itself chimed in. By the late 1980s global temperatures had begun an ascent that hasn't abated since, except for a sharp two-year drop after 1991's eruption of **Mount Pinatubo** (see p.199).

The reports and findings accumulated through the 1980s, but with little fanfare outside of research labs and government hearings. Occasionally the media would take note, as when the London *Times* wrote in 1982 of "the experiment that could become too hot to handle", one that could "change the face of the world within three generations." The focus again shifted to cooling during the Cold War angst of the early 1980s, with the controversial pronouncement from a group of climate modellers that nuclear warfare might produce enough Sun-

SCIENCE

The Cooling World

There are ominous signs that the earth's weather patterns have begun to change dramatically and that these changes may portend a drastic decline in food production—with serious political implications for just about every nation on earth. The drop in food output could begin quite soon, perhaps only ten years from now. The regions destined to feel its impact are the great wheat-producing lands of Canada and the U.S.S.R. in the north, along with a number of marginally self-sufficient tropical areas—parts of India, Pakistan, Bangladesh, Indochina and Indonesia—where the growing season is dependent upon the rains brought by the monsoon.

The evidence in support of these predictions has now begun to accumulate so massively that meteorologists are hard-pressed to keep up with it. In England, farmers have seen their growing season decline by about two weeks since 1950, with a resultant over-all loss in grain production estimated at up to 100,000 tons annually. During the same time, the average temperature around the equator has risen by a fraction of a degree—a fraction that in some areas can mean drought and desolation. Last April, in the most devastating outbreak of tornadoes ever recorded, 148 twisters killed more than 300 people and caused half a billion dollars' worth of damage in thirteen U.S. states.

Trend: To scientists, these seemingly disparate incidents represent the advance signs of fundamental changes in the world's weather. The central fact is that after three quarters of a century of extraordinarily mild conditions, the earth's climate seems to be cooling down. Meteorologists disagree about the cause and extent of the cooling trend, as well as over its specific impact on local weather conditions. But they are almost unanimous in the view that the trend will reduce agricultural productivity for the rest of the century. If the climatic change is as profound as some of the pessimists fear, the resulting famines could be catastrophic. "A major climatic change would force economic and social adjustments on a worldwide scale," warns a recent report by the National Academy of Sciences, "because the global patterns of food production and population that have evolved are implicitly dependent on the climate of the present century."

A survey completed last year by Dr. Murray Mitchell of the National Oceanic and Atmospheric Administration reveals a drop of half a degree in average ground temperatures in the Northern Hemisphere between 1945 and 1968. According to George Kukla of Columbia University, satellite photos indicated a sudden, large increase in Northern Hemisphere snow cover in the winter of 1971-72. And a study released last month by two NOAA scientists notes that the amount of sunshine reaching the ground in the continental U.S. diminished by 1.3 per cent between 1964 and 1972.

To the layman, the relatively small changes in temperature and sunshine can be highly misleading. Reid Bryson of the University of Wisconsin points out that the earth's average temperature during the great Ice Ages was only about 7 degrees lower than during its warmest eras—and that the present decline has taken the planet about a sixth of the way toward the Ice Age average. Others regard the cooling as a reversion to the "little ice age" conditions that brought bitter winters to much of Europe and northern America between 1600 and 1900—years when the Thames used to freeze so solidly that Londoners roasted oxen on the ice and when iceboats sailed the Hudson River almost as far south as New York City.

Just what causes the onset of major and minor ice ages remains a mystery. "Our knowledge of the mechanisms of climatic change is at least as fragmentary as our data," concedes the National Academy of Sciences report. "Not only are the basic scientific questions largely unanswered, but in many cases we do not yet know enough to pose the key questions."

Extremes: Meteorologists think that they can forecast the short-term results of the return to the norm of the last century. They begin by noting the slight drop in over-all temperature that produces large numbers of pressure centers in the upper atmosphere. These break up the smooth flow of westerly winds over temperate areas. The stagnant air produced in this way causes an increase in extremes of local weather such as droughts, floods, extended dry spells, long freezes, delayed monsoons and even local temperature increases—all of which have a direct impact on food supplies.

"The world's food-producing system," warns Dr. James D. McQuigg of NOAA's Center for Climatic and Environmental Assessment, "is much more sensitive to the weather variable than it was even five years ago." Furthermore, the growth of world population and creation of new national boundaries make it impossible for starving peoples to migrate from their devastated fields, as they did during past famines.

Climatologists are pessimistic that political leaders will take any positive action to compensate for the climatic change, or even to allay its effects. They concede that some of the more spectacular solutions proposed, such as melting the arctic ice cap by covering it with black soot or diverting arctic rivers, might create problems far greater than those they solve. But the scientists see few signs that government leaders anywhere are even prepared to take the simple measures of stockpiling food or of introducing the variables of climatic uncertainty into economic projections of future food supplies. The longer the planners delay, the more difficult will they find it to cope with climatic change once the results become grim reality.

64

Newsweek, April 28, 1975

April 28, 1973: *Newsweek*'s report on the fear of a forthcoming ice age blocking particles to cause **nuclear winter**. Meanwhile, theories of global warming remained exotic enough to the public that many journalists kept the term "greenhouse effect" in quotes.

The stunning discovery of the Antarctic **ozone hole** in 1985 was a turning point. Although it fostered long-lived confusion between ozone depletion and global warming (see p.28), the finding was also a new sign of the atmosphere's fragility, borne out by vivid satellite images. The other shoe dropped in the United States during its sizzling, drought-ridden **summer of 1988**. Huge tracts of forest burned in Yellowstone National Park; parts of the Mississippi River ran dry; and on a record-hot June day in Washington DC, NASA scientist **James Hansen** delivered his now-famous testimony before Congress, claiming to be "99% sure" that global warming was upon us, most likely induced by humans. Together, the *New York Times* and *Washington Post* ran more than forty stories on climate change in 1988 after less than two dozen in the preceding four years, according to Katherine McComas and James Shanahan of Cornell University. *Time* magazine named "Endangered Earth" Planet of the Year, in place of its usual Man of the Year.

Even conservative politicians took note. In August, US presidential candidate George Bush (senior) declared, "Those who think we are powerless to do anything about the greenhouse effect forget about the 'White House effect'." And although the meteorological drama of 1988 was focused on North America, the political waves reverberated far and wide. In September, British prime minister Margaret Thatcher warned the Royal Society that "we have unwittingly begun a massive experiment with the system of the planet itself." As Jeremy Leggett recalls in *The Carbon War*, "1988 was the year that broke the mould." Indeed, the events of that year were enough to convince Leggett, who was then teaching at Britain's Royal School of Mines, to join Greenpeace as a science advisor to its climate campaign.

From progress to roadblocks

In the aftermath of 1988, governments began to pour money into global warming research. In 1989 the Intergovernmental Panel on Climate Change was established (see p.287) for channelling research from dozens of nations and thousands of scientists into an internationally recognized consensus. The IPCC produced its first report in 1990, underlining the risks of global warming, and environmental activists did their best to alert journalists and the public to the problem.

Debates among the campaigners

Climate change is now a favoured cause of most environmental groups, but key differences in strategy and ideology remain. One question is whether to embrace or reject certain **aspects of capitalism**. With consumerism driving the world economy, some eco groups are going with the flow – encouraging people to use their spending power to make climate-healthy choices, such as buying a hybrid car or using low-energy light bulbs, and supporting "cap and trade" policies to keep emissions down. But other groups, especially those with roots in 1970s counterculture, retain an abiding suspicion of the corporate world and the governments that support it. Friends of the Earth stresses its anti-globalization work in its climate change publicity materials, and some of its activists claim that emissions trading and other parts of the Kyoto Protocol are rigged against poor countries.

For environmentalists who *do* accept the idea of green commerce, the struggle to isolate genuinely "ethical" choices can be tricky. For instance, Toyota makes the world's most popular hybrid – the Prius – yet it's also one of the plaintiffs in the lawsuit aimed at blocking California's strict new emissions standards (p.304). Hence some groups encourage consumers to push big companies in a climate-friendly direction through shareholder actions or socially responsible investing (see p.337).

Another difference between climate activists is **how to approach fossil fuels**. Some take a pragmatic approach, figuring that oil, gas and coal aren't going away tomorrow. The US Natural Resources Defense Council has thrown its considerable weight behind "cleaner coal", on the condition that it includes CO_2 sequestration. NRDC notes the large number of coal plants already on order in China and elsewhere and stresses the need to develop cleaner technology to install in those plants. Other groups keep the focus on renewables when discussing innovations in energy. In a 2007 statement, Greenpeace International made its position clear: "Clean coal and cheap coal are big industry lies."

Looming in the background is **nuclear energy**, and the question of whether countries should turn to it to bridge the potential gap between fossil fuels and large-scale deployment of renewables. Some influential scientists have lent support to the idea of using nuclear as a stop-gap, including James Lovelock and an interdisciplinary panel of MIT scientists (p.322). However, no major environmental groups had come out in favour of this idea as of early 2006, and many remain adamantly opposed to any expansion of nuclear energy.

A final point of difference is **picking goals**. A concrete target is the best way to motivate volunteers and supporters. The most commonly cited benchmark, consistent with the European Union's goal, is to stabilize the climate at 2°C (3.6°F) above the pre-industrial global temperature (see p.280). Yet the globe has already warmed nearly 0.8°C (1.44°F), and scientists estimate that at least another 0.5°C (0.9°F) will result from the CO_2 we've added to date. Thus, emissions will have to be cut drastically in order to meet the 2°C goal – perhaps by more than 50% by the year 2050. With this in mind, some environmental groups use other types of targets as well – including legislative ones.

By this time, the leading **environmental groups** in North America and Europe were well established, most with fifteen years under their belts. No longer a fringe movement, their cause was now part of the fabric of public life. Years of activism helped slow the growth of nuclear power to a crawl in many countries (with no small assistance from the Chernobyl debacle in 1986). Governments and politicians became so attuned to environmental risk that it took less than three years – lightning speed by diplomatic standards – from the time the ozone hole over Antarctica was discovered in 1985 to the signing of the 1987 Montreal Protocol that's now guiding the planet towards eventual ozone recovery.

For a while, it looked as if the same success might be seen with global warming. In 1992, thousands of activists joined similar numbers of journalists and diplomats in Rio de Janeiro for the United Nations-sponsored **Earth Summit**. The meeting, and the global climate treaty that emerged, kicked off years of negotiations that led to the historic **Kyoto Protocol**, the world's first attempt to come to grips with greenhouse-gas emissions. (For

Greenpeace protesting against "Dinosaur Diplomacy" at the Climate Conference in Kyoto, Japan, 1997
Orban Thierry/Corbis Sygma

more on the history and future of Kyoto, see the following chapter.)

As media attention grew and as the scientific evidence strengthened, people in many countries became more aware about the risks of climate change and the possible solutions. Yet something wasn't quite clicking. Although support for climate-change action was growing steadily, the support was "wide but shallow". In other words, people were concerned, but not sufficiently concerned to force the issue up the political agenda (nor to take personal action to reduce their own greenhouse gas emissions).

"Are existing environmental institutions up to the task of imagining the post-global warming world? Or do we now need a set of new institutions founded around a more expansive vision and set of values?"

Michael Shellenberger and Ted Nordhaus, *The Death of Environmentalism*

The sheer scope of the problem was one factor. Fossil fuels are used in virtually every aspect of modern society, and climate change threatens to affect every country on Earth in one way or another. It's hard to motivate people to grapple with such an immense and seemingly intractable issue, and the many options for political and personal action could be too much to process. Moreover, even more than smog or acid rain, human-induced climate change is a classic "tragedy of the commons" – the benefits of burning fossil fuels accrue to individuals, companies and nations, while the costs accrue to the planet as a whole. The most important cause, carbon dioxide, can't be seen, smelled or touched. And while the activists urged concrete action, the benefits – avoiding a global meltdown – were intangible as well.

Another difficulty for climate campaigners was the fact that global warming hit it big just when many people were getting tired of fretting about the state of the world. From its earliest days, the environmental movement had relied on stark, pseudo-apocalyptic imagery to motivate people. In her 1962 book *Silent Spring*, which set the template for environmental wake-up calls, Rachel Carson labelled pesticides and similar agents "the last and greatest danger to our civilization". In her footsteps came a series of similarly dire scenarios, from Paul Ehrlich's *The Population Bomb* to the notion of nuclear winter. Global warming lends itself especially well to this type of rhetoric. It's no exaggeration to talk about the risk of coastal settlements vanishing and the Arctic's summer ice pack disappearing. Yet when activists do dwell on these points, it sometimes brings to mind other predictions of environmental doomsday that didn't come to pass (partly, of course, because society *did* respond to those earlier threats).

Even the most painless ways to reduce global warming – such as improved **energy efficiency** – came with cultural baggage in some countries. Efficiency measures had swept the US during the oil shocks of the 1970s – the top interstate speed limit was dropped to 55mph (89kph), for instance – but the practice of saving energy never lost its taint of deprivation. As oil prices plummeted in the go-go 1980s, efficiency quickly fell by the wayside and speed limits went back up. When global warming pushed energy efficiency back onto the national agenda, it was a loaded topic. Because so many activists had been proposing sensible energy-saving steps for years, it was easy for critics to paint them as opportunists, happy to use climate change or any other issue in order to advance their ulterior goals.

To top it all, climate activists were up against some very tricky adversaries, who were doing their best to *stop* the public getting too worried about global warming. These included a number of sceptics within the scientific community, their opinions amplified by lobby groups representing powerful business and political interests.

Sceptics and industry fight back

Especially in the United States, a group of prominent climate-change sceptics – perhaps no more than several dozen – have wielded far more influence than their numbers might indicate. Until very recently, many if not most news articles about climate change included a comment from one or more of these contrarians. Their voices have been backed up in many cases by the immense money and influence of the **oil, coal and auto industries** through think-tanks such as the GCC and the CEI (see box opposite). This support has enabled them to exert much sway on the US Congress, the media and, by extension, the global fight against climate change.

"I am convinced that in 15–20 years, we will look back on this period of global warming hysteria as we now look back on so many other popular, and trendy, scientific ideas."

William Gray, Colorado State University, testifying before the US Congress, September 2005

Although a few sceptics are active in climate research, many aren't. Some of the most vocal are scientists with

Industry lobby groups

Like their peers, the few climate sceptics active in research are employed mainly by universities and private labs. Although a few have received grants from oil and coal companies, most rely largely on public funds to carry out their work. However, that work gets an extra dose of clout, especially in the US, thanks to a number of conservative think-tanks and lobby groups which cite their findings widely and use them in an attempt to convince legislators that climate change science is full of unknowns. Such centres are often influential, and many are buoyed by funding from corporations with a lot to lose from carbon restrictions.

One highly visible group throughout the 1990s was the opaquely titled **Global Climate Coalition**, which formed in 1989 as the prospect of global diplomatic action on climate change appeared on the horizon. Based at the US National Association of Manufacturers, the GCC included some of the biggest oil, auto and coal companies in the world, including General Motors, Ford, BP, Shell and Exxon (aka Esso). Along with lobbying at UN meetings, the coalition angled its way into becoming an oft-quoted presence in the media. They also financed Kyoto-related commercials warning that "Americans would pay the price" for the treaty.

The GCC began to fracture with the departure of BP in 1997, Royal Dutch Shell in 1998, and Ford in 1999. By 2001, it was history, though arguably it had served its purpose and was no longer necessary. A 2001 memo written to Exxon by the US under-secretary of state, Paula Dobriansky, and later obtained by Greenpeace, states that George Bush rejected Kyoto "partly based on input from you [the GCC]". In the group's own words, "The Global Climate Coalition has been deactivated. The industry voice on climate change has served its purpose by contributing to a new national approach to global warming."

Since the days of the GCC, most of the world's major oil companies have shifted towards public acknowledgment of climate change (see p.308), but Exxon – the largest of them all – has continued to sow seeds of doubt. From 2000 to 2003, according to an exposé by Chris Mooney in *Mother Jones* magazine, the company poured more than $8 million into more than forty organizations aligned with climate-change scepticism. In a rare move, the UK's Royal Society wrote to Exxon in 2006 asking it to stop funding sceptically aligned groups. The company never responded to the satisfaction of the society, though by 2007 Exxon had cut its CEI support and was inching toward a greener public stance.

One think-tank that's very active at the time of writing is the Washington-based **Competitive Enterprise Institute**. Headed up by Myron Ebell, the CEI's global warming wing has become the leading institutional voice of climate scepticism while other entities have pulled back or lost interest. Ebell was censured by the British House of Commons "in the strongest possible terms" in 2004 after he told BBC's Radio 4 that Sir David King, the chief science advisor to prime minister Tony Blair, "knows nothing about climate science".

On the release of the Al Gore documentary *An Inconvenient Truth* (see p.275), the CEI issued a pair of glossy TV advertisements that noted how fossil fuels have made life more comfortable and convenient. They ended with the tagline "Carbon dioxide: They call it pollution. We call it life."

backgrounds in subjects like solid-state physics or mathematics, sometimes with impressive resumes in their fields but little if any experience working directly on climate-change science. For example, physicist Frederick Seitz, who presided over the US National Academy of Sciences in the 1960s, became an outspoken climate-change sceptic in the 1990s.

Bjorn Lomborg's sceptical environmentalism

Danish political scientist Bjorn Lomborg marshalled a slew of statistics and nearly 3000 footnotes to make his case that, overall, the environment is in better shape than we might think. In his 2001 book *The Sceptical Environmentalist,* Lomborg employs the climate-and-economy models used by the IPCC assessments to argue that major emissions reductions in the short term (*à la* Kyoto) are not only enormously costly but will have little impact on the longer-term climate outcome.

Lomborg's book got rave reviews in *The Economist, Rolling Stone* and elsewhere, but it was panned in other publications and pilloried by some leading scientists. The dust-up got to the point where the official Danish Committee on Scientific Dishonesty labelled Lomborg's book "objectively dishonest" (they later withdrew the finding). The magazine *Scientific American* published "Misleading Math about the Earth", an eleven-page critique of *The Sceptical Environmentalist* by four top researchers, eventually followed by a rebuttal from Lomborg himself and then a re-rebuttal from one of the four critics. Another of the four, climate scientist Stephen Schneider from Stanford University, blasted Lomborg in *Grist* magazine for "selective inattention to inconvenient literature and overemphasis of work that supports his lopsided views."

In his discussion of climate change, Lomborg glides past sea-level rise with little concern for the high-end possibilities. Moreover, Lomborg's economic focus fails to take into account the intrinsic, non-monetary value of protecting particular species and ecosystems. Even so, the book's sunny-side-up view of economic and ecological progress and its critique of environmental doom and gloom has its followers, especially from the sceptical side of the global-warming aisle. Lomborg, in fact, ends his climate-change discussion by claiming that society has the money to control greenhouse emissions if we deemed it a high enough priority. However, he argues that many other issues – such as preventable diseases – deserve to take precedence. In his 2007 follow-up *Cool It: The Sceptical Environmentalist's Guide to Global Warming*, Lomborg stakes out his turf even more firmly, remaining sanguine about such concerns as Antarctic ice (it's growing) and polar bears (their real nemesis is hunting, not warming).

Other dissenters are trained in atmospheric science but have published few peer-reviewed papers on climate change. Others still are retirees, affording them the time and freedom to act as consultants, writers and speakers without having to conduct scientific studies of their own. Of course, there have also been plenty of spokespeople for sceptical positions who aren't scientists at all, just as there are plenty of non-scientists speaking out for climate-change action.

Right from the beginning, uncertainty has been the overriding theme in the arguments of climate-change contrarians. The core of greenhouse science – such as the consensus estimates on how much global temperature rise to expect from a doubling of CO_2 – has held firm for decades. But climate change is such a multifaceted and complicated enterprise that it's easy enough to find minor weaknesses in one study or another. Furthermore, there are always exceptions that prove the rule, such as an expanding glacier or a region that's cooled in recent decades. Sceptics seize such uncertainty and exceptions and – amplified by the PR budgets of corporations heavily invested in fossil-fuel use – give the false impression that the entire edifice of knowledge about climate change might crumble at any moment (or even that the whole thing is a colossal scam, a claim voiced more than a few times).

When contrarians point to a single event or process like this one as a disproof of global climate change – a glacier growing here, a city cooling there – they're often accused by mainstream scientists of **cherry-picking**: selecting a few bits of evidence that seem to prove their point while omitting counter-examples. It's a classic rhetorical technique, one well known to skilled lawyers and politicians.

Arguments and counterarguments

Aside from cherry-picking, sceptics soon developed a more general set of criticisms – many of which you still hear today – to throw at mainstream climate scientists and at the concept of global warming in general. Taken to the extreme, you could sum up the classic sceptical view like this:

> The atmosphere isn't warming; and if it is, then it's due to natural variation; and even if it's not due to natural variation, then the amount of warming is insignificant; and if it becomes significant, then the benefits will outweigh the problems; and even if they don't, technology will come to the rescue; and even if it doesn't, we shouldn't wreck the economy to fix the problem when many parts of the science are uncertain.

Probably no single sceptic would endorse the whole of that rather convoluted statement. Yet each of the points within it has been argued vigorously over the years by various contrarians. Let's look at each point in turn.

▶ **"The atmosphere isn't warming"** This one has been put safely to rest, although as recently as the 1990s some sceptics insisted there was no planet-wide warming at all, and the notion still crops up on the Internet. Fuelling this line of argument was the apparent lack of warming in upper-air temperatures as measured by satellites and radiosondes. But since the year 2000 it has become increasingly clear that average upper-level temperatures are in fact warming at close to the same rate as the surface (p.185).

▶ **"The warming is due to natural variation"** This point is still argued often, even though the IPCC – drawing on many individual studies – has concluded that the warming of the last century, especially since the 1970s, falls outside the bounds of natural variability (see p.8).

One of the most widely publicized studies to downplay the human influence on the rising temperature was a 2003 paper for *Climate Research* by Willie Soon and Sallie Baliunas of the Harvard-Smithsonian Institute for Astrophysics. After surveying other studies that examined the climate of the last thousand years, Soon and Baliunas claimed that "the twentieth century is probably not the warmest nor a uniquely extreme climatic period of the last millennium". Their study was lauded in the US Congress, but its methodology was slammed by a number of climate scientists. Three editors of *Climate Research* resigned amid the fallout. The journal's publisher later acknowledged that the Soon/Baliunas paper failed to back up its own claims and said that the journal "should have been more careful and insisted on solid evidence and cautious formulations before publication".

More recently, a group of sceptics led by Canadian mathematician and former mining executive Stephen McIntyre has taken on a watchdog role, hunting for errors in the massive analyses carried out by climate scientists. In 2005, McIntyre and economist Ross McKitrick published a paper in *Geophysical Research Letters* criticizing the methodology behind the "hockey stick" reconstruction of the last millennium's temperature record (p.226). Their work drew widespread attention from many quarters, including the US Congress. In a 2006 review of hockey-stick findings, the US National Research Council noted the presence of a few statistical flaws in hockey-stick analyses but lent support to the overall picture of substantial twentieth-century warming.

Similarly, in 2007 McIntyre discovered that NASA's analysis of US temperatures in the twentieth century had inadvertently omitted some data. The resulting adjustments, though small, were just enough to juggle the NASA rankings of hottest US years, bumping 1934 slightly ahead of 1998. But NASA's year-by-year rankings of *global* temperature didn't change at all – a point mangled by several sceptical columnists. Still, some observers have supported McIntyre's call for more transparency in the data, procedures and models used in climate-change research, while reiterating that any such adjustments aren't at all likely to change the major themes and conclusions of global warming science.

"...we must carefully examine all claims that, if true, would lead to paradigm shifts like that caused by Galileo, but at the same time, it is wise to note that for every real Galileo or Einstein who radically alters conventional wisdom, there are probably a thousand 'fossil fools.'"

Stephen Schneider, Stanford University

It's often claimed that **solar variations** account for the last century's warmth. It's true that the Sun is producing more sunspots than it did in the early 1800s. However, this mainly reflects an increase in the ultraviolet range of sunlight, which is only a tiny part of the solar spectrum. In fact, the total solar energy reaching Earth changes very little over time. Across the eleven-year solar cycle, it varies by less than 0.1%, and even across the period since the Little Ice Age chill of 1750 solar output climbed no more than about 0.12%, according to the 2007 IPCC report. The IPCC now deems it "very likely" (more than 90% certain) that greenhouse gases wielded more influence than total solar output in driving the last fifty years of warming.

There's still a question mark or two when it comes to ultraviolet radiation, where the lion's share of solar variability occurs. It's possible that UV rays interact with ozone in the stratosphere to change circulation patterns, though more work is needed to clarify how this might occur. UV light also helps shield Earth from **cosmic rays** that bombard and ionize the atmosphere – a point much discussed by sceptics in recent years, thanks to work by Henrik Svensmark (Danish National Space Center) and others. This concept got major play on UK and Australian TV in the 2007 documentary *The Great Global Warming Swindle* and in the popular book *The Chilling Stars*, cowritten by Nigel Calder and

Svensmark. The idea is that highly reflective low-level clouds might form more easily when tiny particles that serve as cloud nuclei are ionized, helping them to clump together more readily. Should this be true, then an active Sun would inhibit low-level clouds, thus allowing more sunlight to reach Earth and fostering warming. In lab work, Svensmark and colleagues found some evidence for the clumping effect, but it's an open question whether these particles actually make low-level clouds more prevalent in the real world, since vast numbers of potential cloud nuclei are normally present anyway. Moreover, there's no clear evidence that more cosmic rays have actually made it into Earth's lower atmosphere over the last several decades. *The Great Global Warming Swindle* was

Michael Crichton gets sceptical

It's not often that a novelist gets the opportunity to speak to members of a science-oriented committee in the US Congress. Michael Crichton, king of the techno-thriller, got his day on Capitol Hill on September 28, 2005, not long after his novel *State of Fear* stormed the bestseller lists. Crichton, a Harvard-trained physician, was invited to discuss the role of science in environmental policy-making. He also scored a private hour-long meeting with US president George Bush, according to the 2006 book *Rebel in Chief* by Fred Barnes.

Crichton first turned his literary attention to the atmosphere when he co-wrote the screenplay for the 1996 film *Twister*. That movie painted a colourful, largely positive portrait of the meteorologists who chase, observe and study tornadoes. Climate scientists would not get such favourable treatment in *State of Fear*, an outcome that careful Crichton-watchers could have predicted. For years he'd dismissed the more grave predictions of environmentalists as fear-mongering. In a 2003 speech at Caltech entitled "Aliens Cause Global Warming", Crichton mocked the projections from global climate models, using one of the classic critiques discussed elsewhere in this chapter: "Nobody believes a weather prediction twelve hours ahead. Now we're asked to believe a prediction that goes out 100 years into the future". He also railed against consensus-driven science such as that practised by the IPCC, calling it "an extremely pernicious development that ought to be stopped cold in its tracks."

In *State of Fear,* Crichton pits a lawyer and a swashbuckling scientist/sceptic against a band of eco-terrorists who plan to create a devastating tsunami in order to drum up support for their cause. (Never mind that tsunamis have nothing to do with climate change; they're actually triggered by earthquakes.) The book alternates between nail-biting action sequences in the usual thriller mould and tutorials in which the lawyer (and by extension the reader) discovers that the consensus on global warming is full of holes. *State of Fear* is surely the first blockbuster novel to include more than 25 actual graphs of long-term temperature trends at stations across the world. Many of these graphs show criticized by many experts not only for downplaying these unknown but also using discredited data and inaccurate graphs.

▶ **"The amount of warming is insignificant"** This claim mingles bona fide uncertainty about the future with a judgment call on how much warming should be labelled as significant. If you were a polar bear, or a Parisian in 2003's heat wave, you might consider the warming we've already experienced in the last few years to be highly significant.

The genuine uncertainty is how much warming we can expect in the coming decades and centuries. As noted in Circuits of Change (p.227), the most widely accepted estimate for the rise in global temperature

cooling trends; others are for urban stations, such as New York, that show dramatic long-term warming, presumably due to the heat-island effect. In Crichton's eyes, these examples weaken the case that greenhouse gases have produced substantial warming to date. In real life, no climate scientist expects that every station on Earth should warm in lockstep. Moreover, urban biases have already been exhaustively studied and corrected (see p.176).

The book veers even further from the thriller template after the action stops. The back part of *State of Fear* includes a summary of Crichton's personal beliefs on climate change and environmentalism as well as a 21-page bibliography. Among his claims is the idea that land use will outweigh fossil fuels in shaping global temperature. He calls the IPCC projections of 21st-century warming "guesses" (since nobody knows how technology will evolve in the next century), and he offers a guess of his own (the sarcastically precise 0.812436°C). Crichton also claims that models vary by 400% in their estimates of warming by 2100, based on the IPCC's range of 1.4–5.8°C (2.5–10.4°F). In fact, as NASA modeller Gavin Schmidt points out on realclimate.org, that range encompasses several different scenarios for how much greenhouse gas we will emit. For any particular scenario, the uncertainty range is more like 60%.

Jim Meddick

from a doubling of atmospheric carbon dioxide in the atmosphere is about 3.0°C (5.4°F) over pre-industrial times, with a range of 2.0–4.5°C (3.6–8.1°F) deemed "likely" (more than 66% chance) in the 2007 IPCC report. A number of sceptics believe that the low end of the range is the most likely outcome, but in fact we've already warmed close to 0.8°C (1.44°F) since pre-industrial times. That puts us almost halfway to the low end of the IPCC range, and that's with carbon dioxide up only about 35% from its pre-industrial value.

The following chapter looks more closely at the levels of warming considered most dangerous (p.278).

▶ **"The benefits will outweigh the problems"** Most of those who stress the bright side of global warming favour the idea that increased carbon dioxide will fertilize greenery of all types, enhancing agriculture and nourishing forests. Overall, some stimulation of plant growth does appear likely, but it's not at all certain that the benefits will be prolonged or planet-wide or that the nutritive value of crops will be sustained amid the growth (see p.163). Moreover, while CO_2 may give forests a boost, the changing climate raises the risk of devastating fires and insect attacks (see p.155).

Against the potential pluses of CO_2 fertilization, and other benefits such as fewer cold-related illnesses and the possibility of sailing through the Arctic in midsummer, we have to balance the various negative

Betting on a changing climate

Frustrated by the prominence of climate sceptics in the media, scientist James Annan set out to see if any of the naysayers were prepared to put their money where their mouth was and bet on the future of climate. After months of searching, he found two Russian solar physicists prepared to take part, and a deal was struck. If the period 2012–2017 ends up warmer than 1998–2003, Annan collects US $10,000 from Galina Mashnich and Vladimir Bashkirtsev; if not, Annan coughs up the ten grand. They'll use numbers from the US National Climatic Data Center to settle the bet.

"A payoff at retirement age would be a nice top-up to my pension", says Annan, who studies climate prediction and probability at Japan's Frontier Research Centre for Global Change. He's confident he'll win this one – although, as he notes, "there are no exemptions for volcanoes, large meteorite strikes or nuclear winter." If the climate does cool, Mashnich and Bashkirtsev may need to plough their winnings into insulation. The Siberian scientists work at the Institute of Solar-Terrestrial Physics in Irkutsk, which gets the coldest winters of any city of its size on Earth.

symptoms discussed elsewhere in this book, such as an increased risk of drought in much of the poorest areas of the world; rising seas that could devastate cities and displace millions; more intense hurricanes and heat waves; massive species loss ... the list goes on.

▶ **"Technology will come to the rescue"** This isn't scepticism about global warming so much as an affirmation of human ingenuity. Some optimists believe that geo-engineering might save us from the clutches of global warming (see p.330). Even if such an approach proves feasible, it would face an uphill trek to gain funding, international approval and public confidence. Still, it's important to keep in mind the possibility and promise of technical innovation, while at the same time recognizing the reality of our present situation and the emissions trajectory we're on.

▶ **"We shouldn't wreck the economy"** For sceptics motivated more by economic than scientific considerations, this is the ultimate bottom line. It was the rationale used by US president George W. Bush in opting out of the Kyoto Protocol. The argument hinges on the uncertainty of the science, as well as a host of economic assumptions. If we don't know with absolute confidence how much it will warm and what the local and regional impacts will be, so the reasoning goes, perhaps we're better off not committing ourselves to costly reductions in greenhouse-gas emissions. However, the eventual costs of environmental remedies have often proved much less than economic models indicated at first. Moreover, it's unclear how much further the scientific uncertainty around specific regional outcomes can be reduced – perhaps not as much as we'd like to think. And many decisions of grave importance to nations and the world at large are made in the absence of ironclad certainty.

Perhaps more importantly, it would be foolish to assume that reducing emissions will cost more than coping with a changing climate – a point made emphatically by the Stern Review (see box overleaf). In some areas, such as energy efficiency, the reductions are likely to be money-saving even in the short to medium term. And there's always the chance that new energy technologies will stimulate job growth that could more than compensate for any jobs lost in the transition. We'll discuss some of those technologies and approaches in the following chapters.

A few other critiques continue to pop up on talk shows or on newspaper op-ed pages, surviving even after they've been debunked time and again. The "**global cooling**" scare of the 1970s is a perennial favourite of sceptics who imply that climate researchers can't really make up their minds. Although media did jump on the cooling bandwagon for a few years, as noted above, there was never widespread consensus that anything like an ice age was imminent.

Another brickbat often lobbed by sceptics, including Michael Crichton (see p.260) is the accusation, "If they can't predict the weather for next month, how can they predict the climate a hundred years from now?" Of course, these are two fundamentally different processes. A weather fore-

How much will climate change cost us?

One of the key points of debate surrounding national and international efforts to tackle climate change has been the potential economic cost. Conservative commentators and politicians, especially those in the US, have frequently argued that emissions cuts could bring major economic downswings. Green-minded writers and legislators, on the other hand, have often suggested that the cost of *not* acting could be even higher.

The terms of the debate shifted sharply in 2006 with the release of the *Stern Review on the Economics of Climate Change*. Commissioned by Gordon Brown, then the head of the UK treasury, this massive study headed by Nicholas Stern (formerly the chief economist for the World Bank) reverberated in financial and political circles around the globe. Like a mini-IPCC report, Stern's study drew on the expertise of a wide range of physical and social scientists to craft a serious, comprehensive look at the potential fiscal toll of global warming.

Stern's conclusions were stark. Our failure to fully account for the cost of greenhouse emissions, he wrote, "is the greatest and widest-ranging market failure ever seen". He estimated that climate change could sap anywhere from 5% to 20% from the global economy by 2100, and that "the benefits of strong, early action considerably outweigh the cost".

While the Stern Review drew massive publicity and galvanized leaders around the world, it wasn't universally praised by economists, many of whom found it overly pessimistic. One of the biggest sticking points was how Stern handled the discount rate – the gradual decline in value of an economic unit over time. In traditional analysis, a unit of spending today is valued more highly than a unit spent tomorrow, because the present is more tangible and the unit will be worth less in the presumably richer future. But climate change is a far more complex matter because of its global reach and the long-term effects of current emissions. In that light, Stern chose not to "discount" tomorrow the way most economists would: "if a future generation will be present, we suppose that it has the same claim on our ethical attention as the current one".

Stern's report also drew fire for its assumptions about weather extremes. In the journal *Global Environment Change*, for example, Roger Pielke Jr of the University cast tracks day-to-day changes at a given point. A climate projection looks at longer-term trends that in turn tell you about the type of weather we might expect. If you live in Germany or Minnesota and it's the first day of January, you can say with some confidence that the first day in July ought to be warmer than today, even if you can't predict whether the high will be 20°C (68°F) or 35°C (95°F).

Two other points of contention are the quality of the global models that project future warming and the data that tell us about past climate. The

of Colorado criticized the report for extrapolating a 2%-per-year rise in disaster-related costs observed in recent decades over and above changes in wealth, inflation and demography. As it happens, 2004 and 2005 brought a number of expensive US hurricanes, and Pielke argues that this happenstance skewed the trend used by Stern. At the same time, Pielke grants that the Stern report "helped to redirect attention away from debates over science and toward debates over the costs and benefits of alternative courses of action".

Long before Stern's report, the insurance industry realized that the climate change threw an uncertain element into the detailed calculations it uses to gauge risk. According to industry giant Swiss Re, the global total of insured losses from natural disasters topped $225 million in 2005. That was nearly twice the constant-dollar record set only a year earlier. Part of this rise was due to steep rises in property prices in hurricane-prone regions such as the US Gulf Coast, but there's no doubt that climate change is a significant factor.

As early as 1989, the Lloyd's of London insurance market began incurring massive extra losses. The head of the American reinsurance association said in 1993 that "changes in the number, the frequency and the severity of natural catastrophes are threatening to bankrupt the industry". Since then, Swiss Re and Munich Re have been among the strongest corporate voices calling for climate protection, issuing reports and raising public awareness. They're also starting to use their leverage as institutional investors (in the UK, they own around a quarter of the stock market) to persuade other companies to take climate change on board.

By comparison, US insurance companies have been rather mute on the issue, despite suffering massive financial hits. As late as 2007, the Institute for Business and Home Safety, an industry group, made no mention of climate change on its website. A former president of the group told climate reporter Ross Gelbspan in 2003 that US insurers are "burying their heads", dropping customers and abandoning high-risk areas. Indeed, hundreds of thousands of Floridians saw their home insurance cancelled in 2005 and 2006, after the state's string of hurricanes. Over two million people along the US Gulf and Atlantic coasts have turned to "insurers of last resort" established by state governments – with the public often paying the tab if major disaster strikes.

models certainly aren't perfect (as discussed in Circuits of Change, p.227), but they've agreed for years that we can expect a significant warming. Likewise, shortfalls do exist in the records of past weather (which weren't really designed to detect climate shifts in the first place), but they aren't enough to rule out the overwhelming evidence of change already under way. It's hard to debate a world full of melting glaciers.

Two sides to *every* story

By the mid-1990s, the media had largely lost interest in global warming. According to the Cornell study cited above, the number of climate-change articles in the *New York Times* and *Washington Post* dropped from more than seventy in 1989 to less than twenty in 1994. Even the release of the second IPCC assessment in 1995 prompted only a slight up-tick in journalist interest.

In part, the drop-off was typical of how news stories come and go – the alarm bells couldn't go on ringing forever, at least not without some major disaster to make climate chaos seem like an imminent threat. But the sceptics and lobby groups undoubtedly played a role, having successfully convinced many journalists – and large swathes of the public – that global warming was at best an unknown quantity and at worst "ideological propaganda … a global fraud" (in the words of UK *Daily Mail* journalist Melanie Phillips).

Even when climate change *did* appear in the news, it suffered from the media paradigm that seeks to give equal weight to both sides of every story, with an advocate from each side of the fence battling it out in a fair fight. It's a time-honoured form of reporting, honed to perfection in political coverage, but in the case of climate change it often conveys a misleading sense of symmetry, as if the sceptic camp represented half of the world's climate scientists rather than a small group of contrarians.

"Although the scientific community has known since 1995 that we are changing our climate, the US press has done a deplorable job in disseminating that information, and all its implications, to the public."

Ross Gelbspan, *Boiling Point*

Despite its faults, the "duelling scientist" mode of coverage soon became the norm, especially in the US. One study of seven leading US papers showed that a group of five top climate scientists from the US and Sweden were quoted in 33 articles in 1990 but in only five articles

You don't need a weatherman...

For a variety of reasons, most TV forecasters keep remarkably quiet about global warming. Obviously, with only a couple of minutes available on the daily weather segment, there's little time to explain the greenhouse effect or other global-warming science. What's striking is that TV producers so seldom turn to their resident weather experts for coverage that might tie global warming to local concerns and conditions. "The last thing any station wants is an activist weatherman", Matthew Felling, a US consultant, told the *Salon* website in 2006.

There are far more weathercasters in the expansive US television market than anywhere else – more than 500 of them. Perhaps half are trained meteorologists, but few of them produce science or environmental stories on top of their regular weather-reporting duties. The American Meteorological Society (AMS) has embarked on a campaign to train weathercasters as "staff scientists", with an eye towards giving them a higher profile in environmental coverage at their stations. The Clinton administration even organized a Washington summit for over a hundred weathercasters in 1997, complete with tutorials on climate science. Yet many forecasters remain dubious. After one CNN report in 2006 on rising temperatures, weathercaster Chad Myers mentioned his concern about how much the heat-island effect – which is already corrected for in the leading global analyses (p.176) – might be skewing the trend. (Myers did add, "I absolutely believe that CO_2 is heating the atmosphere").

Perhaps the most reliable spot for news about global warming in the TV-weather world is The Weather Channel, which reaches most American subscribers to cable TV. In 2003 the network hired Heidi Cullen as its first-ever climate expert. Cullen – a former research scientist – has since produced numerous segments on climate-change science and impacts. The network's position statement on climate change was also beefed up to acknowledge "strong evidence" that a good part of the current global warming is human-induced. There's been some resistance, though. When Cullen suggested in 2006 that TV weatherpeople who can't "speak to the fundamental science of climate change" shouldn't be certified by the AMS, her website received more than 1700 email protests.

The UK doesn't have an exact counterpart to Cullen, but the BBC has encountered its own challenges in covering climate change. In late 2007, the network cancelled a day-long special, *Planet Relief*, that would have focused on the topic. Comments from producers and commentators reflected a fierce internal struggle at BBC over where reporting stops and advocacy begins. "It's absolutely not the BBC's job to save the planet", said *Newsnight* editor Peter Barron, while writer and activist Mark Lynas told *The Independent* that climate change was now "too hot for the BBC to handle".

Heidi Cullen
The Weather Channel

Climate change and the church

Most environmental activists operate from a secular viewpoint, but that's not always the case. In the US, there's a small but growing faction of what one headline writer dubbed "earthy evangelists". They made the news in 2006, when nearly a hundred of them signed a statement in support of the fight against climate change. It was the first salvo in the Evangelical Climate Initiative, along with TV ads that include the tag line, "With God's help, we can stop global warming."

This was hardly the first faith-based action on global warming. Christian charity Tearfund is a founding member of the Stop Climate Chaos coalition, and environmental groups with Jewish and other religious ties have entered the fray over the years (many of them as part of the US National Religious Partnership for the Environment). But for those accustomed to thinking of US evangelicals as moving in lockstep with the nation's far-right wing, the 2006 statement was a startling move. It noted the scientific evidence for climate change and the risks it could pose to the world's poorest residents. The statement added, "Christian moral convictions demand our response to the climate change problem", and went on to endorse federal action to establish emissions cap-and-trade programs (p.292).

There are a lot of evangelical Christians in America – at least thirty million – and not all are on the same wavelength as the earthy evangelists. The activists have spun off from the National Association of Evangelicals, which declined to endorse their project. A rival group quickly sprang up – the Interfaith Stewardship Alliance – featuring some of the nation's best-known conservative Christians, including James Dobson and Charles Colson. They wrote their own statement, claiming "global warming is not a consensus issue", and their positions align much more closely with traditional sceptic fare. A paper by one of their founders, E. Calvin Beisner, went so far as to draw an analogy between coal and Jesus: "Vegetation is sown a natural body. Then, raised from the dead as coal and burned to enhance and safeguard our lives, it becomes a spiritual body – carbon dioxide gas – that gives life to vegetation and, through that, to every other living thing."

The earthy evangelicals and the conservatives come to their vastly different perspectives from a similar starting point. Both subscribe to the Biblical view of humans as stewards of Earth. Both express concern over the fate of Earth's poorest residents, especially in the developing world. But where the activists point to climate change as "the latest evidence of our failure to exercise proper stewardship", Beisner says that "a truly Biblical ethic of creation care simply cannot ignore the Biblical mandate for man to fill, subdue and rule the Earth."

The activists note that "millions of people could die in this century because of climate change, most of them our poorest global neighbours", while the rival group claims that the money presumably required to reduce greenhouse emissions could lift millions out of poverty – a point also made by Bjorn Lomborg (see p.256). Of course, this presumes that a giant pot of money is at hand, ready to be spent on either climate-change protection or poverty relief. In truth, of course, policymakers seldom put such big goals side by side in setting budgets. Indeed, it could end up being the risk to the world's biggest economies, rather than to its poorest people, that motivates real action on climate change.

in 1994. The use of sceptics changed little over the period, so that what were once minority opinions were soon getting roughly equal time.

"Conspicuous by its absence has been any sense of urgency in the British media ... the public has been left uninformed about a serious issue."

The British Medical Journal, 1996

In a more recent study for the journal *Global Environmental Change*, Maxwell Boykoff and Jules Boykoff detailed the sudden shift in US reporting styles at four major newspapers. Most articles in 1988 and 1989 focused on the evidence that human-induced global change is a real concern. But in 1990 these were eclipsed by "balanced" articles that gave more or less equal weight to natural variations and human factors as causes of climate change. This became the standard format through most of the 1990s, according to Boykoff and Boykoff. As they noted, "through the filter of balanced reporting, popular discourse has significantly diverged from the scientific discourse."

How did the sceptics gain such a high profile? One factor was the powerful public-relations machine funded by industry and facilitated by conservative think tanks already mentioned. Several campaigns were designed to highlight uncertainties in the science through news releases, press conferences and direct vehicles such as TV advertisements. In 1998, for example, just after the Kyoto Protocol was drafted, *The New York Times* learned of a proposal by an industry faction to spend $5 million campaigning against the treaty. "If we can show that science does not support the Kyoto treaty – which most true climate scientists believe to be the case – this puts the United States in a stronger moral position," noted the industry document, which called for "an action plan to inform the American public that science does not support the precipitous actions Kyoto would dictate." Victory will be achieved, it said, "when media 'understands' (recognizes) uncertainties in climate science."

Muzzled scientists

It wasn't only lobby groups and think tanks working to calm the public mind about climate change. According to many commentators, certain conservative governments were doing their best to stop – or at least slow – the research and views flowing from their own scientists to the public. In June 2005, *The New York Times* reported that a political appointee of US president George W. Bush (himself a former oil man) had reviewed several climate reports, including the annual overview "Our Changing Planet", with a heavy hand. For example, in the statement "The Earth is undergoing a period of rapid change," the term "is" had been changed to "may be". Within days the Bush appointee – who had earlier been a staffer with the American Petroleum Institute and a lobbyist against the Kyoto Protocol – left the White House and took a job with ExxonMobil. Rick Piltz, longtime editor of "Our Changing Planet", resigned from the US Climate Change Science Program in 2005 before bringing his story to the media. "I decided that continuing to sacrifice the ability to speak freely and publicly in order to attempt to limit damage and win minor victories on the inside was no longer the most appropriate thing for me to be doing", wrote Piltz. He's now running a website called Climate Science Watch that includes ample coverage of how climate-change science is being disseminated (or not).

Similarly, several researchers at NASA and the National Oceanic and Atmospheric Administration have claimed they were stymied by higher-ups in their attempts to speak with journalists on the latest findings in controversial areas, such as the effect of climate change on hurricanes. NASA's James Hansen told *The New York Times* and the BBC in 2006 that he was warned of "dire consequences" should he fail to clear all media requests for interviews with NASA headquarters. One of the political appointees charged with such clearance was a 24-year-old who resigned after it turned out he'd falsely claimed a degree in journalism on his resumé. Both NASA and NOAA reiterated their public commitment to openness and transparency after the stories broke.

Traditionally, US government scientists are allowed to speak their mind as private citizens; in their official capacities, they're allowed to discuss their research with the media but expected to refrain from commenting on federal policy. However, the rules vary widely by agency. The US National Science Board has called for a more uniform federal guideline to encourage open communication and prevent "the intentional or unintentional suppression or distortion of research findings". More broadly, a group of esteemed US scientists – including a dozen Nobel laureates – issued a statement in 2004 claiming that science was being ignored by US politicos. "The Earth system follows laws which scientists strive to understand," said F. Sherwood Rowland, who received the Nobel for his pioneering work on ozone. "The public deserves rational decision-making based on the best scientific advice about what is likely to happen, not what political entities might wish to happen."

The scientist's dilemma: speak out or keep quiet?

Among climate researchers, **Stephen Schneider** is the ultimate media veteran. He burst onto the scene early and dramatically while at the US National Center for Atmospheric Research. In 1976, only five years after completing his doctorate, Schneider co-wrote *The Genesis Strategy,* a book that stressed the need for society to prepare for intense climate shifts. Schneider had studied the powerful effects of polluting aerosols in the early 1970s, and when media interest in global cooling ran high, he was often quoted. But his book acknowledged the eventual risk of greenhouse warming as well as aerosol cooling. Schneider's quick way with words landed him on NBC's *Tonight Show* in 1977. In the early 1980s, Schneider was among the leaders of research into nuclear winter, again an offshoot of his early work on aerosols.

By the late 1970s, it was increasingly apparent that greenhouse warming was going to outweigh aerosol cooling in the long run. Like any good scientist, Schneider learned from the new evidence. In 1988 he wrote *Global Warming,* one of the first lay-oriented books on the topic, and shot to new fame as climate change took on a higher media profile. But Schneider soon learned how a single quote from one interview can have major repercussions. In discussing global change with *Discover* magazine in 1989, Schneider tried to explain how scientists like himself were often plunked into boxes: "climate change is a big problem" or "it's too uncertain to do anything about." He described the tightrope that scientists in the media must walk between conveying complexity and maintaining accessibility. Schneider told *Discover* that "each of us has to decide what the right balance is between being effective and being honest. I hope that means being both." That quote, yanked out of context – and often with only the first of the two sentences included – became prime ammunition for climate sceptics. Many of them have worked long and hard trying to discredit Schneider and his extensive body of research on climate risks and assessments.

Schneider, now at Stanford University, continues to believe in the importance of scientists speaking out through the media. But on a webpage that discusses "mediarology," Schneider recalls the incident above and warns, "This example illustrates the risks of stepping from the academic cloister to the wide world out there." He adds, "A scientist's likelihood of having her/his meaning turned on its head is pretty high – especially with highly politicized topics such as global warming."

Stephen Schneider
Courtesy Stephen Schneider

On the other side of the Pacific, a media storm erupted in 2006 with an investigative report from the Australia Broadcasting Corporation. According to the ABC report, at least three scientists at Australia's Commonwealth Scientific and Industrial Research Organization (CSIRO) were dissuaded from airing their views on climate policy in various reports. The CSIRO discourages its scientists from commenting on policy, as is typical in the US. However, at least one CSIRO scientist claimed that his comments on policy-relevant science – such as sea-level rise, which could affect immigration from Pacific islands to Australia – were getting quashed as well. Among the casualties in this turmoil was a short-lived communications director at CSIRO who had joined the science agency after serving as a spokesperson for Australia's tobacco industry.

Even the UK government, which is participating in the Kyoto Protocol, has found itself under the klieg lights of investigative reporters. Sir David King, the UK's chief science advisor, stated in January 2004 that "climate change is the most severe problem we are facing today – more serious even than the threat of terrorism." A few weeks later, US reporter Mike Martin discovered a floppy disk left in the press room at a Seattle science meeting attended by King. On the disk was a memo from Ivan Rogers, the chief private secretary to Tony Blair, asking King to avoid media interviews. The discovery led to an article by Martin in the journal *Science* and a follow-up from *The Independent* in London that was headlined, "Scientists 'gagged' by No 10 after warning of global warming threat." In the eyes of Martin – the only reporter who saw the actual wording – the memo was more of a suggestion than a gag order. But as US journalism professor George Kennedy opined, "It doesn't seem unreasonable to read a 'request' made by one's superior as an 'order' to be followed on pain of consequences."

The tide turns?

Despite the naysayers and lobby groups, media interest began climbing again in the late 1990s, with the run-up to the Kyoto Protocol and the record-setting 1997–98 El Niño. And coverage continued to increase into the new millennium. In her study of three UK newspapers – the *Times*, *Guardian* and *Independent* – communications scholar Anabela Carvalho found that the number of climate-change stories per year jumped from about fifty per paper in 1996 to around 200–300 each by 2001. Carvalho noted that the articles were increasingly laced with a sense of urgency and more talk of extreme events, including such contemporary happenings as England's severe flooding in 2000 and the European heat wave of 2003. By 2006, front-page climate stories were a frequent sight in Britain. The environment also shot up the political priority list, with the new leader of the opposition – David Cameron – aiming to entice voters to the Conservative Party with the slogan "Vote Blue, Go Green". By 2007 the three major parties were all jockeying to be seen as the most environmentally friendly, with the Liberal Democrats even promising to make the UK carbon neutral by 2050.

Such a shift was slower to arrive in the US, where contrarians have generally been highly visible in the editorials and opinion pages of right-wing papers such as *The Wall Street Journal*. Especially with op-ed pieces, the interwoven connections between sceptical authors and major corporations haven't always been apparent. For instance, *The Washington Post* ran an editorial in 2003 by James Schlesinger, a high-ranking official in several US administrations. In claiming that "the science isn't settled", Schlesinger stressed uncertainty and implied it was too soon to attribute global warming to human-produced greenhouse gases. What the article and byline didn't point out, as environmental writer Chris Mooney has noted, is that Schlesinger was sitting on the board of Peabody Energy, the world's largest coal company.

Still, things have changed in a big way in the US media over the last several years. One of the first concrete signs was a front-page story in *USA TODAY* on June 13, 2005, entitled, "The debate's over: Globe is warming." In the sceptic-friendly US press, this was a banner development. As the article proclaimed, "… the tides of change appear to be moving on." This developed further in 2006, when Democrats took control of both houses of the US Congress, a shift that pushed sceptical voices on climate toward the margins of American politics.

Climate change also started featuring more frequently on those most mainstream of media outlets: TV and film. On TV, Alanis Morrissette gave viewers in the UK, US and

Climate change and the cinema

"Where are the books? The poems? The plays? The goddamn operas?" asks Bill McKibben. In a 2005 essay for the openDemocracy website, the US writer vented his frustration at the fact that global warming hasn't generated the outpouring of creative expression that other issues of key importance to society have in the past.

It's true that theatres aren't exactly packed with films about climate change, but the topic has made a few appearances on screen, both major and minor. Weather, of course, is a perennial mood-setter in films, and sometimes it's a lead character. Filmmakers have used the backdrop of particular climates to great effect ever since the silent-film days. But the abstraction of climate *change* – the evolution of weather over time – is a much trickier concept to put into cinematic terms. Plus, as McKibben points out, the instigators of global warming are typically far removed from the consequences of their actions. In other words, you can't exactly resolve the plot with a thrilling chase.

Some of the first films to address the dystopian prospects of human-induced climate change were in the realm of science fiction. *Zombies of the Stratosphere* (1952) included a young Leonard Nimoy as part of a gang of Martians bent on exploding Earth from its orbit so that Mars can move sunward and benefit from a milder climate. The James Bond spoof *In Like Flint* (1965) featured hero Derek Flint, played by James Coburn, facing off against a sinister organization called Galaxy that plans to flood valleys and send icebergs into the Mediterranean. Then there's *Soylent Green* (1973), the first of the lot to present a scenario based in part on global-warming science. It's set in an overcrowded New York City in the year 2022, with pollution and other environmental ills run amok. ("A heat wave all year long," grouses Charlton Heston. "Greenhouse effect," replies his partner, played by Edward G. Robinson.) With food scarce, residents turn to the concoction that provides the film's name – ostensibly a blend of soya and lentils ("soylent"), but actually something far more gruesome.

Waterworld (1995), with Kevin Costner as co-producer and star, was the first major US film to draw on modern concerns about global warming. The most expensive movie ever made up to that point, at $175 million, it was considered a commercial and critical flop, even though it ended up making a profit overseas. In *Waterworld*, hardscrabble camps of people are left to fight and adapt long after the world has been entirely flooded by the melting of icecaps. (Such a sea-level rise goes far beyond anything possible in the real world, even if every inch of Antarctic and Arctic ice melted.) Global warming isn't so much the topic of *Waterworld* as it is the backdrop for a more conventional action story.

Climate change crept into other films of the period as well. In *The Arrival* (1996), aliens intent on occupying Earth pollute the atmosphere in order to boost the greenhouse effect and melt polar ice. *The March* (1990) depicted Africans migrating to a Europe stressed by global warming, and *Prey* (1997) featured a bioanthropologist who fears that a new über-species, nurtured by global warming, wants to kill off Homo sapiens.

"Great movie ... lousy science."

George Monbiot, writing about *The Day after Tomorrow*

Stephen Spielberg set his futuristic *A.I. Artificial Intelligence* (2001) in a world where New York and other coastal cities have been abandoned due to rising sea levels. High water also figures into *Split Second* (1992), set in London circa 2008. Global warming coupled with "forty days and nights of torrential rain" (as noted in the prologue) has pushed the ocean into the city streets, where a policeman played by Rutger Hauer – who lives on "anxiety, coffee and chocolate" – hunts for a mysterious, lethal creature.

It took until 2004 for Hollywood to produce a big-budget extravaganza with climate change at its centre. The hero of the *The Day after Tomorrow* is Jack Hall (Dennis Quaid), a paleoclimatologist who discovers that global warming has triggered a shutdown of North Atlantic currents. Within days, the Northern Hemisphere is plunged into an ice-age-scale deep freeze, thanks to cold air descending from the stratosphere (and disobeying the laws of physics, which dictate that the descending air would become warmer than the surface air it replaced). Before long, Hall is tramping through snowfields up the East Coast to rescue his son from the icebound New York Public Library.

Well before its release, *The Day after Tomorrow* got the attention of people on both sides of the global-warming debate. The activist group MoveOn.org distributed fliers to people outside theatres showing the film, while sceptics bashed it as a propaganda tool. But whatever potential the movie had to sway real-world debate was compromised by its absurdly telescoped view of how fast climate change might unfold. Although the Atlantic currents that help keep Europe warm for its latitude have shut down before, and they could diminish in the future (p.119), the process should take years to decades to unfold, rather than mere days. Still, many saw the film's success as a sign of growing public interest in climate change.

Al Gore became an unlikely movie star in 2006 with the release of *An Inconvenient Truth*. The Oscar-winning documentary is a glossed-up version of Gore's stump speech, which he claims to have presented over a thousand times. *An Inconvenient Truth* rode the wave of two big trends – interest in global warming and a surge of hugely popular US documentaries that includes Michael Moore's *Fahrenheit 9/11*. Outside of a few nuggets on Gore's life story, the film is what it claims to be: the story of global warming, told in a style that's sober and thoughtful, yet visually rich and emotionally compelling, enhanced by Gore's 30-year storehouse of knowledge on the topic. Expanding on Gore's thesis, *The 11th Hour*, a powerful 2007 documentary narrated by Leonardo DiCaprio, linked climate change with other environmental woes.

Global warming even penetrated the children's film market in 2006 with *Ice Age: The Meltdown*. This sequel to the 2002 animated feature *Ice Age* – featuring the tag line "the chill is gone" – follows a batch of prehistoric animals in a newly warmed world, with species going extinct and floods threatening. Though set in the great meltdown after the last ice age, the film subtly tips its hat to modern-day worries.

Along with serving as a plot device, global climate change may be working its way into the process of movie-making itself. The crew that headed for Calgary, Canada, in late winter to film the comedy *Snow Day* (2000) encountered an unusually mild and dry March. It took 450 truckloads of snow to save the day.

Canada an overview of the "signs and science" of global warming. A few months later, BBC Television announced a series of sixteen programmes on climate impacts, themed to the massive Stop Climate Chaos campaign organized by a group of British non-governmental organizations. Soon after, Al Gore's long-time mission to raise awareness on global warming suddenly went mainstream with the launch of *An Inconvenient Truth*, a feature-length documentary film (and accompanying book) which smashed per-screen box-office records on its launch in 2006. Magazines

The New Great Game – who owns the Arctic?

As the world reaches a near-consensus over the existence of climate change, more of the debate focuses on who will pay the costs – or reap the benefits – of the warming atmosphere. One example is playing out in the Arctic.

Throughout the 1800s, Britain and Russia vied for power in Central Asia in a century-long contest dubbed the Great Game. The thawing Arctic has been described as the "New Great Game", with the erosion of the region's summer-time ice pack opening the door to high-latitude wheeling and dealing. Later this century, shipping routes that now link Europe, Asia and North America through the Arctic could be open months longer than they are now, and huge vaults of oil and gas that are believed to lie beneath the sea may become accessible for the first time.

The five countries with Arctic coastlines – Canada, Denmark (through Greenland), Russia, Sweden and the US (via Alaska) – long paid little mind to each other's business in a region seemingly devoid of usefulness. Intersecting borders across the Arctic were once drawn informally or not at all. Now they're a topic of intense debate, as countries angle for the best shipping routes and undersea resources.

The competition heated up in 2007, as did the Arctic itself. Russia planted a flag on the seafloor beneath the North Pole, which it claims is an extension of the Lomonosov ridge running northward from Russia's Arctic coast. Canada and Denmark, meanwhile, are attempting to claim the ridge as their own. These and other border disputes will be resolved through the Law of the Sea, the United Nations convention ratified by more than 150 nations, including all of those bordering the Arctic except the United States. (Conservatives long blocked US ratification of the Law of the Sea, but bipartisan support is building and the US may end up approving the treaty after all.) The Law of the Sea dictates that each nation has an "exclusive economic zone" – including the right to drill for oil and gas – extending 200 nautical miles (about 370km) poleward from its Arctic coastline. That leaves a sizeable area under contention, from near the North Pole westward toward the Siberian coast. As it happens, the unprecedented melt of summer-time ice in the Arctic has opened much of this area, with prevailing currents pushing much of the remaining ice toward Greenland.

also took note, with titles from *Time* and *Time Out* to *Vanity Fair* releasing special environmental editions. By 2007, climate change was a prominent presence in most media – a situation bolstered in October, when the IPCC and Al Gore were co-awarded the Nobel Peace Prize.

It remains to be seen whether this media focus will be sustained in the years ahead. If it is, it may just generate sufficient political will and public interest to engender the potential political, technological and personal solutions detailed in the following chapters.

Entrepreneurs and companies are also getting involved in the New Great Game. For example, Pat Bode of Denver snapped up the port of Churchill, Manitoba, from the Canadian government for a token $7 US in 1998, as part of an $11 million purchase of train lines running into Canada's heartland. Though it's only fitfully busy now, Churchill's port – which lies on the west side of Hudson's Bay – could become a major transport hub if climate change thaws the region as much as projected.

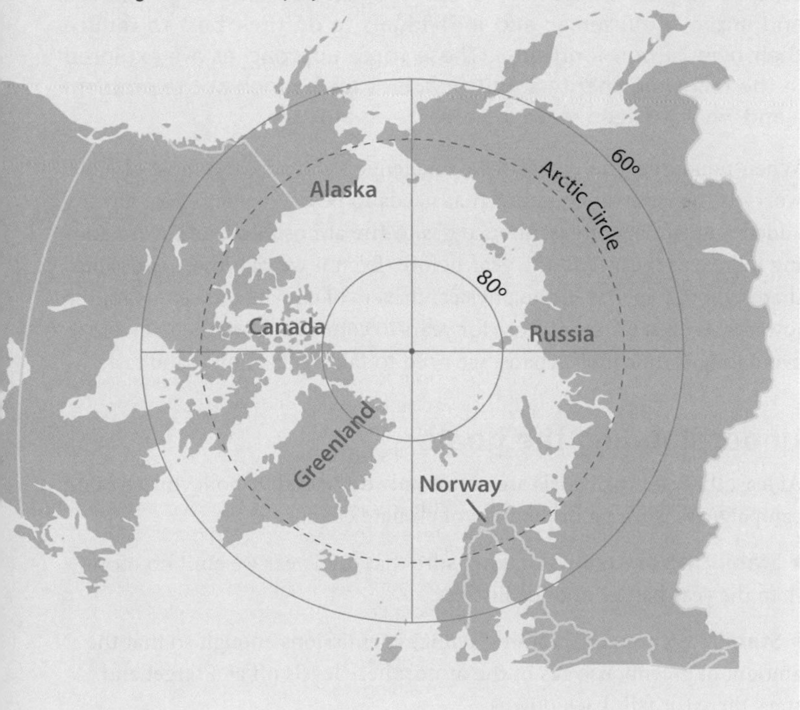

The predicament

Can we solve global warming?

The global-warming problem isn't going to be solved tomorrow, next week, or next year: we're in this one for the long haul. And there clearly isn't one single solution. We need governments to agree and enforce targets; innovators to develop low-carbon energy sources and improve efficiency; and individuals to do their best to reduce their own carbon footprints. These three approaches are explored in the following chapters. But first, let's take a look at the problem – and what we need to do to solve it – as a whole.

When pondering the global-scale challenge before us, the most obvious worry is the sheer momentum that needs to be overcome. We've already added a great deal of greenhouse gas to the atmosphere, and we're adding more each year than the year before. Even if we can lower emissions, there's enough inertia in the physical drivers of climate to keep us rolling towards an even warmer future for years to come. In short, then, if we're to avoid long-term climate chaos, we need to take real action – and fast.

Understanding the goals

At least three types of goals are commonly discussed by policymakers and campaigners working in the field of climate change:

▶ **Stabilizing emissions** Making sure that each year we emit no more than the year before, and ideally less.

▶ **Stabilizing concentrations** Reducing emissions enough so that the amount of greenhouse gas in the atmosphere levels off at a target and stays there (or falls back down).

▶ **Stabilizing temperature** Keeping the atmosphere from warming beyond a certain point.

Obviously, these three types of goals overlap: our emissions build up in the atmosphere, changing the concentration of greenhouse gases, which in turn alters the temperature. However, the relationship between emissions, concentrations and temperature is complicated by a few factors.

First, as discussed in the Circuits of Change chapter (see p.227), there's still some uncertainty about exactly how much hotter Earth will get as greenhouse-gas concentrations rise – the so-called **sensitivity** of the climate.

Second, since some greenhouse gases persist in the atmosphere for years or even centuries, simply levelling off emissions won't be enough to stabilize concentration – and therefore temperature. After all, if water's flowing into a bathtub faster than it can drain out, you need to *reduce* the flow – not just keep it constant. Likewise, though stabilizing global emissions would be an enormous short-term achievement, we need to go much further and make significant *cuts* in emissions in order to keep the world from getting hotter. As we'll see later, some countries have already managed to stabilize their emissions, but globally we haven't come close.

Third, there's the **time-lag** factor to consider. Even if we *do* manage to make significant emissions cuts, it will take decades for the concentration to level off, and many decades more for the temperature to stop rising (see graphs).

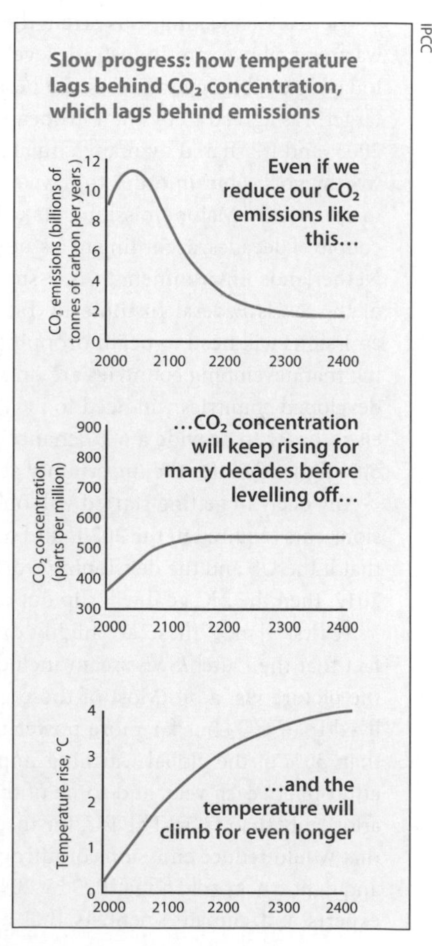

Selecting a target

Perhaps the ultimate question when it comes to defining goals for tackling climate change is this: how much global warming would be truly dangerous? That topic was examined in depth at a 2005 conference hosted by the UK Met Office in Exeter. The preliminary findings were that a rise of 3°C (5.4°C) relative to pre-industrial levels may be well past the edge of the comfort zone. It may be enough to trigger unstoppable melting of the Greenland ice sheet, for example, and any further warming could jeopardize parts of the Antarctic ice sheet or flip land areas from being carbon sinks to carbon sources.

The Exeter meeting thus strengthened a consensus that a **2°C** (3.6°F) warming above pre-industrial levels, or around 1.2°C (2.1°F) above today's global temperature, is the best goal for climate stabilization. This target was agreed to by the European Union in 1996 and reconfirmed in 2005, and it's shared by many climate scientists and activist groups. It's a worthy goal – but in order to have any chance of getting to it, we'll have to act quickly. Major emission cuts will need to take place within the next couple of decades, according to a study in 2005 by Michel den Elzen of the Netherlands Environmental Assessment Agency and Malte Meinshausen of the Swiss Federal Institute of Technology. They estimate that global emissions will need to begin dropping by the year 2015. By 2020, assuming that developing countries are on a slower reduction track, *à la* Kyoto, developed countries will need to have emissions 15–30% below 1990 levels in order to provide a 60% chance that the 2°C goal will be met. (The 60% figure denotes the uncertainty about the climate's sensitivity.)

Any delay in getting started will only add to the sharpness of the emissions cuts required in the 2020s and beyond. Interestingly, the team found that if the US and the developing world both continue to emit freely after 2012, then the 2°C goal will slip out of reach by 2030.

At first glance, these are highly discouraging numbers. However, the fact that the Dutch/Swiss team included gases other than CO_2 brightens the picture just a bit. Most of these gases are less prevalent and shorter-lived than CO_2 but far more powerful. Together, they account for more than 30% of the global-warming impact from gases we're adding to the atmosphere each year, and some of them can be reduced more easily and affordably than CO_2. The EU, for instance, introduced new rules in 2006 that would reduce emissions of nitrogen oxides from diesel-powered cars and light trucks sold after 2009 by 20%. There is some hope among policy experts and climate scientists that a two-step approach – focusing on non-CO_2 greenhouse gases right away and CO_2 in the long run – might prove fruitful.

2°C: can it be done?

Whether the 2°C goal can be met depends on various factors, such as how fast **new technology** is developed and adopted, and how seriously we take **energy efficiency**, which in turn depends partly on the political will to prioritize climate change. Another important factor is the state of the world **economy**. Global emissions of CO_2 from fossil fuels actually fell more than 4% between 1980 and 1983, a period of high oil prices and widespread recession, and they dropped by about 2% in 1992 and 1999, when the economies of eastern Europe and Russia were struggling.

Naturally, all these factors are interrelated in complex ways. For example, high oil prices can simultaneously dampen economic growth, encourage efficiency and stimulate investment in alternative energy sources – all of which are likely to reduce emissions. But if the economy suffers too much, governments may feel pressure from voters to prioritize short-term growth over long-term environmental issues.

Despite these complications, it seems at least theoretically possible that we could manage global emission cuts of at least 10% within a decade or two. And, though doing so could cause some short-term fiscal pain, the long-term gains from energy efficiency and renewable energy are likely to be enormous – economically as well as environmentally.

The wedge strategy

Since almost all human activity contributes to greenhouse gas emissions on some level, the task of reducing global emissions can seem somewhat overwhelming. What if we thought of it as a series of simultaneous smaller goals? That's the philosophy behind Stephen Pacala and Robert Socolow's "wedge" approach to climate protection. The two Princeton University scientists brought the wedge concept to a wide audience through a 2004 article in *Science,* and it's since generated a fair bit of excitement. The idea is to break down the enormous carbon reductions needed by mid-century into more manageable bits, or wedges, each of which can be assigned to a particular technology or strategy.

The wedge concept originates from Pacala and Socolow's projection of historical CO_2 emissions into the future (see graphic). Let's assume that emissions could be instantly stabilized – in other words, the yearly

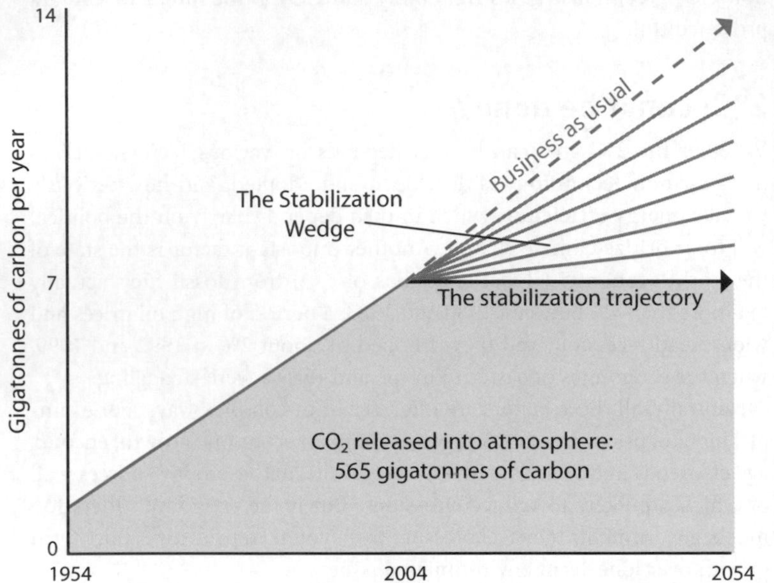

increases in CO_2 all went to zero right now – and remained that way until at least the 2050s. This is represented by the flat, black line on the diagram. That would correspond to an eventual CO_2 concentration of about 500 parts per million and, under the best assumptions of climate sensitivity, a global temperature rise somewhere in the ballpark of 2°C.

However, if emissions continue to increase as they have in the last several decades – at more than 1% per year – then we'll be adding twice as much CO_2 to the atmosphere in 2054 as we are doing now (see the red "business as usual" line on the diagram). The result will be a warming far more severe than 2°C.

The triangle between the black and red lines shows the difference between the desired path of steady emissions and the dangerous uphill path. To get rid of the triangle, we'd need to come up with at least seven wedges, each of which would eliminate a billion tonnes (a gigatonne) of annual carbon emission by 2054. (Further wedges would probably be needed after 2054 to stay below the 2°C target.)

According to Pacala and Socolow, we're almost spoilt for choice on wedges that we can implement promptly. They identify fifteen examples, each of which is already being implemented on a reasonably large scale somewhere in the world. "Scaling up from present capacity to an entire wedge would be challenging but plausible in each of the fifteen cases", says Socolow. Here, then, are the fifteen possibilities. All comparisons below are expressed in terms of the year 2054 versus 2004.

Efficiency and transport

Doubling vehicle fuel economy and **cutting the distance driven per car in half** (one wedge each).

Installing lights and appliances with state-of-the-art energy efficiency in all new and existing residential and commercial buildings, thus reducing emissions by about 25%. This would actually provide two wedges, but business-as-usual scenarios already account for one of those wedges since they assume that recent gradual improvements in energy efficiency will continue. Simply installing compact fluorescent bulbs in all of the world's fifty billion light fixtures would provide a third of a wedge in itself.

Improving the efficiency of coal-fired power plants from 40% to 60%, plus cutting in half the energy lost when fossil fuels are extracted, processed and delivered to those plants.

Renewable energy

A 50-fold expansion in wind energy, or adding about two million wind turbines of the standard one-megawatt size (there are now about 30,000 turbines worldwide), replacing coal. Socolow estimates these turbines would cover an area about the size of Germany, but the land beneath them could be used in other ways (such as grazing or farming – or shipping, for offshore turbines). To hit the fifty-fold goal, wind energy would have to increase an average of only 8% per year, compared to its recent pace of 30% per year. A second wedge could be obtained by adding another four million turbines to generate hydrogen for fuel-cell vehicles (assuming the infrastructure for delivering the hydrogen was in place – see p.324).

A 700-fold expansion of photovoltaic (PV) solar energy, again replacing coal. This would require enough panels to blanket every inch of an area the size of New Jersey. However, many of the panels could be mounted on roofs and walls. To reach this goal, PV solar would have to expand by about 14% per year; it's now growing at about 30% annually.

A 50-fold expansion in ethanol, displacing gasoline. Even with grasses and other plants that yield far more energy per acre than the corn and sugar cane used now for ethanol, these fuel crops would still take far more land than the wind or solar options: a region about the size of India, or more than 15% of all the land on Earth that's now used for agriculture.

A halting of current deforestation, coupled with plantations on non-forested land, together covering a total area about 40% the size of Australia by 2050 (this would require a doubling of the current rate of reforestation).

Employing conservation tillage on all cropland. This is the practice in which farmers avoid ploughing and thus reduce the amount of CO_2 escaping from tilled soil. It's now used on less than 10% of cropland globally.

Replacing coal and reducing coal emissions

Tripling the amount of energy now generated by nuclear sources, adding about 700 one-gigawatt plants as well as maintaining or replacing all nuclear plants now in use.

Quadrupling the use of natural gas in power plants, replacing an equal number of coal-fired plants.

Sequestration – capturing carbon emitted by large fossil-fuel power plants and storing it underground. Socolow believes that up to three billion tonnes of carbon could be sequestered per year by 2054, providing three wedges, one each from standard coal or natural gas plants, synfuel plants (which generate synthetic fuel from coal), and hydrogen plants that draw on coal or natural gas.

It's obvious that some of these potential wedges are far more practical and/or politically palatable than others. And if the history of energy development is any guide, the economies of scale tend to push each sector towards a strongly preferred fuel type – gasoline for vehicles, coal for electric power plants, and so on. This would work against, say, the parallel large-scale growth of electricity generation from wind power, nuclear and cleaner types of coal.

There's also the obvious fact that we can't stabilize emissions today or tomorrow as if we were flipping a switch. Even using current technologies, such massive change would take years to implement. This is why many scenarios constructed by the IPCC and others include a ramp-up phase of continued growth, followed by substantial cuts in emissions – ultimately bringing us well below today's emission levels – in contrast to the flat-line "stabilization trajectory" shown on the preceding graph.

The beauty of the wedge concept, though, is that each wedge is technically feasible – it's simply a matter of society choosing which ones to emphasize. Government subsidies or other incentives could shape this course, and it's possible that different countries will opt for different technologies, thus providing a well-rounded portfolio on the global scale. The wedges also allow one to work backwards and set even shorter-term goals, taking into account that some wedges might grow more rapidly at first while others would take a while to get going. As Socolow puts it, the wedges "decompose a heroic challenge into a limited set of monumental tasks."

We'll be considering the above options in the Technological Solutions and What You Can Do chapters. But first, let's explore what governments are doing – and not doing – to speed us along the path to a low-carbon future.

Political solutions

Kyoto and beyond

Climate change must be dealt with on a global level – and this poses an unparalleled challenge in international diplomacy. For one thing, a whole host of intractable issues, from excessive consumption to South-North inequity, tend to glom onto climate change, as if the latter weren't a daunting enough problem in itself. However, the United Nations has made some headway in the two decades since it began addressing climate change. Nearly all countries are on record as taking the problem seriously, and the UN's 1997 Kyoto Protocol – though weaker than many had hoped – was ratified by countries that together represent the majority of people in the developed world (though only a third of global emissions).

Kyoto isn't the only game in town. Some countries that opted out of Kyoto are exploring other mechanisms for grappling with the carbon problem. For instance, many of the most populous US states and cities are volunteering for reductions that match or exceed Kyoto goals. And a whole smorgasbord of new ideas will be on the table over the next few years as diplomats hash out a global agreement to extend or replace Kyoto after 2012.

The road to Kyoto

In some ways, global warming couldn't have arrived on the diplomatic radar screen at a better point than in 1988. A year earlier the UN had forged the **Montreal Protocol**, which called for the gradual phase-out of the industrial chemicals (chlorofluorocarbons) implicated in the Antarctic ozone hole (an issue separate from global warming, as explained on p.28). The Montreal Protocol came about only after a thorough assessment of the science behind ozone depletion. Previous work had already shown the connection to chlorofluorocarbons, but it was the assessment that gave an international imprimatur to the findings. It showed that researchers from around the world agreed with the science, which in turn provided the green light for global action.

Such success made it natural for the UN to ask for another high-level scientific assessment when global climate change reared its head. In response, the World Meteorological Organization and the UN Environment Programme established a group called the **Intergovernmental Panel on Climate Change** (IPCC). Few if any people at the time could have guessed how influential the IPCC would turn out to be. Its findings have been featured in countless news stories and cited in thousands of reports. The IPCC's four major assessments – released in 1990, 1995, 2001 and 2007 – have become the definitive reference tools on the state of climate science. See overleaf for more about how the IPCC works. Several nations have taken a cue from the IPCC and produced their own comprehensive assessments; several of these are listed in Resources (see p.364).

The IPCC's first assessment in 1990 underscored the seriousness of the climate-change threat, so there was plenty of momentum driving world leaders to the 1992 UN **Earth Summit** in Rio de Janeiro. Global warming was only one of a host of environmental issues on the table at this Woodstock of green politics, but it played a starring role. An agreement called the UN Framework Convention on Climate Change (**UNFCCC**) was introduced at Rio and signed by 166 nations that summer (and a total of 189 to date). Because the US and several other countries were steadfast in their opposition to binding emissions reductions, the UNFCCC doesn't include any targets or timelines, much to the dismay of climate activists. Yet it was, and is still, a landmark agreement. Among its key points:

▶ **The ultimate objective** is to stabilize climate "at a level that would prevent dangerous anthropogenic interference with the climate system" and in a time frame "sufficient to allow ecosystems to adapt naturally to climate change, to ensure that food production is not threatened and to enable economic development to proceed in a sustainable manner."

▶ **International equity** is given a nod with a reference to "common but differentiated responsibilities and respective capabilities", a theme that ended up in the Kyoto Protocol.

▶ **Uncertainty in climate science** – one of the favourite criticisms from sceptics – was acknowledged but put in its place: "Where there are threats of serious or irreversible damage, lack of full scientific certainty should not be used as a reason for postponing such measures."

▶ **Countries agreed to inventory their greenhouse emissions** and publish yearly summaries.

With the UNFCCC in place, the world's industrialized countries plus "economies in transition" (mostly in Central and Eastern Europe) – together referred to as **Annex I** – agreed to reduce their emissions to 1990 levels by the year 2000. This was fairly easy for Russia and the former Soviet republics due to their lacklustre economic performance and the resulting drop in energy use, but many other countries missed the goal. Britain managed it only because of the phase-out of subsidies for coal-fired power plants and a resulting shift to natural gas. As a whole, the Annex I emissions were 6% above 1990 levels by 2002.

Inside the IPCC

The IPCC has only a few permanent staff, but it's a far larger enterprise than the term "panel" might suggest. Indeed, it's one of the biggest science-related endeavours in history. That said, the IPCC doesn't conduct any science of its own. Its role is to evaluate studies carried out by thousands of researchers around the world, then to synthesize the results in a form that helps policymakers decide how to respond to climate change. Each assessment is a bit different, but typically each of several IPCC working groups generates an exhaustive report as well as a summary for policymakers. All of the reports are available online (see p.364). For their 2001 and 2007 reports, the three working groups have dealt with: **the basis in physical science** (how climate change works), **impacts, adaptation and vulnerability** (options for dealing with climate change), and **mitigation** (options for minimizing it).

Every IPCC assessment involves around 100–200 researchers from many dozens of countries, generally nominated by their governments or by a non-governmental organization. Each working group is headed by a pair of scientists, one each from a developed and a developing country. By and large, these scientists volunteer their time to be involved with the IPCC with the blessing of their employers. They survey peer-reviewed science studies and other pertinent materials; meet with peers to gather input as needed; and draft, revise and finalize reports. Several hundred other experts then review each report. Finally, each document is scrutinized by technical reviewers within each government and accepted at a plenary meeting. The policymaker summaries take shape on a parallel track; they're approved by a panel of governments on a word-by-word basis.

With so much riding on its conclusions, it's no wonder that the IPCC has drawn scrutiny from those aiming to discount the risk of climate change. The second assessment provoked an attack from the Global Climate Coalition (p.255) and other sceptics, who criticized the process by which the wording of one of the

It was already apparent by the mid-1990s that some type of **mandatory reductions** were needed to make real progress on emissions. Setting targets, though, was far easier said than done. At a 1995 meeting in Berlin, it was agreed that industrialized countries (the ones that had caused the lion's share of the greenhouse problem thus far) would bear the full brunt of the first round of agreed-upon emissions cuts. This was in keeping with the principles of the UNFCCC. The idea of mandated cuts eventually gained support from US president Bill Clinton (albeit weaker cuts than those promoted by the EU). But in 1997 the US Congress voted 95-0 against any treaty that did not specify "meaningful" emission cuts for developing as well as developed countries.

summaries for policymakers was prepared (including the fateful statement "the balance of evidence suggests that there is a discernible human influence on global climate"). The process was revised for the next assessment in response to this and other concerns, although no IPCC rules had been violated and the panel's earlier conclusions were unchanged. After the third assessment came out, sceptics' focus turned to its "hockey stick" graphic depicting climate over the last thousand years, including the sharp upturn of the twentieth century (the "head" of the stick – see p.226). Controversy aside, the IPCC's pronouncements on the state of global climate resonate worldwide. As noted earlier in this book, the IPCC has grown increasingly emphatic about its conclusions on human-induced climate change.

Putting together an IPCC report is unlike any other job in the science world. Susan Solomon, the NOAA scientist who unravelled the role of polar stratospheric clouds in creating the ozone hole (p.28), served as co-chair of Working Group I for the 2007 assessment. "It's a very intense activity", says Solomon, who estimates she went through more than 17,000 comments from more than 500 reviewers on the first draft of her group's report. "I've learned a tremendous amount about climate, but it is demanding, both personally and professionally." More than anything, she stresses the community aspect of the panel: "It's very important for people to understand that the IPCC is not one scientist's voice."

Susan Solomon with a draft of the 2007 IPCC assessment
Bob Henson

"I told the world I thought that Kyoto was a lousy deal for America."

US president George W. Bush, March 2006

The standoff roiled the diplomatic world right up to the fateful meeting in Kyoto, Japan, late in 1997, where targets would be set. In the end – after virtually non-stop negotiations and some last-minute assistance from US vice president Al Gore – the parties agreed to a Kyoto Protocol that would exempt developing countries and produce an average 5.2% cut among Annex I countries by 2008–2012, compared to 1990 levels.

The US and Australia signed the treaty, but neither country went on to ratify it. However, the Kyoto Protocol was crafted so that it could become international law even without US participation – but just barely. The rules required ratification by at least 55 Annex I countries that together account for at least 55% of the total Annex I emissions in 1990. Thus, if every other major developed nation apart from the US and Australia ratified the protocol, it could still take effect. Most of the industrialized world quickly came on board, but it wasn't clear whether or not Russia would. Years of suspense passed before Russia, after stating its opposition to Kyoto in 2003, ratified the protocol in late 2004. The protocol became international law ninety days later, on 16 February 2005.

How Kyoto works

The meat of the Kyoto Protocol (which is technically an amendment to the UNFCCC) is the requirement for developed nations to cut their emissions of six greenhouse gases: **CO_2**, **nitrogen oxides**, **methane**, **sulphur hexafluoride**, **hydrofluorocarbons** and **perfluorocarbons**. Different countries have different reductions targets and, for various reasons, some countries are actually allowed an *increase* in their emissions over the period:

- **Bulgaria, Czech Republic, Estonia, Latvia, Liechtenstein, Lithuania, Monaco, Romania, Slovakia, Slovenia, Switzerland:** 8% reduction
- **US:** 7% reduction
- **Canada, Hungary, Japan, Poland:** 6% reduction
- **Croatia:** 5% reduction
- **New Zealand, Russia, Ukraine:** no change
- **Norway:** 1% increase
- **Australia:** 8% increase
- **Iceland:** 10% increase
- **All other countries:** no mandated targets

These targets apply to emissions in the period 2008–2012 – the "commitment period" – and are expressed in CO_2 equivalent, which takes into account the varying power of each of the six gases (see p.33). With a few exceptions, the targets are relative to 1990. Those for the US and Australia are hypothetical, since they didn't ratify the treaty.

As you might guess, the targets that emerged at Kyoto are political creatures. Australia is a good case in point: how did this prosperous country win the right to increase its emissions? As in the US, Australia had come up with voluntary reduction plans in the 1990s, hoping to avoid mandated cuts, but these failed to do the trick. Its negotiators then argued that Australia was a special case because it was a net energy producer with a growing population and a huge, transport-dependent land area. They also pointed to reductions in Australian deforestation after 1990. Kyoto diplomats were keen to get all of the industrialized world on board, so in the end they acceded to Australian wishes, but the country still ultimately opted out of the treaty.

Another wrinkle is the so-called "EU bubble". The EU was assigned an 8% target but was allowed the freedom to reallocate targets country by country within its overall goal, given that some EU economies would find the 8% limit much easier to meet than others. The resulting targets range from reductions as big as 29% for Denmark and Germany to increases as high as 19% and 20% for Portugal and Luxembourg, respectively.

The emissions market

Because Kyoto didn't become international law until 2005, many countries are now playing catch-up in order to meet their targets. They're assisted by several market-based mechanisms woven into the protocol, designed to help countries meet their targets at the lowest possible cost. The basic idea is that countries or companies finding it tough to cut their emissions directly can meet their targets by purchasing reductions achieved more cheaply elsewhere. A secondary goal is to promote the export of clean technologies to emerging economies. The mechanisms include the following:

▶ **Emissions trading** is just what its name implies. Annex I countries that exceed their emissions targets can buy allowances from another Annex I country that's doing better than its goal. This idea was inspired by the successful US drive to reduce sulphur dioxide pollution by allowing corporations to buy and sell emission credits from each other under a federally imposed cap on total emissions. Such systems are often labelled "cap and trade" because they operate by capping the total emissions from a group of companies or countries (which assures the overall target will be met) and by allowing the various participants to trade emission credits (which helps prioritize the lowest-cost reductions strategies). Ironically, it was the US that lobbied hardest for emissions trading in the early days of Kyoto negotiations, while Europe balked. Today, the US is out of the Kyoto picture – at least for the initial period – while the EU is not only participating in the Kyoto trading scheme, which began in 2008, but has also created an internal trading system of its own (p.291).

▶ **The Clean Development Mechanism (CDM)**, first proposed by Brazil, allows developed countries to get credit for bankrolling projects such as reforestation or wind farms that reduce emissions in developing countries. The scheme was slow to get off the ground, with many investors waiting to see if Kyoto would become official. By mid-2007, though, more than 700 CDM projects had been approved, with the majority of the emissions savings taking place in Brazil, India and China. Together, these projects are expected to keep more than two billion tonnes of CO_2 equivalent out of the atmosphere by the end of 2012. That's a substantial amount – equal to about 6% of the CO_2 emitted globally per year. However, the CDM is not without its detractors, some of whom object on philosophical and environmental grounds (see box opposite), while others are critical of the methodology. Due to a loophole that was present in the CDM until 2007, more than £3 billion was allocated for reduced emissions of triflouromethane (HFC-23). That's more than ten times the actual cost needed to remove HFCs at the factory, according to US researcher Michael Ware, who argued in a 2007 *Nature* article that CDMs would be far more effective if they were devoted purely to CO_2 than to all six Kyoto gases.

▶ **Joint Implementation projects** allow an Annex I country to earn emission credits by subsidizing an emission-reduction project in another Annex I country. As with the CDM, these projects officially began in 2005 and are only now gathering steam, with thirty countries involved as of late 2007. The favoured location is the former Eastern Bloc, where there are plenty of inefficient technologies ripe for upgrading.

Right now it looks as if some Annex I countries will meet their Kyoto goals with relative ease, while others will almost certainly fall short. The

The pros and cons of clean development

The Clean Development Mechanism seems like a win-win part of the Kyoto Protocol. It allows nations that come up against their emissions targets to pour money into emission-reducing projects in developing countries. But not everyone likes the CDM. Some oppose it on principle, claiming it's simply a tool that legitimizes the polluting legacy of Annex I nations by allowing them to buy their way out of any commitment to actually reduce emissions. Others are more concerned with how the CDM is implemented. The parties who arrange a CDM project have to demonstrate that it produces a reduction in developing-country emissions that wouldn't have happened otherwise. But many of the initial projects got started before they came under the CDM umbrella, thus complicating the task of deciding what would have happened without the CDM.

Some campaign groups, including the WWF, have lobbied to see certain projects ruled off-limits for CDM, such as those involving coal, large-scale hydropower and forest-based carbon sinks. A particular sore spot is the Plantar project in southeast Brazil, funded by the World Bank. Grassland at the Plantar site is being replaced by a eucalyptus plantation spanning about 230 square km (89 square miles). The trees are being harvested and converted to charcoal that goes to local pig-iron smelters, providing them with a less carbon-intensive form of fuel than coal. But activists say that such monoculture plantations have been a problem in Brazil for decades, displacing local residents as well as indigenous ecosystems.

In a more distributed approach, a solar-energy project in Bangladesh is now being expanded through CDM support. Since 1999 the Bangladeshi bank Grameen (winner of the 2006 Nobel Peace Prize, with founder Muhammad Yunus) has provided small, no-collateral loans to more than 70,000 householders – most of them women – which allow them to install solar photovoltaic (PV) systems. Less than 30% of homes in Bangladesh are on the national electric grid. The others typically use kerosene for lighting and cooking and perhaps a lead-acid battery, recharged at a regional centre, to power a TV. Adding a solar system not only reduces greenhouse emissions but relieves women from the drudgery of maintaining kerosene lamps and provides cleaner indoor air and better light for reading. Revenue from the CDM will cut the cost of the household loans, allowing the project to expand to as many as a million homes by 2015.

Children at a solar house in Bangladesh
The Ashden Awards

above mechanisms should enable all countries to make the grade. It's quite possible that the non-performers will make up some of their shortfall by purchasing emission credits from Russia. Because its economy declined through the 1990s, Russia expects to have little trouble meeting its no-change-in-emissions target, thus giving it the right to sell unused emission credits to other nations. This comfortable margin has been derisively termed "hot air", since it arose from the peculiarities of the Kyoto

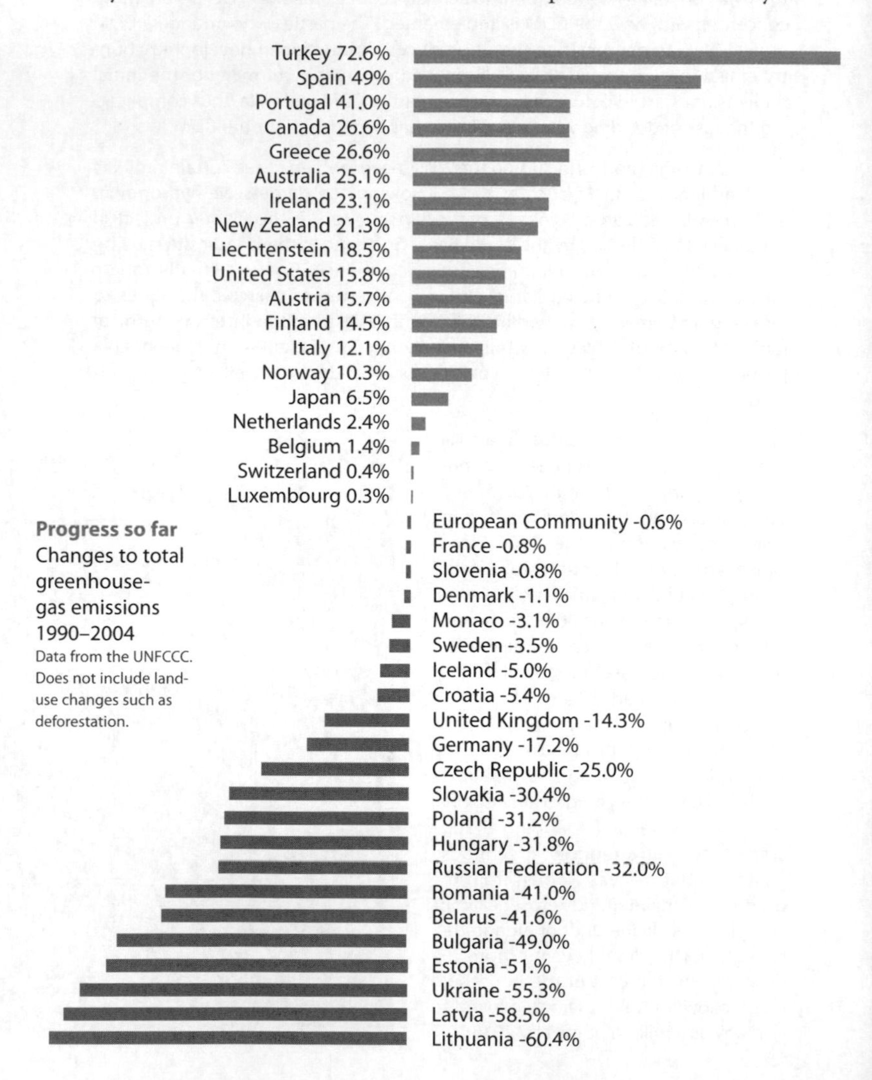

Progress so far
Changes to total greenhouse-gas emissions 1990–2004
Data from the UNFCCC. Does not include land-use changes such as deforestation.

target-setting process and the downfall of the Eastern Bloc economies, rather than from a concerted effort by Russia to clean up its industries.

> **"In effect, Kyoto is learning by doing, even if what is agreed initially is minimal."**
>
> Dieter Helm, *Climate-Change Policy*

Should all countries meet their Kyoto targets through bona fide emission reductions, it won't make a great deal of difference to the atmosphere. Even if the prescribed reductions were maintained for a full century – rather than until 2012 – they would only bring down the year-2100 temperature increase by a few percent, according to a 1998 analysis by Tom Wigley of the US National Center for Atmospheric Research. Critics have cited the weakness of even the potential best-case outcome in order to bash the protocol altogether. But as Wigley noted about his findings, "This does not mean that the actions implied by the Protocol are unnecessary." Indeed, the larger goal of Kyoto was to establish a template through which the world could work for even greater reductions in subsequent agreements.

Chicago's emissions exchange

The US government may have kept out of Kyoto, but many US-based industries – especially multinationals – know that change is in the air. Concerned about their ability to compete in the other Annex I countries, dozens of major firms have made public commitments to reduce their emissions. Perhaps the most tangible evidence of this is the **Chicago Climate Exchange**, the first market for emissions trading in North America. Membership is entirely voluntary; companies that sign on agree to cut their greenhouse emissions 6% by 2010 as compared to 1998–2001. As with the EU scheme, members who can't make the cut can buy allowances through the exchange's electronic trading platform, or they can purchase offsets by funding carbon sequestration or methane collection efforts.

The exchange sells itself as a chance for companies to test the waters, do the ecologically right thing, build their international reputations and familiarize themselves with a form of currency – carbon – that looks set to grow enormously over the coming years. The exchange itself is also expanding its reach. In 2005, New Mexico became the first state to join. ("Markets drive social and environmental change. That's a fact," says governor and 2008 presidential contender Bill Richardson, who's setting out to earn New Mexico a new title: The Clean Energy State.) And a spin-off company, the European Climate Exchange, now operates across the Atlantic, intersecting with the EU's emissions trading scheme.

Beyond Kyoto: what's next?

The next few years could make or break the global mission to deal with climate change. Kyoto's first compliance period expires in 2012, and as of this writing there is no consensus on what to do next. In December 2005 the largest meeting of climate diplomats in nearly a decade took place in Montréal. While many developing countries were ready to start work, the US resisted up to the last minute, and the chief American negotiator walked out of a midnight discussion. In the end, the diplomats managed to eke out an agreement for a two-year round of non-binding talks under the UNFCCC that "will not open any negotiations leading to new commitments" (as the official wording says).

The next phase began with a meeting in Bali in December 2007, with the pressure on to develop firm commitments that pick up where the initial Kyoto period leaves off. Earlier in 2007, G8 members agreed to "consider seriously" the goal of reducing global emissions by at least half by 2050. Meanwhile, US president George Bush promoted an Asian-Pacific consortium that's considering long-term emissions reductions separately from other players (see below). While meeting in Sydney in September 2007, Asian-Pacific leaders endorsed "aspirational goals" to reduce carbon intensity 25% by 2030 – although this wouldn't necessarily imply any drop in actual emissions (see p.42). Critics noted that the plan is based on targets that are entirely voluntary, and it remains unclear how this strategy will mesh or conflict with the larger UN-based approach.

Much will depend on the American political climate as the decade draws to a close. A new president will take office in 2009, and the victor will likely be more amenable to binding greenhouse targets than was the Bush administration. Moreover, the US Congress moved from Republican to Democratic control in 2007 – a key shift that sparked a diverse set of proposed legislation, with goals ranging from modest to ambitious. If these trends in US politics continue, it's possible that the nation may throw considerable momentum behind post-Kyoto planning from 2009 onward.

Over the Atlantic, there's powerful momentum from the EU as a whole – and from industries that expect to be heavily involved in carbon trading – for a post-Kyoto emissions plan. A big question is what shape such a plan might take, and there's been no shortage of suggestions on the table. A 2004 report from the US Pew Center on Global Climate Change summarized more than forty ideas, with names that range from the grandiose

("Orchestra of Treaties" and "Climate Marshall Plan") through the humble ("Broad but Shallow Beginning" and "Soft Landing in Emissions Growth") to the droll ("Keep it Simple, Stupid"). The main points in question include:

- **Should there be a single global plan or an array of decentralized alliances?** As noted above, the US has teamed up with Australia, China, India, Japan and South Korea to form the Asia-Pacific Partnership on Clean Development and Climate. Thus far the alliance – whose members generate half the world's carbon emissions – is focused on non-binding actions such as sharing technologies for renewable and reduced-carbon fuel sources. With an eye toward boosting this technology-driven approach, the US held a meeting of the world's fifteen largest emitters in September 2007. However, most of the participants reiterated their support for binding targets. "I think that the argument that we can do this through voluntary approaches is now pretty much discredited internationally", said UK representative John Ashton.
- **What's the best time frame to consider?** Some plans are focused on the second Kyoto commitment period (2013–2017), while others extend all the way out to 2100.
- **What type of commitments should be specified?** There is a whole array of possibilities, from emission targets by nation or region to non-emission approaches such as technology standards or financial transfers.
- **How should the burden of climate protection be shared among developed and developing countries?** This remains a key sticking point, as was the case from the very beginning.
- **How does the world make sure that commitments are enforced?** It's an issue that many say Kyoto hasn't fully addressed.

Among the most intriguing plans offered to date is the **contraction and convergence** (C&C) model developed by the Global Commons Institute, a British group headed by Aubrey Meyer. It was introduced by the Indian government in 1995 and adopted by the Africa Group of Nations in 1997 during the run-up to Kyoto. The plan has also received votes of support from the European Parliament and several UK and German advisory groups. The two principles at the heart of C&C are:

▶ **Contraction:** the need to reduce overall global emissions in order to reach a target concentration of CO_2, with a commonly cited goal of 450 parts per million.

▶ **Convergence:** the idea that global per-capita emissions, which now vary greatly from country to country, should converge towards a common amount in a process more or less parallel to contraction.

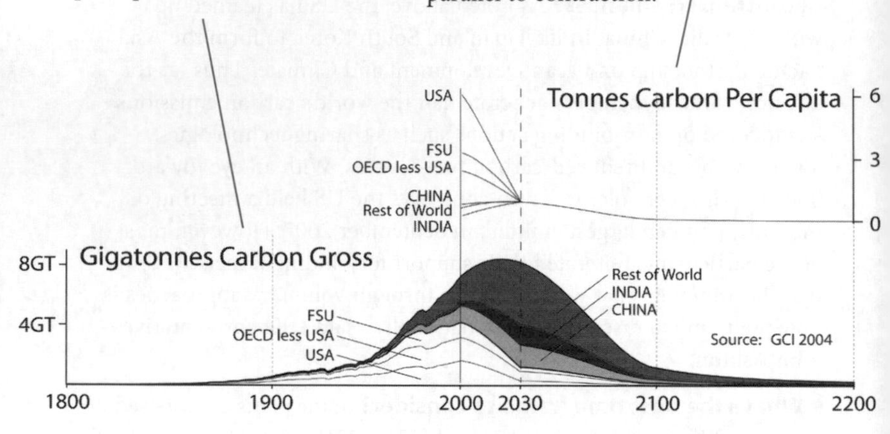

From the Global Commons Institute C&C briefing document, available at at www.gci.org.uk/briefings/ICE.pdf
FSU is the Former Soviet Union

In short, C&C calculates how much carbon the world can emit in order to reach its target, then apportions that total equally among the world's residents based on population. It's an elegant concept that moves the process towards a climate-protected future that virtually everyone recognizes as fair. As ecologist and author Tim Flannery put it in a 2006 speech, "In some ways C&C is an ultra-democratic variant of the Kyoto Protocol."

Some critics of C&C point out that it could provide developing countries with an anti-incentive for birth control, since it allocates emissions rights based on population. And many activists feel that the C&C plan lets developed nations off the hook for their hundred-plus years of creating our current predicament. A competing plan, the **Brazilian Proposal**, uses the historical pattern of emissions as a starting point.

"In the politics of climate change, the Kyoto Protocol is the equivalent of kerb-crawling. It is utterly inadequate and doesn't provide the legal framework we need."

Aubrey Meyer, Global Commons Institute

Adaptation vs mitigation

What might seem like a straightforward response to climate change – adapting to it – is actually fraught with politics. There's nothing especially novel about being prepared for what the atmosphere may bring. As the IPCC notes in its 2007 assessment, "Societies have a long record of adapting to the impacts of weather and climate." There are plenty of sensible steps that vulnerable nations and regions could take right now to reduce their risk of climate-change trauma. Just as energy efficiency makes sense for other reasons than climate protection, many forms of adaptation – such as moving inwards from coastlines – will help keep people and their property safer regardless of the extent of global warming. Yet there's a tension between adaptation and mitigation: to some, the former implies a disregard of the latter, as if society were giving up on trying to reduce greenhouse-gas (GHG) emissions. "There's one way to directly address climate change, and that's reducing the GHG emissions that drive it", writes David Roberts in the online magazine *Grist*. "In the context of the climate-change debate, advocating for adaptation means advocating for a non-response."

Others argue that the problem is so vast that neither adaptation nor mitigation alone can do the trick – we need both. "The UK government is very clear on a twin-track approach," says Chris West, director of the UK Climate Impacts Programme. The UKCIP was founded specifically to help cities, business and other entities determine their climate-change risk so they can better prepare for it. "Regardless of mitigation efforts to reduce greenhouse gas emissions, there are unavoidable impacts already in the system", says West. With the North Sea rising against its dykes, the Netherlands is also considering adaptation in a big way. Among the long-range options being explored are large-scale relocation to the eastern Netherlands and the creation of a "hydrometropole" – in essence, a floating city of fifteen million people.

Speaking on behalf of a wide range of non-governmental organizations, the Climate Action Network (CAN) has put forth another option, informally called the **Global Framework**. It envisions three tracks: a Kyoto-style track for developed countries, a "decarbonization" track for developing nations that sets goals for carbon intensity (see p.42) rather than emissions, and an adaptation track that provides assistance to the poorest and most climate-vulnerable countries. As developing

> **"What is needed above all right now is US leadership, for no country bears greater responsibility for climate change, nor has greater capacity to catalyze a global response."**
>
> Eileen Claussen and Elliot Diringer, Pew Center on Global Climate Change, 2007

countries prosper, they'd move into an emission stabilization mode before graduating to the reduction targets of the Kyoto track. CAN says that any post-Kyoto plan "must be built on core principles of equity and fairness and include an appropriate balance of rights and obligations."

There's no disguising the fact that the above plans, and many others, would require a massive explicit and/or implicit transfer of wealth from rich to poor countries, at least in the beginning, in order to get results. One might argue that such a transfer is unavoidable if the developed world is to take full responsibility for its outsized contribution to the greenhouse effect. However, there's sure to be fierce political resistance in some quarters.

With this in mind, the Pew Center on Global Climate Change brought together a group of 25 experts from 15 countries in 2004 and 2005 for a series of "climate dialogues" that produced a new proposal. The idea is to secure commitments from the world's biggest emitters by acknowledging the tremendous diversity among them and by creating a common but flexible framework that allows a variety of approaches and alliances. For instance, a climate deal spanning the automotive industry could be reached by involving only fifteen countries and a handful of corporations. A series of these kinds of agreements between groups of like-minded nations and companies could be pursued in parallel and tied together under the auspices of the UNFCCC. To make all this happen, the Pew group cited the need for open dialogue, a broader engagement of citizens and the corporate world, and – toughest of all – political willpower.

The next few years should make for fascinating and hugely important politics, as various plans vie for consideration and leaders from around the world jostle for position. In Meyer's words, "The global community continues to generate dangerous climate change faster than it organizes to avoid it. The international diplomatic challenge is to reverse this."

National, regional and local schemes

Governments of all sizes are taking a variety of actions independent of Kyoto and its potential successors. Some of these actions involve setting reduction targets based on much longer time frames than Kyoto specifies, while others focus on specific sectors or market mechanisms.

A big part of the EU's strategy for climate protection is its **emissions trading scheme** (ETS), launched in 2005. It's similar in spirit but separate from the Kyoto emissions trading. Policy analyst Dieter Helm calls it "the most ambitious attempt yet to design an economic instrument to address

What's working in Woking

Woking, a town of 90,000 people just southwest of London, offers an encouraging example of how cities, towns and villages worldwide could respond to climate change. It has a combined heat and power plant, which provides both electricity and (by capturing the generator's waste heat) hot water to the city's car park and town hall, as well as an arcade, conference centre and hotel. The green electricity comes through a public-private venture with a Danish energy firm and gives Woking a measure of independence from the national power grid. All told, the town's civic buildings have slashed energy consumption by nearly one-half and cut CO_2 emissions by an estimated 77% over 1990 levels.

Woking started early – the town's first report on climate change was issued in 1990 – and it looks far ahead. Its long-term plan is believed to be the first in Britain aimed at meeting the UK emission goals of 60% cuts by 2050 and 80% by 2080. Woking's land-use guidelines call for any new development to generate emissions that are at least 80% lower than those from the same land area in 1990. One secret to the town's long-term planning success is an energy efficiency fund, by which savings brought by green practices are ploughed into future climate-protection projects.

climate change." When the EU's scheme began, each nation assigned each of its largest industries the right to emit a certain amount of carbon. The scheme covers a variety of activities, including oil refining, cement production, iron and steel manufacture, glass and ceramics works, and paper and pulp production, as well as fossil-fuel-based power generators of more than twenty megawatts. Together, these companies emit almost half of the EU's carbon – more than two billion tonnes of CO_2 per year. In order to meet its own targets, each of these big EU companies can either cut its emissions or "pay to pollute" by purchasing credits from another company that's done better than targeted. The credits can even be purchased from a country outside Annex I, via the Kyoto mechanisms discussed previously. Some campaign groups found the initial targets assigned by EU governments to their biggest companies too weak. "They need to be strengthened considerably", said Climate Action Network on behalf of its major NGO members in early 2006. "With investments in the covered sectors being made for decades to come, the ETS needs to send the clear signal that cleaner technology and processes will be rewarded." Largely because the initial targets were so lenient, there were far more sellers than buyers for carbon credits in the first ETS phase (2005–07), and the price plummeted from a peak of near €30 per tonne of CO_2 in 2006 to only a few cents per tonne by late 2007. Prices are expected to hold more firmly in the second phase (2008–2012), which will expand from CO_2 to include all six of the greenhouse gases covered by Kyoto.

Across the Annex I countries, there's considerable variety in how the instruments of government are applied to make emissions cuts happen. The result is a patchwork that, for any one country, might include a blend of taxes, emissions trading, support for low-carbon fuel sources, campaigns to raise public awareness about energy efficiency, and environmental aid to developing countries. The **UK** is a good example, having implemented a range of measures including:

▶ **The Renewable Transport Fuels Obligation**, which specifies that at least 5% of petrol and diesel be in the form of biofuels.

▶ **The Climate Change Levy**, designed to turn businesses towards greater efficiency through a tax on their energy bills of around 15%. The levy proceeds support a Carbon Trust to help foster lower-carbon technologies.

▶ **The Renewables Obligation**, which requires that at least 10% of electricity comes from wind power or other renewable sources by 2010, with an "aspiration" to increase this proportion to 20% by 2020.

▶ **A range of directives** dealing with everything from co-generation to energy efficiency in buildings (see p.327). In July 2006, Secretary of State for the Environment David Miliband floated ideas for banning inefficient light bulbs and electrical goods and even introducing a citizen-level carbon-trading system. If the latter idea became reality, each UK citizen would be allocated an allowance of "carbon points", which would be carried "on our bank cards in the same way as we carry pounds". These points would be used – in addition to normal money – when purchasing electricity, gas, fuel and air tickets. People exceeding their quotas would have to pay for extra points; people using less would be able to trade in their excess points for cash. So far, however, the government has never talked seriously about implementing such a plan.

With a raft of initiatives such as these in place, the UK aims to reduce its overall emissions (relative to 1997 levels) by 20% by 2010, 60% by 2050 and 80% by 2100. However, progress so far is less than encouraging. The decade up to 2007 has seen UK emissions *rise* rather than fall – even when international aviation and goods manufactured abroad are excluded. Indeed, the country is only on target to meets its Kyoto goals due to a widespread switch from coal to gas power plants in the early 1990s – a

A leader in mandating renewable energy is the German government, whose remodelled Reichstag reduces emissions by exploiting natural light and a ground-source heat pump
Bob Henson

switch motivated primarily by economics and energy security rather than concern for the environment.

In contrast to the UK, the **US** has put only a smattering of climate-protection policies in place, and some American laws seem to encourage more greenhouse gas rather than less. For instance, the hulking sport utility vehicles (SUVs) that dominated the US auto market in the 1990s were required to meet only minimal fuel-economy standards, thanks to a loophole in the original 1970s legislation that allowed them to be considered "light trucks". (This was partially remedied in 2006, and in 2007 the new US Congress moved towards the passage of the first major increase in fuel efficiency standards in more than twenty years.) On the greener side, the US offers a tax credit for ethanol producers and another for consumers who purchase super-efficient hybrid vehicles. Businesses and individuals can also get tax credits for installing solar panels. All told, US incentives for alternative-fuel vehicles should total at least $1 billion by 2015. The US Environmental Protection Agency (EPA) has also made gains with its Energy Star efficiency labelling scheme (see p.343). And in a landmark 2007 ruling, the US Supreme Court ordered the EPA to begin regulating carbon dioxide as a pollutant, paving the way for future nationwide legislation on climate change.

The biggest action in the US, though, is happening on the state level, where there's more of a chance for a few motivated legislators and policy analysts to make a real difference. **California**, as usual, is a bellwether. Its greenhouse emissions are comparable to those of Australia or Brazil, so its actions could have a real impact in themselves, as well as through their influence on other US states. In 2005, California's governor and erstwhile movie star Arnold Schwarzenegger signed an executive order calling for massive emission reductions in the long term – a reduction to 1990 emission levels by 2020 and a further reduction of 80% by 2050 – as well as a variety of policy moves, including a proposed gasoline tax. California is also seeking the right to enforce a 2002 law making it the nation's first state to mandate reductions in greenhouse emissions from cars and trucks. A lawsuit from a group of automakers – including virtually all of the big firms from the US and Japan – has kept this rule in limbo for years, but this could change by the end of the decade, especially given the Supreme Court ruling mentioned above.

Meanwhile, a group of northeastern states extending from Maryland to Maine is putting its own emissions trading system in place, with a goal of stabilizing emissions from power plants by 2015 and cutting them by 10% between 2015 and 2020. Six western states and two nearby Canadian provinces have also teamed up, calling for emission cuts of 15% below 2005 levels by 2020. Even the conservative bastion of **Texas** has taken steps towards a lower-carbon future. Against the grain of its oil-and-gas legacy, the state has ruled that 2.2% of its energy must be drawn from renewable sources by 2009. The target is extremely humble by European standards, but with Texas ranking higher than the UK, Canada, France or Italy in its greenhouse emissions, even a small reduction is better than nothing.

Across the world, cities and towns are also getting involved. More than six hundred US mayors have signed pledges to meet Kyoto targets (all told, at least a quarter of Americans live in cities, counties or states that are doing so). In the UK, Newcastle – a cradle of coal mining for centuries – made a splash in 2002 when it set out to become the world's first carbon-neutral city. It's still aiming for that goal, along with a growing number of other municipalities.

One such city is taking shape on an island just northeast of Shanghai, at the mouth of the Yangtze River. In stark contrast to the well-worn cities of Europe, the new city of **Dongtan** will be built from scratch on Chongming, China's third-largest island. The population of up to 10,000 expected by 2010 (with another 70,000 in later phases) will drive hybrid

An architect's image of Dongtan, soon to be China's greenest city
Arup

cars or pedal bikes along a network of paths. Renewable power for homes and businesses will flow from a centralized energy centre being developed in partnership with the University of East Anglia. It will draw in part on an array of wind turbines on the island's west edge. More than half the island will stay agricultural, to minimize unnecessary food transportation. The British firm Arup, which is designing Dongtan, may extend the concept to several other Chinese cities.

Technological solutions

Energy, engineering and efficiency

Human ingenuity got us into our greenhouse mess, and we'll need to call on it again in order to find our way out. Reducing the emissions from electricity generation and vehicles is absolutely essential to solving the climate-change problem. Indeed, these two areas account for most of the options for filling the seven "wedges" in the approach suggested by Robert Socolow and colleagues (p.281).

Of course, it's not just where we get our energy that counts. Equally important is how *efficiently* we use it. Hence this chapter also covers technologies and policies designed to raise efficiency and reduce energy waste. Finally, we look at geo-engineering plans – schemes to pull greenhouse gases directly from the atmosphere or reduce the amount of sunlight reaching Earth. No matter what technologies come down the pipeline, the economics and politics discussed in the last chapter will play a huge role in shaping their cost-effectiveness and thus how quickly they're adopted.

The future of fossil fuels

For the time being, fossil fuels remain at the core of the global energy supply. Of course, these fuels are all finite resources, which, if we continue to use them, will cease to be available at some stage. Indeed, many commentators believe that our currently abundant supplies of **oil and gas** will start to dry up very soon (see box). But even if they're correct, that won't be enough to stop climate change. There's certainly enough remaining oil and gas to push us deep into the greenhouse danger zone. And even if

When will the oil run dry?

One key question in predicting the energy mix of the future is the amount of oil left in the ground. A flurry of recent books has publicized the concept of "**peak oil**" – the point at which we've extracted half of the world's accessible oil reserves. The claim isn't that oil will suddenly become scarce, but that production will start to drop, demand will outstrip supply and prices will rocket, causing massive economic dislocations. There's intense disagreement on whether the peak-oil point is at hand or decades away, but one thing is certain: since the 1980s, the world has consumed more oil each year than it discovers. In fact, the ratio has been around two to one since the late 1990s.

The US offers an example of how an oil peak can sneak up on people who want to deny its existence. In 1956, M. King Hubbert, a geologist at Shell, told colleagues that American oil production would peak in 1971. Few people took his warning seriously – after all, production was climbing year upon year through the 1960s – but sure enough, the peak arrived in 1970, just one year from Hubbert's forecast. Annual US oil production has fallen ever since and now stands at about half of that 1970 peak. Pundits vary in their predictions about the global oil peak, but some believe that it may arrive within the next few years (or even that we're already there but don't realize it yet). It's impossible to know for sure, because the data on the vast bulk of oil supplies are held by members of the Organization of the Petroleum Exporting Companies (OPEC), which keeps a tight lid on numbers and methods. Oil companies tend to put an optimistic face on the situation, as recent scandals at Shell have made clear.

So-called unconventional oil sources, such as **tar sands** in Canada and **oil shale** in the US, could provide as much oil as the world's current reserves, but the current methods of extraction are hugely greenhouse-intensive and environmentally problematic – not to mention expensive. Thus, it seems quite possible that global oil supplies *will* gradually tighten in the coming decades.

Natural gas supplies don't appear to be quite as imperilled as oil reserves, but they're already beginning to dwindle across some countries that are traditional exporters, including Canada. The world's largest stocks of natural gas are in many of the same locations as oil reserves, including Russia and the Middle East, and transport is a big issue. Liquefied natural gas is fiercely explosive, and few coastal cities are eager to play host to LNG tankers.

Any drop in supplies of either oil or natural gas would have massive effects on the economics of other energy sources. Oil is used above all as a portable fuel, while much gas is burnt for domestic heating and cooking. The alternatives – from hydrogen cars to electric heating – would require massive increases in electricity generation. For this, the world would need to turn to a mix of coal (clean or dirty), renewables and/or nuclear. The relative investment in each of these options could be the single greatest factor in our success or failure in combating climate change.

both fuels ran out tomorrow, there would still be vast reserves of **coal** to consider. Currently, some two-thirds of the world's electricity comes from coal-fired power plants, and coal is responsible for more than a quarter of global CO_2 emissions.

What's scary from a climate perspective is that there are virtually no brakes on coal emissions. Many of the world's most extensive coal reserves are in countries that fall outside the Kyoto Protocol's emissions mandates – particularly the US, Australia and China. The numbers of new coal-fired plants in the works are astonishing: more than fifty in the US, more than two hundred in India, and many hundreds more in China. Once built, these alone would generate power – and emissions – equal to more than half of today's coal-fired plants around the world.

Is Big Oil really turning green?

Climate change policy affects all kinds of companies, but none more so than Big Oil, whose stock in trade is a key driver of global warming. The major oil firms are some of the largest corporations in the world, and they realized early in the climate-change debate that mandatory emissions cuts could seriously hurt their profits. In the mid-1990s, they collectively worked to derail – or at least defer – serious action on climate change. Their primary means of doing so was the enormously influential Global Climate Coalition, which at its peak included BP, Exxon (Esso in the UK), Royal Dutch/Shell and Texaco (now part of Chevron) and which lobbied hard to stop the US joining Kyoto (see p.251)

All this is a far cry from recent moves by **BP** and others to brand themselves environmentally progressive corporations. The first sign of change was BP's decision to drop out of the GCC in 1997. The same year, chief executive John Browne made waves in a speech at Stanford University, claiming that "the time to consider the policy dimensions of climate change is not when the link between greenhouse gases and climate change is conclusively proven ... but when the possibility cannot be discounted and is taken seriously by the society of which we are part. We in BP have reached that point." Since then, the company has made various investments in greener energy, expanding its solar division and spending $500 million on developing the world's first biofuels research lab. Overall, the company is pouring $8 billion into alternative energies in its stated aim of becoming "a world-leading low-carbon power developer by 2015."

These are certainly significant investments, though renewable energy is still only a tiny proportion of BP's business. Whether this justifies the company's ethical makeover – complete with green-flower logo and "Beyond Petroleum" slogan – is a hotly debated topic. The same question applies to **Shell**, which has, among other things, teamed up with the government of Iceland on hydrogen projects (see p.318), but has continued to be dogged by allegations of unethical prac-

Adding to these hair-raising statistics is the fact that the typical coal-fired plant is in operation for sixty years or more. As David Hawkins of the Natural Resources Defense Council points out, fully two-thirds of the expected global supply of coal-based electricity in the year 2040 will be drawn from plants that are not yet constructed. Hawkins told the US Congress in 2003 that if such plants are built without technology to reduce greenhouse emissions, "we will be creating a 'carbon shadow' that will darken the lives of those who follow us."

The spectre of hundreds of new, carbon-belching coal-fired power plants has added urgency to the task of figuring out how coal can be burned more cleanly. The same challenge applies to reducing the emissions of power stations that burn oil, gas or even biomass (wood). One

tices (not helped by the recent scandal in which executives lied to shareholders over the size of the company's oil reserves). Shell and BP have also joined other energy-sector partners teaming with the UK government on the £1 billion Energy Technologies Institute, designed to advance the commercialization of low-carbon energy sources.

The big US-based oil companies have lagged behind their peers in addressing climate change, but there are some signs of progress. **Chevron** now pours $100 million per year into research on greener power, and it came in fifth in a 2006 report by investor-responsibility group Ceres on the commitments of twenty oil firms to tackling climate change. (BP and Royal Dutch/Shell came in first and second.) **ExxonMobil**, meanwhile, has been pilloried by environmentalists over the years, with human rights issues and the horrific *Exxon Valdez* oil spill of 1989 among the triggers. On climate change, the Texas-based firm has arguably been the most recalcitrant among its major peers, continuing to fund climate-sceptic think-tanks and lobby groups long after BP and Shell had moved on. Exxon now acknowledges the existence of climate change, but extremely cautiously – its European website ventured in 2007 that risks from carbon emissions "could prove to be significant." To its credit, the company has pledged $100 million towards the Stanford-based Global Climate and Energy Project, whose research runs the gamut of renewables and low-carbon fuels. Of course, this is less than 0.3% of its staggering 2006 profits ($39 billion).

Profits and ethics aside, even the smallest greenward movements of Big Oil could make a big difference. As pointed out by George Monbiot and other observers, the biggest oil firms are among the few companies large enough to make costly new technologies affordable through economies of scale. It remains to be seen how quickly these green technologies evolve next to the growing threat of climate change.

straightforward approach is to increase the efficiency of such plants. To an extent, this has already been happening for some time. For example, incremental improvements in technology over the years have raised the efficiency with which coal-fired plants can pulverise and burn their fuel by a percent or so each year. This is no small achievement, but even in the most modern plants only about 35–40% of the energy in the burned coal is actually translated into electricity.

Another approach to cleaning up coal is **gasification**. Instead of burning the coal directly, a gasification plant heats the fuel to a high temperature with just a dash of oxygen. The resulting gases are either burned or further refined into **synfuels** – synthetic versions of oil-based fuels such as petrol or diesel. This isn't a new idea; Hitler was driven to synfuels when Germany's oil supplies ran low during World War II. On the face of it, gasification might seem like an inefficient way to use an already problematic fuel, but there are benefits to the process. Chief among them, in terms of greenhouse gases, is that CO_2 can be easily separated out of the stew of gases after combustion.

Sequestration: capturing carbon

The attraction of isolating CO_2 created in a coal gasification plant – or any other power station – is the possibility of capturing the gas before it has a chance to escape into the atmosphere. This process is known as **carbon capture and storage** (CCS), or **sequestration**. There was no economic or regulatory motive for most of the world to explore sequestration until the Kyoto Protocol kicked in. Now there are pilot projects across Europe, with a few scattered elsewhere. Most involve large power plants, since these are the biggest point sources of greenhouse gases. Many campaigners view CCS as an extension of the throwaway culture that's led to climate trouble in the first place, but it may become an important tool nevertheless.

The idea behind CCS is to put the captured carbon in some sort of underground location where it will remain stable for a long time. Some proposals have explored injecting CO_2 into the sea – at shallow or intermediate depths, where it would dissolve into the water, or at the ocean bottom, where the high pressure would keep it liquid for long periods before it eventually dissolved. These options have yet to overcome a serious environmental objection: we know that CO_2 is making the oceans more acidic (p.124).

A more promising solution is to put the carbon underground. If it's pumped to depths of more than 800m (2600ft) below ground, CO_2 takes on a liquid-like form that greatly reduces the chance of it rising to the surface. Given the right geological formations, the gas could remain underground almost indefinitely. Among the possible locations being explored are:

- **Depleted oil and gas wells** This technique has been used on a limited basis for decades in the practice called enhanced oil recovery, which has the economic – if not climatic – benefit of prolonging production from ageing reserves (as the CO_2 is pumped down into the wells, it helps push out the remaining oil or gas).
- **Saline aquifers** Found in basaltic formations and other types of rock, these typically hold salty water that's separate from the freshwater found in other aquifers. There may be enough saline aquifers to hold centuries' worth of CO_2.

In a special 2005 report on carbon capture and storage, the IPCC found that the technique holds real potential as part of the world's climate-protection package. They estimate that many of the largest point sources of CO_2 probably lie within 300km (180 miles) of potential underground storage sites. According to the IPCC, these sites would likely hold 99% or more of the CO_2 added to them for more than a thousand years. The IPCC also found that the most expensive part of CCS isn't likely to be the storage but the capture and compression of CO_2 – although these costs could drop by 20–30% by the mid-2010s.

There are two pieces to the CCS puzzle that haven't yet come together but should very soon. One is the industrial-scale testing of sequestra-

An artist's impression of the FutureGen cleaner-coal plant, currently in development
US Department of Energy

tion. To date, there have been only a handful of such projects – one in the North Sea, where millions of tons of CO_2 have been injected into an undersea aquifer from a natural-gas drilling platform, and another in North America, where CO_2 from a synfuel plant in North Dakota is being piped to Saskatchewan and used to enhance recovery at an ageing oil field.

The other puzzle piece is next-generation coal plants that are designed from the start to make CCS a real option. The leader in this realm is a type of coal plant dubbed **integrated gasification and combined cycle** (IGCC). The technology has been explored since the 1980s, but as of 2007 there were only two full-sized commercial IGCC plants up and running in Europe and two in the US, with several smaller demonstration units there and elsewhere. IGCC plants employ gasification but also use the waste heat to make steam, which drives a second turbine. Although it would be a fairly simple matter to capture the CO_2 from an IGCC plant, it hasn't happened yet. There's now a $1 billion initiative under way in the US to do just this. The goal of FutureGen is to build an IGCC facility by 2011, to be placed in either Illinois or Texas, that would be the world's first electrical plant of any type to capture and store carbon. If FutureGen fulfils its promise, it'll bring coal far closer to being carbon neutral. The question that remains is whether the fruits of FutureGen could be harvested quickly enough to prevent a climate-wrenching set of old-school coal plants from being built.

Beyond fossil fuels: renewables and nuclear

Cleaning up fossil-fuel-based electricity sources is all well and good, but it's equally important to consider the alternatives. These fall into two camps: **renewables**, such as wind, solar, hydro, tidal, geothermal and biomass, and **nuclear**. The degree to which each of these sources can meet our demand for low-carbon energy is among the most important and fiercely debated issues in climate-change policy. It's an immensely complex area – an intimidating mix of economics, technology and politics.

For one thing, it's not just the *amount* of energy each source can generate that counts. Equally important are various other factors, including:

▶ **Generating time** Demand for electricity varies across the day and storage is difficult, which can be a problem for solar, for example.

▶ **Reliability** Wind power, for instance, has enormous potential but needs to be backed up by other sources during calm periods.

Making sense of power

As you read through this chapter, you may find the following comparisons useful. Watts and their derivative terms (kilowatts, megawatts) are measures of the rate of power being delivered or consumed; a watt is one joule of energy per second.

100 watts = the power used by a strong incandescent (non-energy-saving) light bulb

1000 watts = one kilowatt = close to the average electricity demand of a typical household in the developed world (the actual demand goes up and down depending on the time of day)

1,000,000 watts = one megawatt = the amount of power delivered by a typical wind turbine (some modern ones provide 2–3 MW) = a thousand households' worth of electricity

1,000,000,000 watts = one gigawatt = the amount of power generated by a large coal-fired power plant = a million households' worth of electricity

Kilowatt-hours (kWh) and similar terms describe the amount of power delivered or consumed over a period of time. The world's total electricity usage in 2005, including both residential and commercial needs, was close to 17 trillion kWh. If consumed at a steady rate, that would represent about 1900 gigawatts at any one time.

▶ **Public safety** Nuclear being a case in point.

▶ **Roll-out time** In an era of proliferating coal power, every year wasted is bad news for the climate.

As for **cost**, this is complicated by subsidies and tax regimes and the fact that each source becomes cheaper and more efficient as it becomes more widespread. Things are further clouded by the lack of definitive statistics on the total greenhouse emissions or maximum potential contributions of each energy source. As UK commentator Simon Jenkins put it, "Nobody agrees about figures ... energy policy is like Victorian medicine, at the mercy of quack remedies and snake-oil salesmen." And there's also the fact that some amount of fossil fuel will be needed to build, deploy, and maintain renewables such as wind and solar on the mammoth scales needed.

With so much to consider, a full assessment of non-fossil energy is beyond the scope of this book, but following is a brief description of the pros and cons of each major source.

Solar power

Nearly all our energy ultimately comes from the Sun. Wind, for example, is kicked up by the temperature contrasts that result when sunlight heats Earth unevenly. And you can't have biofuels (or even fossil fuels) without the sunshine that makes plants grow. However, when people refer to **solar power**, they're usually talking about devices that convert sunlight directly into electricity or heat.

Solar energy took off in the oil-shocked 1970s, but its momentum sagged as the limits of that era's solar technology became clear and as oil prices dropped. Today, with the risks of climate change looming, the Sun is rising once more on solar power. Because solar power is so decentralized, exact numbers are hard to come by, but for electricity alone, the International Energy Agency estimates that almost four gigawatts of solar capacity was available at the end of 2005, more than half of it in Germany and Japan. That's double the amount in place only two years earlier, with an annual growth rate of more than 40%. Overall, solar power currently costs up to ten times more than electricity from a standard coal-fired power plant, mainly because of the expensive raw materials that go into solar cells. In some locations, though, tax structures and other factors make solar power roughly competitive with other sources. And for people who want to go off the grid – or who have no grid to draw from – solar power ranks near the top of the list.

Some solar technologies have been used for centuries, such as orienting windows towards the south (in the Northern Hemisphere) to allow warming sunlight in. These and other **passive** solar techniques are now commonly used in home and office construction. The higher-tech, "active" approaches fall into two camps:

▶ **Solar cells (also known as photovoltaic or PV systems)**, use sunlight to split off electrons from atoms and thus generate electricity. A typical PV cell is about 10 x 10cm (4 x 4") and a standard PV "array" consists of 400 or more such cells. In small numbers, PV cells are ideal for small-scale needs such as patio lights. With 10–20 PV arrays and a sunny climate, you can power a entire household; a few hundred arrays can be linked for large industries or utilities. Despite its northerly latitude, Europe has embraced PV power in a big way. The world's largest PV plant in 2007 was a twenty-megawatt facility in Spain, and facilities in the works will be many times larger than that. A 40-megawatt plant, spanning a lot the size of 200 soccer fields, will open in 2009 in Germany's Saxony region. And a BP solar plant near Madrid is expanding its capacity to 300 MW – roughly a third the size of a typical coal-fired power plant. (BP also plans to build a similarly sized plant in Bangalore, India.)

▶ **Thermal collectors** use sunlight to create heat rather than electricity. These are most commonly roof-mounted plates used to provide domestic hot water (see p.340). Tubes that carry water (or, in some cold climates, antifreeze) are sandwiched between a glass top and a back plate, which is painted black to maximize heat absorption. The fluid heats up as it flows through the collector and then proceeds to the hot-water tank. The whole process is typically driven by a pump, though there are also passive systems in which the heated water produces its own circulation.

Either approach can be scaled up with **concentrating systems** – which use curved reflectors to focus sunlight onto PV arrays or tubes filled with a fluid such as oil or molten salt (which in turn heats water, to produce steam to turn turbines).

Solar power is both limited and underused – limited because of the fixed distribution of sunlight around the world and around the clock, and underused because far more buildings could benefit from strategically placed solar systems than now have them. The developing world is a particular point of interest, because even a small PV array can power an entire household if the demand for electricity is modest (see p.283). At the other extreme, the world's most advanced showcases for energy efficiency also make good use of passive and active solar systems.

The world's subtropical deserts are potential gold mines of solar energy. (It's been estimated that a Sahara entirely covered with solar panels would produce many times more power than the world's present consumption of all types of energy, including oil and coal.) The problem is that storing and transporting the resulting energy is expensive and inefficient. Hydrogen might help with this at some stage, but there are hurdles to overcome (see p.324).

Wind power

Although it's growing at a somewhat less rapid pace than solar power (around 25% a year), **wind power** represents a much bigger slice of the planet's renewable energy profile. As of 2006, wind accounted for about 75 gigawatts of electric capacity. Factoring in the variability of wind speeds, this represents about 1% of the electricity generated globally. Over half of the capacity is in Europe, where 10–20% of electricity is expected to come from wind turbines by 2010. Wind power is especially popular in Germany and Denmark, and it's coming on quickly in Spain as well. The US now has only about a quarter as much installed capacity of wind power as Europe, but the potential for growth is impressive, especially across the perpetually windy Great Plains from Texas to North Dakota (and on to Saskatchewan in Canada). It's been estimated that the US could generate up to eleven trillion kilowatt-hours of wind power per year, about three times the total from all US power plants.

Once in place, wind turbines produce no greenhouse gases at all – the only petrochemicals involved are the small amounts used to manufacture and lubricate the units. Another bonus is that the giant blades of wind turbines typically spin at heights of 30m (100ft) or more, leaving the land below free for other uses. Many farmers and ranchers rake in substantial royalties for allowing turbines to be placed atop fields where crops grow and cattle graze.

Based on all this, wind power seems like a no-brainer, yet some commentators claim that wind is neither as cheap as its fans claim nor quite as green, since the possibility of a lack of wind means that another power station – probably one burning fossil fuels – has to be kept running as a

Cows graze under a wind turbine near Clear Lake, Iowa
Alliant Energy

backup. Wind farms have also come up against local opposition, especially in the UK and US, where some activists have fought wind farms with a vehemence approaching that of the anti-nuclear movement that peaked in the 1970s and 80s. Mostly, the opposition isn't based on ecological or economic concerns but on aesthetics (some Brits have referred to wind turbines as "lavatory brushes in the sky") and noise (the distinctive whoosh-whoosh of the blades and the creaking that sometimes results when the blades are moved to take advantage of a wind shift). Another allegation is that wind farms are a danger to bird life.

These are all contended claims. Many people enjoy the motion and symbolic cleanliness of the structures, and wind-power proponents claim that you can hold a normal conversation while standing right underneath a modern turbine. As for bird deaths, it's hard to measure accurately, but a 2005 report by the UK's Royal Society for the Protection of Birds claims that, as long as the farms are positioned appropriately, they "do not pose a significant hazard for birds". Adding fuel to the argument is the suggestion that some "local" opposition has been orchestrated behind the scenes by those with vested interests in other energy sources.

Placing wind farms offshore is one way to reduce their visual impact, but it's not enough to placate all protesters. A heated battle has been under way on the coast of New England, where people living near Nantucket Sound are fighting plans for the first US offshore wind farm, which will include 130 turbines, each taller than the Statue of Liberty. Meanwhile, a group called Country Guardian leads the charge against wind power in the UK. It calls wind-power farms "environmental disasters" and even employs a bit of climate-change scepticism, decrying the "unsubstantiated predictions of global warming".

Outside of the local controversies, there is strong overall public support for wind power – typically running around 80% in UK and US polls – so at least for now, the tugs-of-war at specific sites appear unlikely to throw wind power off its course of vigorous expansion.

Geothermal energy

The same forces that can blow the top off a volcanic mountainside can also generate electricity. **Geothermal power** taps into the hot water produced where rain and snowmelt percolate underground to form pools that are heated by adjacent magma and hot rocks. Some of these power plants draw hot water and steam from more than a mile below ground, while others harvest steam that emerges directly at the surface. Either way, a small amount of naturally produced carbon dioxide escapes into the air – but it's less than 10% of the amount emitted by a standard coal-fired power plant of the same capacity. **Binary geothermal** plants avoid even these minimal emissions of CO_2 by using water from underground to heat pipes that carry a separate fluid. The geothermal water returns to Earth without ever being exposed to the atmosphere.

More than nine gigawatts of electric capacity is available from geothermal sources worldwide each year, and geothermal plants can run close to their capacity almost 24/7, since their power source doesn't vary like sunshine and winds do. The best sites for geothermal plants are where continental plates separate or grind against each other, especially around

Iceland's low-carbon lifestyle

Thanks largely to its fortuitous location atop a geological cauldron, Iceland may point the way towards a cooler and cleaner future for the rest of the world. This island nation of about 300,000 people draws on a unique portfolio of energy that's already low in carbon emissions. By 2050 Iceland could be the world's first hydrogen-based, carbon-neutral economy.

The two continental plates that play host to North America and Europe are separating beneath Iceland, which leads to the world's most concentrated zone of **geothermal** energy. The very name of the capital city, Reykjavik, derives from *reyka* – the plumes of steam visible from many points around the country. Steam and hot water from underground furnishes heat for some 90% of Iceland's buildings. Although depletion of geothermal energy is a concern in some other parts of the world, it's less so here: the magma, rocks and steam beneath Iceland are estimated to hold many centuries' worth of energy, even without factoring in the potential for their recharge through future volcanism. Although some of Iceland's electricity comes from geothermal power plants, most of it is produced by **hydroelectric** plants. They're a reliable source of power in the island's wet climate, but also a source of friction from environmentalists who want to limit the intrusion of dams and reservoirs into wilderness areas.

There's far more potential for geothermal power in Iceland than the country can use, but at the same time the nation still imports oil for its cars and ships. These intertwining factors make Iceland a perfect test bed for **hydrogen**, a long-term project being carried out by a public-private partnership that includes DaimlerChrysler, Shell and NorskHydro. The first step was a mini-fleet of three fuel-cell-powered municipal buses in Reykjavik, which drew power from the world's first public hydrogen filling station. Passenger vehicles are now being tested, with at least thirty expected to run on hydrogen by 2009. By mid-century, Iceland hopes to be exporting hydrogen to the rest of Europe.

the "ring of fire" that surrounds the Pacific Ocean. A third of the global total is generated in the US, where the world's biggest plant – the Geysers, in northern California – provides more than 700 megawatts of capacity. In Kenya and the Philippines, geothermal sources provide about 20% of all electricity. They're also part of the palette of carbon-free energy in Iceland (see box on previous page), where it provides direct heating as well as electricity.

On a smaller scale, buildings in many locations can draw on the steady temperature of near-surface soil through **geothermal heat pump** systems, though whether they're cost-effective depends on climate, household size and other factors. These systems rely on air flowing through pipes buried just underground. A heat exchanger pulls up air from the pipes in the winter, when the ground is relatively warm compared to the air above ground. In the summer, the same system pumps warm air downwards and replaces it with relatively cool air.

Hydroelectric, tidal and wave power

Hydropower plants use dams to channel water through electricity-generating turbines. This approach currently provides far more energy than any other renewable source – more than 15% of global electricity capacity – while producing minimal greenhouse emissions. In theory, this capacity could be scaled up several-fold, especially in Africa and Asia, but actual growth is expected to be much more modest. One reason is the high start-up costs (many of the more easy-to-engineer sites have already been used up). Another is that the construction of dams can wreak havoc with river-based ecosystems far up- and downstream. There are also social issues, as large dams can flood huge areas and displace many residents (China's Three Gorges Dam is forcing more than a million people to resettle). For such reasons, the Climate Action Network has requested that hydropower projects be disallowed from the Clean Development Mechanism through which developed countries can earn emission credits by subsidizing carbon-reduction projects in developing countries (see p.292).

Perhaps more promising for the long term are other ways to generate clean electricity from the movement of water. Turbines driven by the **tide** and **waves** are both being developed, but various technological and economic barriers need to be overcome before either source could meet a significant proportion of the world's energy demands.

Nuclear power

Just when it looked as if it might keel over from the weight of negative public opinion and high start-up costs, nuclear power has got a new lease of life thanks to climate change. It's hard to dismiss a proven technology that produces massive amounts of power with hardly any carbon emissions. Yet it's also impossible to ignore the political, economic and environmental concerns that made nuclear power so contentious in the first place.

Though it's taken a somewhat lower profile since the 1990s, nuclear power never went away. The world's 400-plus nuclear plants provide more than 15% of electricity globally, and a much higher fraction than that in some countries – particularly France, which embraced nuclear power in the 1950s and never looked back. France now gets close to 80% of its electricity from 59 nuclear plants operated through the state-owned Electricité de France and a private partner, Areva. The US has even more nuclear plants – 104 – and though none has been completed since 1996, General Electric and Hitachi are seeking approval for two new plants in Texas that would open in 2014.

Global warming has already changed the nuclear equation in tangible ways. In 2003, Finland's legislature voted narrowly in favour of building its fifth nuclear plant – one of the world's largest, to be completed in 2011 – partly because of the hope that it might help Finland meet its Kyoto goals. In 2006, France and the US agreed to give nuclear technology to India, which now has six reactors on order, and Tony Blair announced that he'd "changed his mind" on nuclear, paving the way for likely approval of a new generation of British atomic-energy plants.

"The only technology ready to fill the gap and stop the carbon dioxide loading of the atmosphere is nuclear power."

Stewart Brand, "Environmental Heresies," *Technology Review*

"Nuclear power has already died of an incurable attack of market forces, with no credible prospect of revival."

Amory Lovins, Rocky Mountain Institute

Though a certain amount of CO_2 is released in the process of building a nuclear plant (plus mining and processing its fuel), there's no question that nuclear power is far kinder to the climate than fossil fuels. That simple fact has engendered some powerful crosscurrents. Environmentalists who oppose an expansion of nuclear power run the risk of appearing callous about climate change. Thus, the anti-nuclear arguments have shifted somewhat in recent years. The unsolved problem of disposing of nuclear **waste** remains a key point, as does the risk of **accidents** like Chernobyl in 1986. (Industry advocates claim that such risks are overstated, pointing out that Chernobyl was a low-budget, Soviet-era construction quite unlike modern reactors, which have never resulted in a single death.) And recent heat waves have underscored how nuclear plants themselves can be vulnerable to a warming climate. Plants were closed in Germany in 2003, Michigan and Spain in 2006 and Tennessee in 2007, in each case because lake or river water had warmed too much to cool down the reactors.

More than before, however, opponents are turning to **economic arguments**, claiming that the massive costs of building and decommissioning nuclear plants – which are typically subsidized and insured in whole or in part by governments – drain money that's urgently needed to fund research on renewables and energy efficiency. According to Amory Lovins, nuclear's "higher cost ... per unit of net CO_2 displaced, means that every dollar invested in nuclear expansion will worsen climate change by buying less solution per dollar." Terrorism also figures into the mix: the more nuclear plants we have, say the nuclear opponents, the more risk there is of radioactive material slipping into the wrong hands, or of an attack on a power plant that could kill many thousands. And uranium

Dirty power

San Francisco, one of the more pet-friendly cities in the US, is leveraging that quality to become more climate-friendly. In 2006 a local sanitation firm launched a pilot program to collect droppings from across the city, including collection carts at a local dog park, and use them to fuel a methane digester. It's the same technology used in hundreds of European farms and in a few US operations as well. Bacteria feed on the animal waste and generate methane – ie natural gas – that can be burned directly or used to generate electricity.

In theory, human waste could be used in exactly the same way, though it's hard to envisage the idea taking off. However, a growing number of start-ups in the US and elsewhere are taking manure from chickens and other farm animals and processing it anaerobically to yield methane. Engineer George Chan folded the concept into a template for a sustainable agricultural system called Dream Farm, and several UK planners are working on a follow-up idea. According to Mae-Wan Ho of London's Institute of Science in Society, Dream Farm 2 will be "a microcosm of a different way of being and becoming in the world, and in that respect, nothing short of a social revolution".

supplies are in question beyond the next few decades, unless the world turns to breeder reactors, which pose their own set of problems.

On the other side of the coin, some influential figures have spoken out in favour of increasing the world's nuclear power capacity in the light of the climate-change threat. These include prominent greens such as Gaia theorist James Lovelock and Steward Brand, founder of the Whole Earth Catalog. Along with other nuclear advocates, they argue that nuclear is the *only* low-carbon energy source that's sufficiently developed and reliable that it could be rolled out on a big enough scale to take a meaningful chunk out of the world's greenhouse emissions in the short and medium term.

A blue-ribbon panel assembled in 2003 by the Massachusetts Institute of Technology gave ammunition to both sides of the nuclear debate. The report recommends keeping open the option of a tripling of current global nuclear capacity by 2050 – which alone could reduce greenhouse emissions by a massive 25% – but it also cites several major obstacles, with the risk of nuclear-weapon proliferation perhaps the most daunting. "Nuclear power should not expand," the report warns, "unless the risk of proliferation from operation of the commercial nuclear fuel cycle is made acceptably small." Whether the risk of proliferation can be compared to the risk of climate change is a debate that's bound to run and run.

Making transport greener

Transporting people and goods by road, sea and air is responsible for around a quarter of global greenhouse emissions, and this figure is rising fast. So if we're to tackle global warming, we'll have to develop lower-emissions vehicles – not to mention promoting public transport and, some would argue, reducing the flow of people and goods around the world.

The easiest gains could come from simply improving the **efficiency** of conventional vehicles, something that is already happening but which could go much further, especially in the US, where the average vehicle achieves roughly half the efficiency of those in Europe and Japan. One positive development is the growing popularity of cars with super-efficient **hybrid** engines (see p.347), which pair a standard engine with an electric one and a battery and can manage as much as 70mpg. Plug-in hybrids, which would draw power from the grid or from renewable home energy systems, could provide even greater efficiency. Ironically, the hybrid family is growing to include hulking SUVs and executive cars, which improve upon their non-hybrid counterparts but still require plenty of fuel per mile. For a car to be truly efficient, it also has to be lightweight.

In the longer term, however, it's changes to the way we fuel our vehicles that holds most promise. Within a decade or so, some new cars, trucks and even planes may be powered by **hydrogen** (see box overleaf). In the mean time, engineers are refining ways to power vehicles through so-called **biofuels** – such as ethanol and biodiesel – which are made from plants. The primary attraction of biofuels is simple: though they release plenty of carbon when burned, much of this will be soaked up by the next round of fuel crops (after all, plants grow by absorbing CO_2 from the atmosphere). How climate-friendly a biofuel is, then, depends on factors such as the amount of fossil-based energy used to fertilize, harvest, process and distribute the crops.

The other major issue for biofuels is the demand for land. It would take *huge* areas of cropland to turn biofuels into a major part of the global energy picture. This could squeeze food production or require the conversion of virgin land into farms, something which has implications not just for wildlife habitat but for the greenhouse effect itself, depending on where these land changes occur. Advocates of James Lovelock's Gaia theory (see p.209) see further changes in land use as one of the greatest threats to Earth's ability to maintain a stable temperature. All this aside, biofuels could certainly fill an important niche in the green energy gap.

Ethanol

Ethanol (ethyl alcohol) is a plant-based substitute for petrol/gasoline that's already quite common in some parts of the world. Across the States, it's usually sold in a mixture of 10% ethanol and 90% gasoline (E10). If people want to use higher proportions of ethanol, such as 85/15 (E85), they need a flexible fuel vehicle (FFV) that can handle any blend of the two. FFVs are rare in the US but widespread in Brazil, which leads the world in ethanol usage. Ethanol made from sugar cane provides more than 25% of Brazil's vehicular fuel (all gasoline there is at least one-quarter ethanol) and the country also exports to Asia and elsewhere. A plus for ethanol use in Brazil is that its mild climate enables motorists to avoid the cold-start problems that can otherwise occur with high-ethanol blends.

As of 2006, ethanol production in America was on par with Brazil's – close to eighteen billion litres (five billion US gallons) – but that equalled less than 4% of the fuel swallowed each year by America's cars and trucks. New tax breaks and incentives could nearly double that percentage by

How much will hydrogen help?

Hydrogen, the simplest and most abundant element in the universe, has been hyped as a solution to the greenhouse crisis in recent years, especially in the US. In truth, hydrogen is no panacea for global warming. Indeed, hydrogen isn't really even an energy *source*, as we have to use power from other sources to produce the pure hydrogen gas that's useful as a fuel. That said, hydrogen does hold great potential as a clean and portable energy *store*, so if we can generate enough low-carbon electricity to manufacture hydrogen on a massive scale, it could be the perfect solution for fuelling tomorrow's vehicles – and for other uses such as domestic heating.

On our planet, hydrogen is usually bound to other molecules – as in water (H_20). But in its pure form, H_2, hydrogen is a lightweight gas (or a liquid at extremely low temperatures) that contains a great deal of energy. Hydrogen gas can be burned directly as a fuel, but the preferred means of exploiting it is through **fuel cells**, which convert hydrogen into electricity. This process produces zero emissions of carbon dioxide and other pollutants – the only byproduct is steam (and though water vapour is a greenhouse gas, the amount emitted is insignificant).

The potential for zero-emissions vehicles has generated the recent buzz and drawn some big-time investment. The EU's European Hydrogen and Fuel Cell Technology Platform is part of a decade-long, multi-billion-euro effort. In the US, the Bush administration launched a $1.7-billion program in 2004 aimed at bringing hydrogen-fuelled vehicles into the marketplace by 2020. General Motors has set itself an even more ambitious target of selling hydrogen cars by 2010.

However, it will take years more – if not decades – to bring hydrogen vehicles into common use. Before hydrogen can provide energy to fuel cells, the gas first has to be broken off from other molecules, stored and distributed. Not only is each of these steps a daunting technical challenge in itself, but each one also consumes energy.

▶ **Production** Currently, the cheapest technique for producing hydrogen involves breaking down natural gas into H_2 and carbon dioxide. This is far from being a carbon-neutral process, and it's expensive, too: hydrogen is still more than twice as costly as the same energy in the form of petrol. It's possible to generate hydrogen by splitting up water molecules, but right now that process is even more expensive than the alternatives. Improvements in technology should gradually bring down production costs, but it may be a decade or more before hydrogen can compete on economic terms with petrol. Of course, that picture could change if more stringent emission targets are put in place – or if oil supplies diminish faster than expected. Equally, developments in capturing and storing carbon could make hydrogen creation more carbon neutral.

▶ **Storage** Assuming that hydrogen will someday be produced as cheaply as oil can be drilled – which could be the 2010s or later – we'll still need a way to store it. That's not easy, because hydrogen needs lots of elbow room. In order to hold the same amount of energy that's in a typical tank of petrol, a car would need a hydrogen tank several times bigger than the car itself. The obvious solution is to pressurize or chill the hydrogen so it occupies less space. The few fuel-cell vehicles already in existence draw on hydrogen that's been compacted to about seventy times normal atmospheric pressure. Even so, the tanks need to be several times larger than those of regular cars in order to provide the same amount of power. To bring down the storage space even more, you could liquefy the hydrogen, but that requires chilling it to an almost inconceivable -253°C (-423°F). In the process, you'd consume nearly a third of the energy that the hydrogen itself would provide. Engineers are working on other storage techniques that may prove useful – including metal hydrides that trap hydrogen gas in their lattice-like molecular structure – but we're a long way from solving the storage problem.

▶ **Distribution** There's also the matter of getting hydrogen from where it's produced to where it's used. Because it needs to be kept either compressed or chilled, the prospect of hauling hydrogen long distances doesn't make much sense, at least with today's technologies. One Swiss study found that to haul a truck full of hydrogen 500km (310 miles), roughly the driving distance from London to the Scottish border, would consume an amount of energy equal to almost half of the cargo. Pipeline delivery might be more efficient, at least over relatively modest distances.

Because of the complications in storage and transport, many hydrogen visionaries foresee a distributed production setup in which hydrogen is produced locally – perhaps even in a home unit connected to a supply of natural gas or electricity. But it could be decades before such small-scale generation is feasible, and even then it would signal a continued reliance on fossil fuels. In fact, hydrogen-fuelled cars may not be the best way to think about emissions reductions over the next several decades, according to some observers, including long-time researcher Joseph Romm (author of *The Hype about Hydrogen*). In their view, the natural gas or renewables used to produce hydrogen would make a bigger dent in the global greenhouse picture if they were instead used to replace coal-fired power plants – especially because those plants have lifetimes on the order of fifty years.

As research continues to make hydrogen fuel cells more efficient, they could become affordable and practical in large buildings fairly soon, whereas it may be decades more before the infrastructure (eg filling stations) is in place for widespread use of fuel-cell vehicles. It's also possible that a new generation of internal combustion engines built for hydrogen could eventually leapfrog the fuel cell. In the meantime, emissions from cars and trucks could be cut dramatically through improving efficiency with hybrid engines and other existing technologies.

A prototype DaimlerChrysler fuel-cell car
www.blueclic.com

2012. One challenge will be getting major oil and gas companies to allow higher proportions of ethanol to be sold at US pumps. Right now many oil companies prohibit it, claiming a lack of quality control.

There's been much debate over whether ethanol really helps protect the climate. Most of America's ethanol to date has been derived from corn, which is grown through farming that makes fairly extensive use of petrochemicals. All in all, corn-based ethanol seems to provide some benefit, but not much. The University of California at Berkeley estimates that corn-based ethanol uses about 74 units of fossil-fuel energy to produce 100 units of ethanol energy. That's somewhat like a charity spending $74 to raise $100. As for emissions, a vehicle running on ethanol does produce greenhouse gases, but when you factor out the CO_2 absorbed by the plants it's made from, corn-based ethanol blends add up to roughly 30% less carbon to the atmosphere than standard gas/petrol does.

A more promising biofuel is on the horizon: **cellulosic ethanol**. Standard ethanol comes from fermenting the sweet starches in corn kernels and sugar, but ethanol can also be brewed from the sturdy material called cellulose (as in celery). Cellulose is found in corn husks as well as a variety of other plants, such as fast-growing grasses, and even in human and animal waste (see box on p.321). Since many plants high in cellulose can thrive on land that's marginal for regular farming, they're less likely to displace food crops, and the yield per acre can be twice or more than of corn. Deep-rooted perennials like switchgrass also help stabilize soil and need little if any help from petroleum-based farming techniques. As a result, 100 units of cellulosic ethanol require only about 20 units of fossil energy to produce, according to the US Department of Energy.

In time, cellulosic ethanol could end up being carbon neutral or even a net remover of carbon from the atmosphere, according to the Rocky Mountain Institute. The main catch is that cellulosic materials don't break down very easily, so the cost of producing cellulosic ethanol is still about double that of the corn-based variety. Chemists have been working on methods to bioengineer enzymes that could be produced en masse to break down the cellulosic materials more affordably. That might level the playing field with corn-based ethanol over the next decade.

Biodiesel

Diesel cars – which are very popular in Europe, though far less common in the US – are typically a bit more climate-friendly than the petrol-based alternatives, since they burn their petroleum-based fuel about 40%

more efficiently. (The flip side is higher emissions of nitrogen oxides and particulates, though in both the US and Europe these pollutants will be reduced by strict new emission standards.) But diesel cars can be greener still when powered by **biodiesel** made from sustainably grown crops. This isn't a new concept. The term "diesel" originally described the engine type rather than the fuel, and its inventor, Rudolph Diesel, used peanut oil to drive the prototype that won him top honours at the 1900 World's Fair in Paris.

Biodiesel has received a boost from tax breaks on both sides of the Atlantic. As with ethanol, it's often added to fossil fuel in small percentages and sold as a way of helping engines run more smoothly while emitting fewer pollutants. But you can also buy 100% biodiesel in some parts of the UK and US. Country singer Willie Nelson even peddles his own blend, called BioWillie, across several Southern states.

At its most climate friendly, biodiesel can be a convenient product of recycling – as in the UK, where it's often derived from used vegetable oil donated by chip shops and other commercial kitchens. In the US and mainland Europe, however, it's produced commercially from canola, soybean and other inexpensive vegetable oils. This allows for a far greater level of supply, but also raises questions about the wider impacts of the burgeoning international market for oil crops. Huge areas of tropical rainforest have already been cleared to make way for soybean and palm-oil cultivation, with disastrous impacts on both biodiversity and the climate. For this reason, biodiesel has become something of a controversial subject, with environmentalists divided over its potential benefits and costs.

Making buildings more efficient

Buildings and their electrical contents are responsible for almost half of all energy consumption. So no matter what energy source we use to power, heat and cool them, it makes sense to do so as efficiently as possible. The main challenge here isn't developing new technologies – there are plenty to choose from already, with more on the way – but getting consumers, manufacturers, architects and governments to choose (or require) efficiency.

The outsized role of buildings as contributors to the greenhouse effect is largely a legacy of the way offices and homes have been designed over the last few decades. With fuel for heating and cooling assumed to be cheap and plentiful, aesthetics and short-term costs have often trumped energy efficiency and long-term savings. "Commercial and public buildings in

Low-power towers

Skyscrapers have always been more than mere office buildings. The best rank among the world's greatest architecture, and they often serve as icons for the companies they house. Now a new breed of eco-oriented towers is going up, aiming to impress with their low-carbon credentials as well as with their high-rise aesthetics.

Swiss Re's award-winning, pickle-shaped 2004 tower, located at 30 St Mary Axe in London's financial district, was dubbed the "**Erotic Gherkin**" years before it was even built. The design can be seen as an elongated riff on Buckminster Fuller's geodesic dome, with the curvature provided by gentle turns in the frame. Love it or hate it, the building's provocative shape has a point. Wind flows easily around it, reducing the amount of support structure needed. Less obvious is a latticework of tunnels through which air can flow into and through the Gherkin's inner reaches, cutting heating and cooling costs by up to half. The tunnels also bring natural light to the building's interior.

London's "Erotic Gherkin"

In New York there's **Four Times Square**, also known as the Condé Nast Building. From a distance it's not much different than any other square-sided tower in Manhattan's clotted skyline, but the 48-storey building stands out in energy efficiency. Completed in 2000, it was the first major US skyscraper built with green credentials in mind. The building draws on a wide array of technologies, including fuel cells that kick in at night, state-of-the-art windows, plentiful solar panels and a ventilation system that pulls in 50% more fresh air than comparable towers.

New York also awaits the **Bank of America Building**. Due to be completed in 2009, it's already being billed as the world's most eco-friendly skyscraper. Its sides will be subtly canted to provide some of the streamlining effects of the Gherkin. Windows will reflect ultraviolet light while letting in visible rays. The main power source will be a five-megawatt co-generator drawing on natural gas, and a heat pump will help keep the building cooler in summer and warmer in winter.

the developed world have generally become sealed, artificially-lit containers," says architect Richard Rogers of the UK design firm Arup. "The increasingly evident threat to the global environment posed by buildings of this sort cannot be ignored."

Efficiency will be key if the UK is to stand any chance of meeting its overall goal of a 60% emissions cuts by 2050. A 2005 report by the UK's Environmental Change Institute (ECI) suggested that such a reduction in emissions from buildings *is* possible, even with an expected 33% increase in the number of households by 2050. Among other measures, this would require:

▶ **Installing high-performance windows** and fully insulating wall spaces and lofts in both existing and newly built homes.

▶ **Demolishing about 14% of current houses** by 2050 (four times the current rate) in a targeted strategy to eliminate the worst energy offenders.

▶ **Employing low- or no-carbon technologies** to heat most homes, including heat pumps, solar cells and solar hot water, combined heat and power and wind turbines.

The biggest challenge may be the profusion of electricity-guzzling equip ment, from fridges and lights to big-screen TVs. Indeed, a 2006 report by the Energy Saving Trust suggested that at current rates of increase, the amount of energy consumed in UK homes by electrical goods is set to double by 2010. Unless this trend can be reversed, warns the ECI, "the 60% carbon reduction in the residential sector will be all but impossible".

The US is unlikely to adopt such a grand goal any time soon. That's a serious concern, because American homes and offices are massive consumers of energy. Since the 1970s, the size of the average new US home has grown by more than 50%, and the fraction of homes with air conditioning has risen from a third to more than four-fifths. Amazingly, per capita energy use in US homes has dropped by more than 15% in this time, but countering that trend is the nation's relentless growth in population – now just over 300 million, and projected to top 400 million by 2050. This implies a great opportunity to bolster America's stock of energy-efficient homes, but that goal is impeded by the fragmented nature of the US construction industry and the lack of a federal mandate. (The American Institute of Architects recently endorsed a proposal to halve the energy consumption of the typical new US house by 2030.) Progress may be more rapid inside US homes: the Energy Star program (p.343), which identifies and certifies the nation's most efficient appliances and electronics, saved more than 100 billion kilowatt-hours between 1992 and 2002.

The quest to make large buildings – such as schools, factories and offices – more energy efficient differs somewhat from the domestic strategy. In the US, lighting and office equipment make up almost half of the energy consumed by commercial buildings. All those lights and computers produce lots of warmth, which means commercial buildings typically demand less heating but more air conditioning than homes. Given this energy profile, some of the most constructive steps to cut energy use in big buildings involve illumination. **LEDs** – the light-emitting diodes used in many traffic lights and stadium signs – may soon become even more efficient than compact fluorescent bulbs for everyday use. The top floors of large buildings could also make use of **hybrid solar lighting**, in which sunlight is "piped in" through optical cables feeding from a roof-mounted solar collector. Green roofs – gardens that spread across much or most of a large building – help keep the building cooler (and, as a bonus, whatever's planted absorbs a bit of carbon dioxide). And several high-profile skyscrapers have become showcases for a variety of energy-saving features (see box on p.328).

Rating and assessment systems are a powerful force for greener building. The BRE Environmental Assessment Method (BREEAM) has certified more than 600 major office buildings since being established in the UK in 1990. Even more influential globally is the Leadership in Energy and Environmental Design (LEED) program, which was founded in the US. Since 1998, it has recognized more than 14,000 projects in 31 countries at levels ranging from "certified" to "platinum".

Geo-engineering: practical solutions or pie in the sky?

It's possible that the various strategies discussed so far in this chapter won't be implemented widely enough, or soon enough, to prevent the risk of dangerous climate change. But a few visionaries think that there's another way: using technology to tackle climate change head-on, either by sucking greenhouse gases from the air or changing the amount of sunlight that reaches Earth. These ambitious plans are often lumped under the heading of **geo-engineering** – global-scale attempts to reshape Earth's environment.

Humans have long dreamed of controlling weather and climate. In the 1950s, the USSR pondered the notion of damming the Bering Strait that separates Alaska from Russia. The idea was to pump icy water out of the Arctic and make room for warmer Atlantic currents, thus softening the nation's climate and easing the way for ships. Famed computer scientist John von Neumann reflected the era's slightly wild-eyed optimism when he wrote in 1956 of spreading dark material over snow and ice to hasten its melting and warm the climate: "What power over our environment, over all nature, is implied!" Today's geo-engineering schemes operate in a more sober realm, the idea being not to create the perfect climate but simply to stop the existing climate from changing too much.

Some geo-engineering plans contemplate reducing the amount of carbon in the atmosphere. One idea is to move the carbon from the air to the oceans. Experiments hint that spreading iron over parts of the ocean where it's limited could produce vast fields of carbon-absorbing phytoplankton. One problem with these and other ocean-storage ideas is that they'd promote the acidification that's already a concern for marine life (p.124). The many geo-political challenges of geo-engineering came to the fore in 2007 when the private US firm Plantos launched plans to dump one hundred tonnes of iron near the Galapagos, ten times more than previous iron-fertilization studies. The plan, which could lead to profit-making projects designed to offset emissions elsewhere, drew scrutiny from the International Maritime Organization and the global London Convention. In another approach, Gaia theorist James Lovelock and Chris Rapley (director of London's Science Museum) have proposed dotting the ocean with a set of vertical pipes that would allow wave action to pull up rich, deep water and promote algae blooms.

Another approach is to expose the atmosphere to chemicals – perhaps in conjunction with certain types of rock – that would react with the air-borne CO_2 and convert it into some solid form. In itself, this isn't technologically very difficult: the main problem is that the production and preparation of the chemicals and rock is likely to be very energy intensive (not to mention expensive), which could mean releasing more greenhouse gases than the chemicals can suck out of the atmosphere. Moreover, if such a scheme could be made to work on a scale that would make even a small dent in the greenhouse effect, there would be the question of how to deal with the massive quantities of carbon-based solids generated. One idea is to use it as a building material, but transporting it would be another drain on energy.

An alternative approach is to try and reduce the amount of sunlight reaching Earth. This could involve adding massive amounts of sulphates to the stratosphere, or deploying a colossal array of mirrors or lenses far out in space that would deflect or refract sunlight before it reached our atmosphere. Astronomers Roger Angel and Pete Worden of the University of Arizona recently proposed an array of ultrathin lenses spanning an area the size of Western Europe that would sit about 1.6 million km (1 million miles) from Earth towards the Sun.

Such ideas have simmered on the back burner for years, but they're now getting a bit more attention. A growing number of top scientists are beginning to study and discuss geo-engineering more openly, if only as a last-ditch solution to keep in our back pockets should the alternative look even more threatening and uncertain. The Nobel Prize–winning atmospheric chemist Paul Crutzen wrote a high-profile article in 2006 for the journal *Climatic Change* exploring the type of sulphate-injection scheme noted above, which would mimic the sunlight-blocking effects of volcanic eruptions. "The very best would be if emissions of the greenhouse gases could be reduced so much that the stratospheric sulfur release experiment would not need to take place," wrote Crutzen. "Currently, this looks like a pious wish."

Even as research gins up, some observers worry that geo-engineering is a less-than-ideal way to deal with the root problem. Climatologist Stephen Schneider at Stanford University has likened it to giving methadone to a drug addict. There's also the classic risk of unintended consequences. Even if enough sunlight could be reflected to compensate for the extra CO_2 in the atmosphere, those two factors (less sunlight, more carbon) could produce effects quite different than our present climate, especially for vegetation and ocean life.

Perhaps the ultimate geo-engineering fix would be to remove carbon dioxide directly from the atmosphere, without adding any disagreeable by-products. "Could machines, located wherever we wish, remove CO_2 from the atmosphere as fast as we put it in, or maybe even faster?" asks a 2003 report by the US National Academy of Engineering. One person who believes the answer may be yes is physicist Klaus Lackner of Columbia University. Lackner's "synthetic trees", through which CO_2 in the air reacts with recyclable chemicals, are already being prototyped, though it remains to be seen how much money and, crucially, energy they'll take to produce, operate and maintain. There's also the issue of how to safely store vast quantities of captured CO_2. Until these and other questions are answered, technological fixes will remain a purely theoretical approach to tackling climate change.

PART 5

WHAT YOU CAN DO

Reducing your footprint & lobbying for action

Getting started

Where to begin

The gravity of our greenhouse predicament is enough to weigh anyone down. But the smart way to deal with climate change is to channel your angst and frustration into constructive action. In this chapter you'll find a wealth of ways to get started. Not all of them will suit your particular situation, but some probably will.

The most obvious way to take individual action on climate change is to reduce the size of your **carbon footprint** – the total amount of greenhouse emissions that result directly and indirectly from your lifestyle. Since this will include increasing your energy efficiency at home and on the road, you may even save some money in the process. The following chapters take a brief look at various ways to reduce your footprint. For more information about low-carbon living, see *The Rough Guide to Ethical Living* (UK) or *The Rough Guide to Shopping with a Conscience* (US).

Another, equally valid approach is to **get involved in the debate**: raising the climate-change issue with your political representatives, employer or local community, as described on p.337.

Measuring your carbon footprint

Just as new dieters often keep a food diary, an excellent way to start reducing your emissions is by using carbon calculators. These simple tools, available online or in book form (see p.363), allow you to calculate how much carbon each activity in your life generates and how your total compares to those of the people around you and elsewhere in the world. Carbon calculators vary by country, reflecting differences in the way energy is generated, priced and taxed. They also vary among themselves in how they organize and categorize activities and how they handle uncertainty about the exact greenhouse impact of particular activities, such as flying. Sometimes the calculations in each step are explained in detail; in other cases they're not, which makes the site simpler but not necessarily more user-friendly. American-based calculators often use short tons, about 10% less than a metric tonne.

The carbon-offset companies (see p.356) all offer calculators for specific activities, but to quickly assess your overall carbon footprint, visit:

Ecological Footprint Quiz www.myfootprint.org
BP www.bp.com/environment

These sites, though useful, tend to focus on actions we're each directly responsible for, such as heating and driving. Of course, in reality many of our emissions result from less obvious sectors such as the construction of our homes, offices and roads, and the manufacture of the goods we buy. The following charts, based on figures from Best Foot Forward (see www.bestfootforward.com), break down the carbon footprint of the average UK and US resident. They include aviation and imported goods, which are often excluded in official statistics.

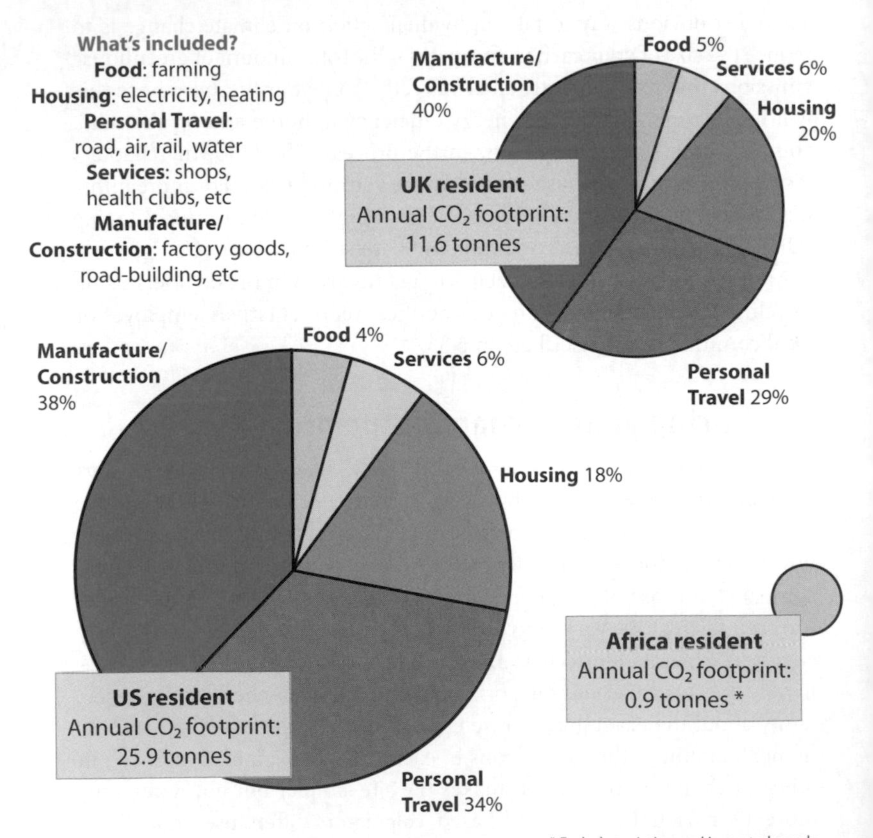

Making things happen

In addition to reducing your own carbon footprint, you can help move society as a whole towards a lower-carbon future. Joining with other like-minded citizens will enhance your own power.

▶ **Work and school** Find out what policies exist on energy efficiency at your workplace or school. If there aren't any, or they seem half-hearted or inadequate, work for something better by relaying your suggestions to the powers that be. Encourage co-workers or fellow students to do the same.

▶ **Community** Cities, towns and councils vary hugely in how committed they are to solving the greenhouse problem. Contact your local government and ask what climate-change measures they've adopted. Then see what you can do to raise awareness and make change happen – by contacting local media, for example, or attending local-government open meetings.

▶ **Finances** If you've got savings or investments, consider moving them towards a bank or fund that supports action on climate change. Generally this will be one that engages in "socially responsible investing" – that is, it considers global warming as well as other social and environmental issues when deciding which companies to include in its portfolio. If you hold stock directly in a company that's behaving in a less-than-ideal way on climate change, look into introducing or joining a **shareholder resolution**. Typically these require companies to report on their actions and plans regarding climate change and its impact on the company's bottom line. At least a dozen major US firms faced such resolutions in 2006 alone.

▶ **Politics** Climate change won't be solved without political will, and there won't be political will without pressure from voters. The Internet makes it easier than ever to contact your political representatives. Simply drop into one of the following websites, enter your postal or zip code and away you go. A good starting point is to ask what emission targets (if any) your legislators support and what they're doing to help ensure these targets are met.

Congress www.congress.org (US)
They Work for You www.theyworkforyou.com (UK)
UFCW www.unionvoice.org (US)
Write to Them www.writetothem.com (UK)

▶ **Pressure groups** You could also consider joining one of the campaign groups working for political change around the climate issue. The following vary in their policy focus and attitudes towards issues such as globalization, but they're all campaigning for major emissions cuts:

Environmental Defense www.environmentaldefense.org
Friends of the Earth International www.foe.co.uk
Greenpeace www.greenpeace.org
Natural Resources Defense Council www.nrdc.org
Sierra Club www.sierraclub.org
WWF www.wwf.org

Home energy

How to save money while saving the planet

Saving energy at home will provide a dual satisfaction: less greenhouse gas in the atmosphere and, sooner or later, more money in your pocket. The following tips should help you streamline your household energy use. You may also want to contact your local government office to see whether it offers free home-energy audits or energy-efficiency grants.

The Web is a good place to find out more. For example, Home Energy Saver allows US residents to carry out a simple online audit of their household energy use, while the UK's Energy Saving Trust provides a wealth of useful information to help you make your home less carbon hungry.

Home Energy Saver (US) hes.lbl.gov
Energy Saving Trust (UK) www.est.org.uk/housingbuildings

Heating and hot water

Heating and cooling is a key area, accounting for a staggering 82% of household energy use in the UK. Older houses are especially inefficient in this regard, though there's room for improvement in nearly all homes.

Insulation

Start small with weather-stripping – sealing up cracks around doors and windows – but be sure to consider beefing up your loft and wall insulation (the attic is a good place to start, since it's easier to access and less expensive to insulate). If you live in a hot climate, you can save on air-con by using bright, reflective window drapes and shades wherever sunlight enters, and by using light colours or a reflective coating on your roof.

Turn down the dial

Reducing your heating and hot-water temperatures by just a small amount can make a disproportionate difference to your energy consumption. You may find you sleep better, too. Try 16–18°C (61–64°F) and throw on a sweater. As for hot water, aim for 50°C (122°F). If you use air conditioning, shoot for 25°C (77°F), or, if you're in a warm, dry climate, investigate swamp coolers, which can take the edge off summer heat while humidifying the air and using far less energy than an air conditioner.

Boilers and heating controls

An efficient boiler (hot-water heater) would make a sensible long-term investment for many households. Modern condensing boilers produce more than 10% extra heat and hot water per unit of energy than a typical boiler from the mid-1990s. Efficient heating controls – especially those that let you specify the temperature of individual rooms, or program different temperature for different times of day – can also take a significant chunk out of your energy demands, in return for a comparatively small investment.

If you have a hot-water tank, be sure to insulate it. Purpose-built blankets are inexpensive and can save 25–45% of the energy required to heat the water. (Some newer hot-water tanks have insulation built-in.)

Pick your fuel

In general, natural gas is a more climate-friendly fuel for home heating and hot water than oil, electricity or coal (assuming you can't access wind power – see overleaf). If you're not connected to the gas network, wood could be your best bet, as long as the fuel is coming from forests that are being replenished as fast as they're being harvested. The UK government offers grants for efficient wood-burning boilers and stoves.

Showers and baths

Everyone knows that showers use less energy than baths, and that shorter showers use less energy than longer ones. Less widely known is that a low-flow shower nozzle, which mixes air with the water flow, can reduce the amount of hot water needed by half. Look for nozzles rated at less than 10 litres or 2.5 gallons per minute.

Electricity supply

Renewable electricity

Many utility companies allow you to specify that some or all of the power you use is generated from renewable sources, most often wind farms. Typically with such plans (labelled "green tariffs" in the UK), you're not literally getting the power generated by the renewables. Instead, the power company agrees to puts renewable energy into the grid in an amount equal to your own consumption. One point to note: in countries such as the UK and certain US states, electricity companies are already mandated to buy a certain percentage of their power from renewable sources, and the companies that exceed this quota can sell credits to those which fail to meet it. The upshot is that your fee won't necessarily increase the overall amount of green power generated. Quiz local suppliers or green groups for more information.

Generating your own power

There are various ways to generate energy at home – but don't expect to be self-sufficient without massive up-front investment. Perhaps the best starting point is a rooftop solar panel to heat your hot water. Unlike the

Starting at the top

Besides hosting solar panels and micro wind turbines, roofs offer various opportunities for combating global warming. One option is to make your roof more reflective. A study by the Earth Institute at Columbia University estimated that a world full of entirely white roofs could add a full 1% to Earth's reflectivity. But you don't necessarily have to paint your roof white to make it more reflective. Several types of tile, metal and shingle roofs can boost reflectivity while maintaining a splash of colour. For large buildings in urban areas, another good option is a "green roof" with plants that can not only help with cooling but also absorb excess stormwater (and CO_2).

Bob Henson

A green roof in Washington DC

photovoltaic (PV) systems that generate electricity, these **collector** panels funnel heat directly into your hot-water system. Even in cloudy Britain, a collector panel – typically spanning about 2 x 2m (6 x 6ft) – can provide up to 70% of the hot water needs of a house with a south-facing roof and a hot-water tank (they won't work with most combination boilers). Such units typically cost less than £5000/$9000US.

Fully fledged **PV systems** can provide homeowners with half or more of their total power needs. However, these systems are expensive – upwards of £10,000/$18,000US, which may take more than a decade to recoup in energy savings, depending on your location and whether you use most of your electricity in the day, when the panels are working hardest. That said, there are government grants and tax breaks available in many countries and US states. And in the UK and parts of the US, it's possible to export any excess power back to the grid, offsetting the cost of power that you draw from it when your solar roof isn't active.

For people who live in a breezy enough spot, domestic **wind turbines** that feed power into batteries or the grid are another option. However, these are also fairly pricey at present (on par with equivalently powerful PV systems), and the towers can run afoul of local building codes – and even jeopardize your home's structural integrity.

For much more information about generating your own electricity, see *The Rough Guide to Ethical Living*.

Appliances and gadgets

Lighting

Close to 95% of the power that drives an old-fashioned incandescent light bulb goes to produce heat instead of light (which explains why those bulbs are so hot to the touch). **Compact fluorescent** bulbs generate far less heat, enabling them to produce almost four times more light per unit energy (the replacement for a standard 100-watt bulb typically uses about 18 watts). These efficient bulbs also last more than ten times longer, meaning that they earn back their higher initial cost several times over and save you time and hassle in buying and replacing bulbs (not to mention the emissions involved in making and shipping them).

Compact fluorescents often take a few seconds to reach full brightness, but they no longer produce unattractive light. When asked to compare the light without knowing the source, the participants of a study by US magazine *Popular Mechanics* actually preferred the compact fluorescent bulbs.

In 2007 Australia became the first nation to ban incandescent bulbs, effective in 2010, and major retailers in the UK have agreed to phase out their sale by 2011. Note that compact fluorescent bulbs do contain tiny amounts of mercury, so when they eventually die take them to a municipal dump to be disposed of properly.

What about **halogen** lights? These are a subset of incandescents and tend to be middle-range performers in the greenhouse stakes. High-quality halogen bulbs are around twice as efficient as typical incandescents. However, they're still half as efficient as compact fluorescents, and halogen light fittings often take multiple bulbs, raising their overall energy consumption. Moreover, they burn much hotter than other bulbs, making them more likely to damage paintwork. If you have halogen fittings, look out for compact fluorescent or LED replacements, available from specialists such as the UK's GoGreenLights.co.uk.

LEDs (light-emitting diodes) may be the bulbs of the future. Already widely used in traffic lights and outdoor displays, LEDs are now being adapted for a broader range of uses. They're already as efficient as compact fluorescents, and further gains are expected over the next few years.

Whatever kind of lights you have, turning them off when you leave the room is an obvious energy-saver. Organizing your home to maximize the use of natural light – putting desks next to windows, for instance – can also help.

Invisible power drains

The proliferation of remote controls and consumer gadgets has come at a surprisingly high energy cost. In the typical modern home, 5–15% of all electricity is consumed needlessly by TV sets, stereos and other devices that are supposedly turned off but are actually on **standby**. Indeed, some devices use almost as much energy in standby mode as when they're in use. To stop the waste, unplug items – or switch them off at the socket – when they aren't in use. A power strip with a switch can help if you don't have switches on your sockets.

Computers vary widely in terms of the energy they consume in standby and screen-saver modes. Turn them off when not in use, or dedicate their downtime to the fight against climate change (see p.236).

Look out for labels

Electrical devices vary enormously in their energy consumption, so when shopping for such items, be sure to look out for the various labels that give you the information you need to make a low-carbon choice. These include:

UK/Europe

▶ **EU Energy Label** Appears by law on a variety of home appliances, light bulbs and air conditioners, rating products from A (efficient) to G (inefficient).

▶ **Energy Efficiency Recommended** This UK label adorns best performers within the EU Energy Label rating system. For example, refrigerators must rank as A+ or A++ and washing machines must be AAA.

▶ **EU Ecolabel** Not yet very common in the UK, the Ecolabel tags products that pass numerous environmental criteria, including energy efficiency as well as expected lifespan and ease of disposability.

North America

▶ **EnergyGuide (US)** Standard on large appliances, this label shows the typical amount of energy consumed per year and how that product compares to the best and worst performers in its category, as well as the expected annual energy cost.

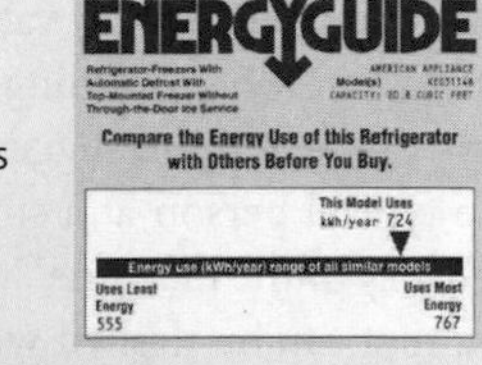

▶ **Energy Star (US)** Identifies products that meet various criteria for efficiency in more than forty categories – including entire homes.

▶ **EnerGuide (Canada)** Similar to the US EnergyGuide, this label shows the amount of energy consumed per year and a comparison to similar models (but no energy cost estimate).

Australia

▶ **Energy Rating Label** Shows annual energy consumption (or, for air conditioners, the rate of consumption) and an efficiency rating from one to six stars. Even the best appliances only rate three or four stars, to leave room for future improvements.

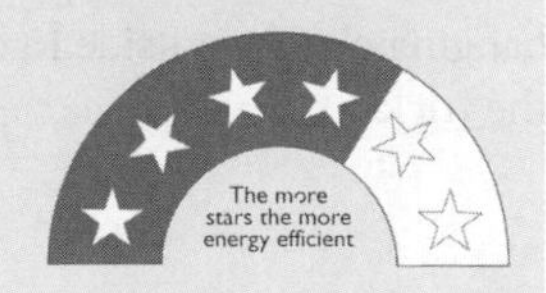

Refrigerators and freezers

These are among the biggest users of energy at home – not least because they're running 24/7. Models produced since the 1990s are far more efficient than their predecessors, so if yours is old, consider taking the plunge: a modern replacement could recoup its costs within several years and save plenty of energy from the start. Be sure to dispose of the old one properly, as many older refrigerators contain ozone-depleting chemicals. Check with your local government or a home-appliance dealer for more information.

If possible, locate your fridge or freezer as far away as possible from hot-running items such as ovens and dishwashers, and make sure it's set no lower than the recommended temperature, which is typically -18°C (0°F) for freezers and 1–4°C (34–39°F) for fridges.

Doing the dishes

Like fridges, dishwashers vary widely in their energy consumption, so be sure to consider efficiency when purchasing. As for how they compare with washing up by hand, this depends on the individual machine (some models use less than fifteen litres or four gallons of hot water per load), the efficiency of your hot-water heater and, most importantly, how economical you are when washing up by hand. A much-cited 2004 study from the University of Bonn suggests that dishwashers use less energy overall than the typical person at a sink, but this doesn't include the production and delivery of the machine. Moreover, the study gives figures for hand-washing that can be slashed with just a bit of care, and assumes you run your machine with full loads, skipping extra features such as "pre-rinse".

Washing clothes

To save energy (and minimize colour fade) keep washing temperatures as low as possible to get the job done well. Even more worthwhile is letting clothes dry on a line or rack rather than using an energy-hungry dryer. In the US, many suburban condominium and housing associations – fearful that clothes-lines will tag a neighbourhood as déclassé – actually prohibit hanging clothes outside to dry. In such situations, an indoor rack may do the trick.

Transport & travel

Trains, planes and automobiles

As we've seen, personal transport accounts for nearly a third of the typical person's carbon footprint in countries such as the UK and US. Moreover, transport as a whole is the fasting growing major source of greenhouse gases. Thankfully, in the case of road transport at least, there's much you can do to make your comings and goings less carbon-intensive. As we'll see, air travel is rather more problematic.

The car and what to do about it

Compared to alternatives such as train and bicycle, most cars are very energy inefficient – especially when carrying only one or two passengers. If you can't live without a car, you can at least reduce its greenhouse emissions by changing your driving habits (see p.347). Making fewer and shorter journeys also helps, of course, as can carpools and liftshare schemes. Depending on where you live, you may even be able to run your present vehicle on ethanol or biodiesel (see p.327).

If you're due for a new vehicle, be sure to opt for the most efficient model that fits your needs. This may be a **hybrid**, such as the Toyota Prius or Honda's Civic Hybrid. These look, work and fuel just like regular cars, but they feature an extra electric motor that charges up when you apply the brakes. Hybrids can achieve up to seventy miles per gallon, though SUVs and executive cars with hybrid engines are far less efficient because big, heavy, high-powered cars eat fuel much faster than smaller, medium-powered ones. If you're concerned about the **safety** of a smaller car, check the official safety data for each model – bigger isn't always better in this regard.

A Toyota Prius hybrid – one of the greenest cars in its class

Even greener than hybrids are **electric cars**, which you charge up via a standard household power socket. At present, electric cars are very limited in terms of top speed (around 72kph/45mph) and battery capacity (you need to recharge every fifty miles or so). However, they can be ideal for city dwellers and they're extremely green: even when charged with electricity generated from fossil fuels they result in far less CO_2 per mile than the most efficient hybrids. The G-Wiz is one such car that's quickly gaining popularity in the UK, with more than 900 on the streets of London (where they're exempt from congestion charges):

G-Wiz www.goingreen.co.uk

In the US, General Motors has pledged to build a commercially viable electric car by the year 2010, with its Chevy E-Volt prototype now in testing. Activists are watching closely in the wake of the cinema exposé *Who Killed the Electric Car?* It profiled GM's ill-fated EV1, which the company torpedoed in 2003 despite fanatically loyal owners.

Air travel

Aviation is problematic from a climate-change perspective. It takes extra fuel to move people at the altitudes and speeds of jet aircraft, and people typically fly much farther than they would ever travel by train, boat or car. The contrails produced by aircraft also have potentially important impacts (p.188). This means that a plane's impact on the climate may be far greater than its already significant CO_2 emissions would suggest. All told, two people flying a round trip between Europe and the US produce the equivalent impact of at least four tonnes of CO_2 – more than half of the average UK household's yearly output and comparable to what a typical US car generates each year.

Low-carbon motoring

Even if your vehicle's a gas-guzzler, the following tips could cut its fuel usage by as much as 30%. Some of them are now incorporated in the tests that new drivers take in the UK, the Netherlands and several other countries.

▶ **Drive at the right speed** Most cars are most efficient at speeds of 45–80kph (30–50mph). Above 90kph (55mph), cars gulp more fuel to travel the same distance – as much as 15% more for every additional 16kph (10mph).

▶ **Lighten the load** Keep heavy items out of your car unless you need them – you'll typically lose a percent or two in efficiency for every 45kg (100lb) you haul. While you're at it, check the tyre pressure often: rolling resistance goes up and efficiency goes down by as much as 1% for every PSI (pound per square inch) below the recommended pressure range. However, there's no benefit, and some risk, to driving with over-inflated tyres.

▶ **Avoid idling** Except when it's required (such as in stop-and-go traffic), idling is a wasteful practice, and it doesn't benefit your car, except perhaps in extremely cold conditions. Even five minutes of idling can throw half a kilo (1.1lb) of greenhouse gas into the air. Anything more than about ten seconds of idling generates more global-warming pollution than stopping and restarting would.

▶ **Use the air-con sparingly** As you'd expect, air conditioning normally saps energy and cuts down on vehicle efficiency by a few percent. However, if you're on a long road trip and it's a choice between driving with the windows down and running the A/C, there may be little difference in fuel usage, according to some studies. That's because wide-open windows can increase the car's aerodynamic drag, especially at high speeds. If outside temperatures are comfortable, try using the vents and fan but leaving the A/C off.

▶ **Starting and stopping** Jack-rabbit starts and stops not only put wear and tear on your car, but they also drain fuel economy. Accelerate gradually, and anticipate stops by starting to brake well in advance. If you have a manual transmission, the best time to change gears is between 1500 and 2500 rpm.

The rules are a bit different for **hybrids**, which has led to some confusion. In 2005, *The New York Times* and *Consumer Reports* magazine declared that hybrids fall far short of their advertised miles per gallon. However, according to energy expert Amory Lovins of the Rocky Mountain Institute, this is because the tests employed by both publications didn't take into account the different driving style needed for hybrids, which typically call on the electric motor at low speeds and the gasoline engine at higher speeds. With a standard engine, you're best off accelerating lightly no matter what the speed. In a hybrid, you should accelerate briskly until you get to the optimal in-between speed, around 45–60kph (30–40mph), where the car is at its most efficient. A technique called "pulse and glide" driving – hovering in that optimal speed range through small accelerations and decelerations – can boost your hybrid's efficiency. When it's time to slow down, brake slowly at first, then increase the pressure: this ensures that the maximum energy goes into recharging the battery versus creating unusable heat in the brakes themselves.

Planes are slowly gaining in efficiency, with improvements of a percent or so each year owing to technological improvements as well as fuller flights. The problem is that these gains are mostly translating into cheaper and more popular air travel, so that the sector's overall greenhouse output continues to grow rapidly. More substantial gains in efficiency are on the drawing board – for example, in 2007 Air New Zealand and Virgin Atlantic announced upcoming tests of biofuels. However, even the most optimistic commentators don't expect such developments to offset the massive growth in the numbers of flights (see p.37).

With all this in mind, cutting back on air travel is one of the most obvious steps for people concerned about global warming. This might mean taking fewer flights and making up for it by staying longer each time. It might mean favouring holiday destinations that are closer to home. Or it might mean considering alternative ways of travelling. With two or more people on the same itinerary, it can even be more climate friendly to drive than to fly – especially for short distances, where planes are most inefficient per mile travelled. Better still are **trains and boats**, which are typically responsible for many times fewer emissions per passenger mile than either cars or planes. To find out how to travel almost anywhere in the world by rail and sea, visit:

The Man in Seat 61 www.seat61.com

When you do fly, consider offsetting your emissions (see p.353) and use the airline's website to see what model of plane you're booking yourself on. Sites such as seatguru.com can help you figure out whether your flight will be on a newer (and thus likely more efficient) aircraft. Especially on long-haul flights, avoid the perks of luxury seating; thanks to the extra room they occupy, business-class passengers are responsible for about 40—50% more emissions than those in economy, and first-class travel may generate up to six times more carbon. Also, try to favour flights in the **daytime** and during the brighter times of the year; this can make a surprisingly big difference. A 2006 study of flights across southeast England, published in the journal *Nature*, points out that plane contrails reflect no sunlight during night-time and wintertime, thus reducing their ability to compensate for the climate-warming gases spewed by aircraft. For the study area, flights between December and February (less than a quarter of the annual total) caused half of the climate-warming effect, and night-time flights – only about a fifth of the total in Britain – produce 60–80% of the effect.

Another trick is to travel light. Two heavy suitcases nearly double the on-board weight of a passenger, adding to fuel consumption.

Shopping

Food, drink and other purchases

It's impossible to know the exact carbon footprint of all the items you buy, but there are a few rules of thumb. First, you can consider the energy used in transporting an item. Heavy or bulky goods manufactured far away inevitably result in substantial carbon emissions, especially if they've travelled long distances by road (shipping is comparatively efficient, though not negligible). On the other hand, smaller items transported by air – such as perishable, high-value fruit and vegetables, and cut flowers – are *far* more greenhouse-intensive than their weight would suggest.

Another thing to consider is the material that something is made from. Steel, aluminium, concrete and precious metals all require large amounts of energy to create. By contrast, wood from sustainable sources – such as local softwoods and anything certified by the Forest Stewardship Council – is practically carbon neutral. The same can't be said for tropical hardwoods such as mahogany or teak, the demand for which is one driver of the tropical deforestation that accounts for a large slice of the world's greenhouse emissions.

The three Rs: reduce, reuse, recycle

The old green mantra holds true for combating climate change. It's heresy in a capitalist society to suggest that downsizing might have its pluses, but a big part of reducing global emissions is taking a hard look at global consumption. So try your best not to make unnecessary purchases, especially of products that take a lot of energy to manufacture and distribute.

Equally, try make use of the recycling facilities offered in your area: almost all recycling helps to reduce energy consumption to some extent. For food waste, use a composter instead of the trash whenever possible. When food is buried in a commercial landfill, it decomposes anaerobically and generates the potent greenhouse gas methane. Landfills are, in fact, the source of a third of US methane emissions.

Weaning yourself from bottled water

Bottled water has made a tremendous splash among health-conscious consumers over the last twenty years. More than 150 billion litres are sold worldwide each year – more than ten times the amount in 1990. All told, the US consumes twice as much bottled water as any other nation, but the UK is close behind in per-capita use. Italy leads the world in that category, though much of its consumption is from glass bottles at sit-down meals rather than from plastic.

Drinking more water is certainly a good move for one's health, but it would be hard to imagine a more gratuitously wasteful product than bottled water, in terms of both cost (up to a thousand times more expensive than tap water) and greenhouse emissions. For starters, all those plastic bottles are made from petroleum, most likely at a factory that burns fossil fuels. Then there are the emissions involved in shipping the bottles long distances (water's quite heavy), keeping them refrigerated and, finally, transporting them for recycling or landfill (yet another ecological impact).

Fortunately, there's a marvellous, time-tested alternative: the tap. There's no denying that a bottle of water can be convenient in certain places and at certain times. However, it's practically as easy and far less costly to buy a sturdy, refillable, washable bottle and keep it with you. Avoid reusing bottled-water bottles, though; they may leach out harmful chemicals after only a few refillings.

As for quality, the supposed superiority of bottled water has been vastly overstated. In most developed countries, bottled water is inspected less rigorously than tap water. In fact, many bottled-water brands are simply tap water accompanied by a label that shows a mountain stream. Aquafina (marketed by PepsiCo) and Dasani (Coca-Cola), which together make up the majority of the US market, are both drawn from city sources. (The US Food and Drug Administration now requires such brands to indicate their municipal origins, but there's a loophole if the water has been "purified" or otherwise treated.) As for taste, blind tests on the US TV program "Good Morning America" found New York City tap water beating out each of the three bottled alternatives by a clear margin. If water taste and/or quality is a concern in your own community, then try installing a filter on your tap or keeping a filtered jug in your fridge.

In the broader picture, many activists worry that the soaring popularity of bottled water – most of it distributed by a few huge corporations, including Nestlé and Danone – points to an increasing privatization of the world's water supply. In developing countries, the advent of bottled water as an alternative to poor public supplies only exacerbates the divide between haves and have-nots. With climate change expected to make water supplies more variable in many parts of the globe, there's all the more incentive to make sure that everyone has access to water that's clean as well as climate friendly.

Food & drink

Cutting back on beef

Perhaps the single best way to reduce the emissions of your diet is to cut back on meat and dairy – and especially beef. Cattle belch and excrete a substantial fraction (perhaps 20%) of the world's methane emissions, and much of Earth's rainforest destruction is driven by the clearing of land for grazing livestock or growing their feed. In fact, a much-cited 2006 study by the United Nations Food and Agriculture Organization found that emissions associated with livestock – from deforestation to fertilizer and cow flatulence – are responsible for an astounding 18% of the current total human impact on the climate.

Food miles

In general, food that's travelled short distances is best for you and the climate. So if there's a farmers' market or local delivery scheme in your vicinity, try it out, or consider starting your own vegetable garden. Keep in mind, though, that "food miles" alone aren't the whole story, especially for food that doesn't have to be shipped fresh. A 2006 study at Lincoln University found that, for UK eaters, New Zealand lamb has only a quarter of the total carbon footprint of British-raised lamb, mainly because the free-ranging Kiwi sheep require far less fuel to raise and are transported overseas using relatively efficient ships. The study omits trucking within national borders, and there are other hard-to-quantify elements in the comparison (which was, unsurprisingly, carried out by agricultural specialists in New Zealand).

In a few years, we might see carbon-footprint labels that make it easier to choose the climate-friendliest foods regardless of where they're grown. The impending shift towards alternative fuels will no doubt tip the balance in some cases. In the meantime, when you do buy imports, such as coffee, choosing items marked with a **fair-trade** label will reduce the risk that rainforests are being chopped down to support your tastes.

Organics

Organic food eschews farming techniques that rely on petrochemicals and tends to result in slightly lower emissions per unit of food – though not all studies agree on this point. Whatever the truth, organic produce is rarely low-carbon enough to justify its being shipped (not to mention flown) over vast distances, so check the country of origin and favour local food over organics if it's a choice between the two.

Buy basics – and in bulk

Some food-processing tasks require a surprising amount of energy, so buying ingredients, rather than ready-meals, is usually a good idea from a carbon perspective. Buying in bulk is a good idea, too, as it minimizes packaging. For fluids such as shampoo and washing up liquid, look out for shops that let you bring in and refill your own bottles.

Bring your own bag

Shopkeepers practically force store-branded plastic bags on us, but each of the estimated 500 billion plastic bags that are used and tossed away each year carries a small carbon pricetag. Where possible, then, take a decent reusable bag with you. Aside from anything else, they're more comfortable to carry and less likely to break.

Offsetting

Paying a carbon cleaner

Carbon offsetting involves paying an organization to neutralize the climate impact of your own activities, thereby making those activities "carbon neutral" or "climate neutral". Some offset schemes focus on reducing future emissions by, for example, giving out low-energy light bulbs in the developing world or buying renewable electricity credits. Others focus on sucking CO_2 directly out of the atmosphere – usually by planting trees.

Offsetting has become very popular in the last decade, not just with individuals but also with global corporations (HSBC and other office-based giants have gone climate-neutral), celebrities (Pink Floyd, Pulp and the Pet Shop Boys have all neutralized their tours), and even publishers (the production of the book you are reading was offset by Rough Guides). In 2007, Vatican City became the world's first climate-neutral state by planting trees in a Hungarian preserve.

Offsetting is also increasingly popping up in point-of-purchase transactions. For example, some online travel services, and even a few airlines, now make it possible for consumers to offset their emissions from air travel when they buy their ticket.

How offsetting works

Whether you want to cancel the carbon footprint of a single long-haul flight, a year of car journeys or your entire existence, the process is the same. First, you visit the website of an offsetting organization and use their **carbon calculators** to work out the emissions related to whatever activity you want to offset. This will be translated into a fee which the offsetting organization will use to soak up – or remove the demand for – a matching amount of greenhouse gas.

Offsetting fees vary by organization and over time, but in the next several years many firms will likely settle around the carbon cost now being established in the second round of European emissions trading – in the neighbourhood of £15/$30 for a tonne of CO_2 at the time of going to press. At this price, a seat on a round trip from London to New York would cost around £25/$50 to neutralize, while a typical year of driving in a typically efficient car would clock in at around £42/$84. Currently, though, many popular schemes charge around half that amount.

If you pick a firm that charges an especially high rate per tonne of CO_2, it doesn't necessarily mean your money is doing more good; as the Tufts Climate Initiative notes, "It is more important to invest in high quality offsets than to buy as many offsets as possible."

Tree planting – help or hindrance?

One of the more clichéd approaches to tackling global warming, tree-planting has also become one of the more controversial – especially in the context of carbon offsetting. It's true that trees soak up CO_2 as they grow. Two or three dozen can be enough to absorb an entire household's emissions. However, at snow-prone higher latitudes, trees can actually *accelerate* climate change, according to a 2006 study led by Govindasamy Bala of the US Lawrence Livermore National Laboratory. That's because their CO_2-absorbing benefit is outweighed by the impact of their dark colour, which in wintertime absorbs sunlight that might otherwise be reflected to space by bright snow cover atop barren ground.

In snow-free warmer climates, and especially in the tropics and subtropics, the dark colour of trees isn't an problem. But there are other catches. First, when a tree dies, much of its stored carbon returns to the atmosphere. So the offset will only be permanent if each tree planted is replaced by another – and so on. Second, there are quicker, cheaper and longer lasting ways to fight climate change, such as distributing low-energy technologies to displace fossil fuels.

For all these reasons, many offset schemes have switched from tree-planting to energy-saving. (The Carbon Neutral Company, for example, was the result of a rebranding of a scheme previously called Future Forests.) That said, unless you live in a snowy region, planting a few trees in your garden is still likely to be beneficial for the climate – at least for the crucial coming decades.

A separate, and altogether more pressing, approach is protecting the forests that are already standing. That's not so much because of the CO_2 that mature forests absorb. Rather, it's because deforestation, and in particular the destruction of tropical rainforest, is one of the very largest sources of greenhouse emissions (see p.11). To help limit these emissions, the charity Cool Earth enables individuals and companies to sponsor areas of critically endangered rainforest in order to "keep the carbon where it belongs". The forest is purchased and given to a local trust, which protects it but allows sustainable harvesting of rubber, nuts and other forest crops. You can even make sure your sponsored area is still tree-covered thanks to satellite photos from Google.

Cool Earth www.coolearth.org

Offset debates

Despite their popularity, carbon-offset schemes are not without their critics. One argument levelled against offsetting is that it's just a plaster on the wound – a guilt-assuaging exercise that hides the inherent unsustainability of carbon-intensive Western lifestyles. Many observers have likened offsets to the medieval Catholic practice of selling indulgences to "offset" sinful behaviour, while the spoof website cheatneutral.com ridicules the offset concept by offering a service for neutralizing infidelity. "When you cheat on your partner, you add to the heartbreak, pain and jealousy in the atmosphere ... CheatNeutral offsets your cheating by paying someone else to be faithful and *not* cheat."

Even the offset schemes themselves tend to agree that offsetting emissions isn't as good as not causing them in the first place. That said, there's no reason why people can't buy offsets *and* make efforts to reduce their emissions directly. Indeed, the very act of digging into pockets for an offset fee may make consumers or businesses more conscious of carbon's larger cost.

A separate criticism is that offset projects may not make the swift, long-term carbon savings that are claimed of them. It's true that some of the projects – most notably tree planting (see box, opposite) – may take decades to soak up the carbon you've paid to offset, which is one reason why many offsetting groups are moving towards sustainable energy projects instead of trees. As for whether the carbon savings are real, the better offsetting services are externally audited (see below) to address just this question.

Still another point of contention is whether offsetting services ought to make a profit. According to the Tufts Climate Initiative, some offsetting firms use as much as 85% of their revenue for expenses before actually funding any emission reduction work. Some of the commercial firms that provide offsets don't reveal how much profit they earn, arguing that the good work they do justifies an unspecified return on their labour and risk. Other offsetters are bona fide charities, which means all of their income goes toward operations and the offset projects themselves.

Ultimately, the benefits of offsetting are open to debate. If you pick the right scheme, it's likely that your contribution *will* help reduce emissions (especially if you pay to offset, say, two or three times the emissions of the activity in question). On the other hand, critics such as George Monbiot argue that, if we're to stand any chance of tackling climate change, then the kinds of projects invested in by offset schemes need to happen anyway, funded by governments, in addition to individuals and companies slashing their carbon footprints directly.

How to pick an offsetter

With dozens of offset schemes out there and no real regulatory framework, it can be hard to know which one to choose. One of the most useful consumer-oriented guides to date was first published in 2006 by Clean Air – Cool Planet.

Clean Air – Cool Planet www.cleanair-coolplanet.org

The report compares offset schemes according to various criteria, including: transparency; the cost of overheads; whether the scheme complies with any voluntary standards; the quality of the online carbon calculators; the type of projects favoured (avoiding future emissions is more effective than tree-planting); and additionality (if and how the scheme ensures that its emissions reductions wouldn't have occurred anyway). Eight companies earned the highest marks in the report:

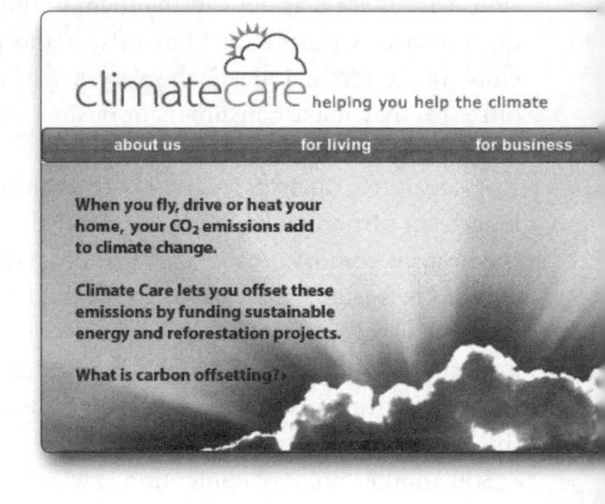

AgCert/DrivingGreen (Ireland) www.agcert.com
atmosfair (Germany) www.atmosfair.de
Carbon Neutral Company (UK) www.carbonneutral.com
Climate Care (UK) www.climatecare.org
Climate Trust (US) www.climatetrust.org
co2balance (UK) www.co2balance.com
NativeEnergy (US) www.nativeenergy.com
Sustainable Travel/MyClimate (US) www.my-climate.com

Another report dealing specifically with aviation offsets, issued by the Tufts Climate Institute (www.tufts.edu/tie/tci), recommends ten schemes, with the topmost ratings going to atmosfair and NativeEnergy, as well as:

Climate Friendly www.climatefriendly.com

PART 6

RESOURCES

Climate change books and websites

Books

Many excellent books touch on climate change as part of more general narratives about the planetary environment. We've reserved this list – which, of course, is only a starting point – for books that have global warming at or near their centre.

General

Tim Flannery, *The Weather Makers* (2005). Flannery, an Australian zoologist, museum director and writer, exploits his gift for pithy explanations and metaphor in this fine climate-change overview. The short, cleverly titled chapters walk readers through climate history, global-warming science, political and technological options and individual action.

Ross Gelbspan, *The Heat Is On* (1997) and *Boiling Point* (2004). Gelbspan is the consummate climate-change muckraker. The sceptics that shaped 1990s climate debate are the main target of *The Heat Is On*, which includes revealing detail from congressional testimony. *Boiling Point* lambasts a broader set of actors, including "criminals against humanity" (big oil and coal), "bad media" and "compromised activists".

Al Gore, *Earth in the Balance* (1992) and *An Inconvenient Truth* (2006). Gore's 1992 book established its writer as the US leader most committed to solving climate change. Though the science has evolved greatly since, Gore's writing remains powerful and surprisingly relevant. The newer release, *An Inconvenient Truth*, is a handsome summary of the documentary of the same name. If you saw the film, you won't learn a great deal more from the book, though it will enable you to linger over the stunning graphics.

John Houghton, *Global Warming: The Complete Briefing* (Third Edition, 2004). Readers who crave scientific meat but don't have the time or inclination to pore over IPCC reports will enjoy this primer. Houghton led two of the IPCC's major assessments and speaks with authority as well as humanity (there's a thoughtful discussion of climate change from ethical and theological points of view).

Elizabeth Kolbert, *Field Notes from a Catastrophe* (2006). Based on a multipart series in *New Yorker* magazine. The influence of Rachel Carson's classic *Silent Spring* (which helped kick start US environmentalism) is evident in Kolbert's dry, crisp prose and her one-step-back perspective on how climate change is already affecting places like Shishmaref, Alaska, and how it's altered past civilizations.

Eugene Linden, *The Winds of Change* (2006). Focuses on the risk of rapid climate change – how it's apparently happened in the past, how it could happen again, and what that would mean. Linden is a skilled writer who draws on extensive experience reporting on the global environment, which lends a peripatetic feel to the narrative and allows for first-person asides.

James Lovelock, *The Revenge of Gaia* (2006). The man who told us Earth is a self-sustaining system now worries that we're pushing that system beyond its capacity to heal itself. In this slender and sometimes gloomy book, Lovelock argues we may need nuclear power to avoid what he sees as the truly catastrophic implications of "global heating". *Revenge* also provides a concise overview of Gaia theory for newcomers.

Mark Lynas, *High Tide* (2004) and *Six Degrees* (2007). One of the best "you are there" books on the impacts of global warming, *High Tide* see Lynas hopping from Britain to China to Peru, describing how climate change is affecting people and landscapes. An engaging travelogue and an approachable introduction to climate-change science. With *Six Degrees*, Lynas ups the ante, extending his on-the-scene approach to how each added degree of warming might affect the planet. His portrayals of the type of dire events that may be in store, such as a Superhurricane Ophelia ravaging Houston in 2045, manage to be both sober and hair-raising. The last chapters tiptoe toward doomsday, with methane release triggering mass extinctions, but as always Lynas is clear and well referenced.

Mark Maslin, *A Very Short Introduction to Global Warming* (2004). Maslin zips through climate-change science and policy in this extended essay. Depth is limited, given the compact format, and graphics suffer in size and consistency. But Maslin touches on areas that others omit, such the worldviews that inform people's positions on the topic.

George Ochoa, Jennifer Hoffman and Tina Tin, *Climate* (2005). Produced in a hardback coffee-table format, this book pairs beautiful photography and clear graphics with text that's very accessible yet comprehensive. It covers all bases, with especially strong sections on ecosystems and climate history.

James Gustave Speth, *Red Sky at Morning* (2004). A US-oriented discussion of climate change in the larger environmental context by a co-founder of the Natural Resources Defense Council and founder of the World Resources Institute. The book benefits greatly from Speth's first-hand knowledge of environmental work, and avoids the potential tangles of policy jargon.

Arctic and glaciers

Richard B. Alley, *The Two-Mile Time Machine* (2000). A personable book written by a geologist who's participated in ice-core expeditions in Greenland and Antarctica. Alley explains in lively, conversational prose how ice cores are retrieved and what they tell us.

Mark Bowen, *Thin Ice* (2005). Bowen's masterful, in-depth look at tropical glaciers and those who study them fills an important niche. The focus is on glaciologist Lonnie Thompson and his exploits in Peru's highest mountains, with more general background on climate change provided along the way. Himself a climber and physicist, Bowen digs deep into the science, but also brings tales of high-altitude exploits that will resonate with outdoor enthusiasts.

Andrew C. Revkin, *The North Pole was Here (2006)*. This book for young readers blends lavish photography with clear, engaging writing. It focuses on a three-day visit to the high Arctic by Revkin, a veteran reporter on climate change at *The New York Times*. Full-page excerpts from the *Times* go back as far as a century.

Charles Wohlforth, *The Whale and the Supercomputer* (2004). The best book to date on what climate change is doing to the people, creatures and landscape of Alaska – and, by extension, high latitudes around the world. Wohlforth's narrative has the broad sweep of an epic, plus plenty of insightful details on researchers and everyday Alaskans pondering the future and grappling with the enormity of the changes already under way.

Historical

John Imbrie and Katharine Palmer Imbrie, *Ice Ages: Solving the Mystery* (1979). Recently reprinted, this classic book explains how our modern understanding of ice ages emerged. Though it's not the place to turn for the latest cutting-edge work in glaciology, its accounts of early figures such as Milankovitch are packed with detail.

Jeremy Leggett, *The Carbon War* (1999). As a former oil geologist and then a science advisor to Greenpeace, Leggett got a close-up look at the negotiations leading up to Kyoto. He shares the victories and agonies of climate-change activism and sheds light on many of the key players and processes, making no secret of whom he considers to be the good guys and bad guys. A book of great historical value.

William K. Stevens, *The Change in the Weather* (1999). One of the best global-warming overviews of its time, this is worth seeking out for its thorough coverage of climate research in the 1990s. A long-time science writer for *The New York Times*, Stevens provides illuminating profiles of scientists (including sceptics) and how they arrived at their positions on climate change.

Spencer Weart, *The Discovery of Global Warming* (2004). A superb chronicle of how fragments of theories and observations gradually coalesced to form a new scientific discipline. The sparkling prose is concise and precise, giving a sense of the full story quickly and pleasurably. Just as impressive is the adjunct website, www.aip.org/history/climate, which includes all the material from the book and far more besides.

Paleoclimate

John Cox, *Climate Crash* (2005). A wonderfully written report on the science of abrupt climate change and its roots in clues such as glaciers and fossils. Cox conveys the excitement of the research chase, the difficulty of the work and the mixed signals it provides about our future. *Climate Crash* is one of the few places to find easily digestible treatments of such complex phenomena as Dansgaard–Oeschger events.

Brian Fagan, *The Little Ice Age* (2000) and *The Long Summer* (2003). Literate, highly entertaining accounts of the first two great climate epochs of the last millennium (the third being the warm-up since the mid-1800s). They're not so much books about science as vivid historical portraits of climate's impact on humanity.

William Ruddiman, *Plows, Plagues and Petroleum* (2005) and *Earth's Climate* (2001). An accomplished paleoclimatologist, Ruddiman shook up his field by speculating that agriculture may have postponed the next ice age. *Plows, Plagues and Petroleum* outlines the hypothesis in lay-oriented language. It's a satisfying read and a great case study of a scientist creating and defending a theory. *Earth's Climate*, designed as a textbook to introduce students to climate history, is clear, colourful and accessible, free from equations and jargon but packed with helpful diagrams.

Policy

Aubrey Meyer, *Contraction and Convergence* (2001). The full story about one of the leading candidates for a post-Kyoto system of controlling greenhouse emissions. Meyer developed C&C more than a decade ago and makes the case for it with passion and conviction.

George Monbiot, *Heat: How to Stop the Planet from Burning* (2006). Meticulously researched and convincingly argued, *Heat* lays out plans for reducing UK emissions to what Monbiot argues is a safe and equitable level (just 13% of what they are now) by 2030. Particularly valuable is the book's thorough analysis of the electricity grid and other parts of the energy infrastructure.

Andrew L. Dessler and Edward A. Parson, *The Science and Politics of Global Climate Change* (2006). This soberly written guide covers the basics of climate change, the main points of contention and the approaches of policy experts to the topic. Striking a middle ground between the comprehensiveness of an IPCC report and the immediacy of *An Inconvenient Truth*, it's especially good for students and scholars needing a one-stop reference.

Barry George Rabe, *Statehouse and Greenhouse* (2004). Many US states have leapt ahead of the federal government in developing responses to climate change. Rabe, a policy analyst, tells the story in a colourful and intriguing way, showing how developments in California and elsewhere are consistent with a great federalist tradition: states serving as laboratories for policies that could later go nationwide.

Energy and climate

Richard Heinberg, *Powerdown* (2004). An expert on sustainability, Heinberg has long been concerned about the potential for oil depletion. In this book he explores four possible scenarios that could unfold as oil and gas resources draw scarce. Overall, Heinberg manages to be both pessimistic and pragmatic, envisioning big trouble ahead while lobbying for action that would at least stanch the bleeding.

James Howard Kuntsler, *The Long Emergency* (2005). A fascinating if depressing read, this book places climate change in the context of other global issues such as oil depletion, terrorism and disease. Kuntsler explains why he believes that renewables, nuclear, tar sands and oil shale will fail to protect us from the massive impacts of diminished oil production.

Jeremy Leggett, *Half Gone* (2005), also published as *The Empty Tank*. An incisive book that explores how a drop in oil production (which Leggett sees as being imminent) could affect climate change. It's full of sharp observations on the workings of the fossil-fuel industry, benefiting from Leggett's own experience as an oil geologist. The book is let down only by the cloying opening and closing parables about "The Blue Pearl".

Joseph J. Romm, *The Hype about Hydrogen* (2005) and *Hell and High Water* (2006). Romm, who worked for years at the US Department of Energy, deftly punctures the bubble of optimism surrounding "the hydrogen economy". While acknowledging the potential of fuel cells for powering large buildings, he explains clearly and succinctly why it could be decades before hydrogen becomes common in mass-market vehicles. In *Hell and High Water*, Romm broadens his view to include the politics swirling around climate change in general (including the role of sceptics and vested interests in thwarting action), as well as energy policies in particular.

Personal action

Duncan Clark, *The Rough Guide to Ethical Living* (2006). Covering personal climate action as well as other forms of ethical consumerism, this UK-focused book provides useful, digestible advice on everything from home energy generation through to greener cars.

Mayer Hillman, *How We Can Save the Planet* (2004). A lucid summary of the threat posed by climate change and the various proposed solutions. Hillman keeps the narrative lively while making it crystal clear what he expects us – as individuals and societies – to do. The last part of the book introduces the notion of personal carbon rations and offers suggestions on how to calculate and live within your own.

Chris Goodall, *How to Live a Low-carbon Life: The Individual's Guide to Stopping Climate Change* (2007). If you're truly passionate about reducing your carbon footprint, this is the ultimate guide. In 326 pages, Goodall walks you through each lifestyle aspect in exhaustive detail – including a full page on how best to heat a teacup's worth of water – with an exacting eye for the subtleties that shorter guides miss or omit. No doubt too rigorous for some readers, but unsurpassed in its depth and commitment and quite readable to boot.

Mark Lynas, *Carbon Counter* (2007). In this compact, inexpensive volume, Lynas provides a handy step-by-step guide to calculating your carbon footprint, in addition to lots of advice on reducing it.

Dave Reay, *Climate Change Begins at Home* (2005). Reay brings a laser-like focus and a light touch to the task of reducing your household's carbon emissions, as brought to life through a fictional family called the Carbones. What could have been a dour series of orders is instead a creative, thought-provoking guide to tackling climate change on a personal level.

Fiction

John Barnes, *Mother of Storms* (1995). In this sci-fi thriller, set in 2028, nuclear warheads plough into the Pacific, destabilizing colossal amounts of methane clathrates. The greenhouse effect then runs amok, spawning cataclysmic hurricanes around the world, and the future rests on an uncommonly smart astronaut and his plan to screen out sunlight.

Michael Crichton, *State of Fear* (2004). The bestselling thriller that cheered climate-change sceptics across the world is an undeniably gripping tale in the Crichton mould, but hardly the place to turn for a well-rounded exploration of global warming science. See p.260 for more on Crichton and the *State of Fear*.

Kim Stanley Robinson, *Forty Signs of Rain* (2004), *Fifty Degrees Below* (2005) and *Sixty Days and Counting* (2007). In this sci-fi trilogy, the collapse of warm North Atlantic currents results in an ice blitz. *Forty Signs of Rain* chronicles a cast of characters coming to grips with worsening climate, while *Fifty Degrees Below* focuses on one man living in snowbound Washington, DC, and a grand scheme to jump-start the Gulf Stream. The action – both interpersonal and meteorological – reaches a peak in *Sixty Days and Counting.*

Bruce Sterling, *Heavy Weather (1995).* This landmark blend of climate change and cyberpunk follows a group of rebellious, tech-savvy storm chasers facing off with epic tornadoes across the US Great Plains – just one manifestation of global warming circa 2031.

Websites

Given the enormous scope of the Web and the equally vast realm of climate-change science and debate, the sampling below is only the barest sliver of what's available online. We've focused on websites with a rich array of climate change content. For sites covering carbon offsetting and home energy, see What You Can Do (p.333).

Scientific assessments

If you want undiluted, in-depth scientific facts about climate change, you can't do better than the assessment publications of relevant scientific institutions. The full text of each report since 1998 from the **IPCC** (see p.287) can be viewed online; for recent reports, full PDFs are also available, typically broken into chapters. The UK Climate Impacts Programme (**UKCIP**) produces scenarios every few years and sends a monthly e-bulletin. The **US National Assessment**, produced in 2000, remains the most exhaustive American study of national and regional impacts, while the ongoing US Climate Change Science Program (**CCSP**) is conducting a series of targeted assessments. National-scale scenarios are also provided by **CSIRO** (Australia) and Environment Canada. The **ENSEMBLES** page isn't fancy but offers plenty of background on this massive EU modelling project. Detailed projections for Europe in the 2071–2100 time range are being created through **PRUDENCE**, and **RAPID** is a UK-funded programme on rapid climate change, including the possible slowdown of North Atlantic currents.

IPCC www.ipcc.ch
UKCIP www.ukcip.org.uk
USNA www.usgcrp.gov/usgcrp/nacc
CCSP www.climatescience.gov
CSIRO (Australian scenarios) www.climatechangeinaustralia.gov.au
Environment Canada www.ccsn.ca/index-e.html
ENSEMBLES www.ensembles-eu.org
PRUDENCE prudence.dmi.dk
RAPID www.soc.soton.ac.uk/rapid

Research centres

Some of the world's focal points for global warming research also provide extensive materials designed for the interested public. These labs include the UK's **Hadley Centre** and the US National Center for Atmospheric Research (**NCAR**), which is operated by the University Corporation for Atmospheric Research (**UCAR**). UCAR's member universities and its international affiliates include many of the world's top academic centres studying global climate. Several branches of **NASA** and the National Oceanic and Atmospheric Administration (**NOAA**) focus on climate change and offer lay-friendly summaries of their work. The non-profit **Pew Center on Global Climate Change** works with leading experts to produce comprehensive, readable analyses tailored for policy makers and the public.

Hadley Centre www.metoffice.com/research/hadleycentre
UCAR/NCAR www.ucar.edu
UCAR members www.ucar.edu/governance/members/institutions.shtml
NASA Goddard Space Flight Center www.giss.nasa.gov
NOAA Climate Dynamics and Prediction Group
www.gfdl.noaa.gov/research/climate
NOAA Paleoclimatology Program www.ncdc.noaa.gov/paleo
Pew Center on Global Climate Change www.pewclimate.org

Climate change news

The following sites are among the best for keeping up with news on global warming and the environment in general. **Grist** is a standout, with content that's light-hearted but razor-sharp. You'll find a full grab bag of recent press clips at **The Heat Is Online**, run by premier climate muckraker Ross Gelbspan. Among UK newspapers, **The Independent** has been relentless in covering climate change in recent years, with **The Guardian** close behind. **The New York Times** leads the US pack in the depth and frequency of its climate-change reporting, with the **Los Angeles Times** also excellent.

Grist www.grist.org
The Heat Is Online www.heatisonline.org/main.cfm
The Independent www.independent.co.uk
The Guardian www.guardian.co.uk
The New York Times www.nytimes.com
Los Angeles Times www.latimes.com

Blogs

Though some should be taken with a grain of salt, blogs can be invaluable for sorting through the ever-increasing stacks of global warming research and news. **Real Climate** is the real thing: run by an international team of top climate scientists, it addresses burning questions in a style aimed at motivated laypeople, with plenty of cross-references. **Climate Feedback**, a new entry from the journal *Nature*, includes a varied cast of top-notch contributors. **Stephen Schneider**'s site isn't a blog so much as a mini-encyclopedia on climate change. Among sceptics, **Pat Michaels** is among the most blog-savvy, with frequent commentary on a wide range of relevant research. **Roger Pielke Jr.** maintains a thought-provoking blog focused on climate policy, while climate whistleblower **Rick Piltz** keeps an eye on US government doings. Several other good sites maintained by individual scientists and journalists are listed below.

Real Climate www.realclimate.org
Climate Feedback (*Nature*) blogs.nature.com/climatefeedback
Stephen Schneider stephenschneider.stanford.edu
World Climate Report (Pat Michaels) www.worldclimatereport.com
Promethus (Roger Pielke Jr) sciencepolicy.colorado.edu/prometheus
Climate Science Watch (Rick Piltz) www.climatesciencewatch.org
James Annan julesandjames.blogspot.com
Mark Lynas www.marklynas.org
George Monbiot www.monbiot.com
Chris Mooney www.scienceblogs.com/intersection
John Fleck www.inkstain.net/fleck

Index

D

E

F

G

S

T

U

V

W

Y

Z